THE PRACTICE
AND MANAGEMENT
OF INDUSTRIAL ERGONOMICS

PRENTICE-HALL INTERNATIONAL SERIES
IN INDUSTRIAL AND SYSTEMS ENGINEERING

W. J. Fabrycky and J. H. Mize, Editors

ALEXANDER, *The Practice and Management of Industrial Ergonomics*
AMOS AND SARCHET *Management for Engineers*
BANKS AND CARSON *Discrete-Event System Simulation*
BEIGHTLER, PHILLIPS, AND WILDE *Foundations of Optimization,* 2/E
BLANCHARD *Logistics Engineering and Management,* 2/E
BLANCHARD AND FABRYCKY *Systems Engineering and Analysis*
BROWN *Systems Analysis and Design for Safety*
BUSSEY *The Economic Analysis of Industrial Projects*
CHANG AND WYSK *An Introduction to Automated Process Planning Systems*
ELSAYED AND BOUCHER *Analysis and Control of Production Systems*
FABRYCKY, GHARE, AND TORGERSEN *Applied Operations Research and Management Science*
FRANCES AND WHITE *Facility Layout and Location: An Analytical Approach*
GOTTFRIED AND WEISMAN *Introduction to Optimization Theory*
HAMMER *Occupational Safety Management and Engineering,* 3/E
HAMMER *Product Safety Management and Engineering*
IGNIZIO *Linear Programming in Single and Multiobjective Systems*
MIZE, WHITE, AND BROOKS *Operations Planning and Control*
MUNDEL *Improving Productivity and Effectiveness*
MUNDEL *Motion and Time Study: Improving Productivity,* 6/E
OSTWALD *Cost Estimating,* 2/E
PHILLIPS AND GARCIA-DIAZ *Fundamentals of Network Analysis*
SANDQUIST *Introduction to System Science*
SMALLEY *Hospital Management Engineering*
THUESEN AND FABRYCKY *Engineering Economy,* 6/E
TURNER, MIZE, AND CASE *Introduction to Industrial and Systems Engineering*
WHITEHOUSE *Systems Analysis and Design Using Network Techniques*

THE PRACTICE AND MANAGEMENT OF INDUSTRIAL ERGONOMICS

David C. Alexander, P. E.

PRENTICE-HALL, INC., Englewood Cliffs, NJ 07632

Library of Congress Cataloging-in-Publication Data

ALEXANDER, DAVID C.
  The practice and management of industrial ergonomics.
  Includes bibliographies and index.
  1. Human engineering. I. Title.
T59.7.A45 1986   620.8′2   85-25593
ISBN 0-13-693649-0

Editorial/production supervision: Colleen Brosnan
Cover Design: Wanda Lubelska Design
Manufacturing buyer: Rhett Conklin

© 1986 by Prentice-Hall, Inc.
A Division of Simon & Schuster
Englewood Cliffs, New Jersey 07632

All rights reserved. No part of this book may be
reproduced, in any form or by any means,
without permission in writing from the publisher.

Printed in the United States of America

10 9 8 7 6 5 4 3 2 1

ISBN 0-13-693649-0 025

PRENTICE-HALL INTERNATIONAL (UK) LIMITED, *London*
PRENTICE-HALL OF AUSTRALIA PTY. LIMITED, *Sydney*
PRENTICE-HALL CANADA INC., *Toronto*
PRENTICE-HALL HISPANOAMERICANA, S. A., *Mexico*
PRENTICE-HALL OF INDIA PRIVATE LIMITED, *New Delhi*
PRENTICE-HALL OF JAPAN, INC., *Tokyo*
PRENTICE-HALL OF SOUTHEAST ASIA PTE. LTD., *Singapore*
EDITORA PRENTICE-HALL DO BRASIL, LTDA., *Rio de Janeiro*
WHITEHALL BOOKS LIMITED, *Wellington, New Zealand*

To Janie and Lucas

CONTENTS

PREFACE xiii

Part 1 Industrial Ergonomics 1

1 INDUSTRIAL ERGONOMICS 1

- 1-1 Goals 1
- 1-2 Definition of Industrial Ergonomics 2
- 1-3 Ergonomics in the Industrial and Service Sectors 3
- 1-4 The Philosophy of Applications 4
- 1-5 The Use of This Book 6
- 1-6 Summary 7
  - Questions 8

2 THE APPLICATION OF ERGONOMICS IN INDUSTRY 9

- 2-1 Introduction 9
- 2-2 Prevention Versus Correction 10
- 2-3 General Areas for Improvements 11
- 2-4 Types of Problems 12
- 2-5 Depth of Knowledge 16

2-6   The Role of the Ergonomist   17
2-7   Summary   18
      Questions   18

Part 2   Starting the Ergonomics Effort   20

3   PLANNING AN INDUSTRIAL ERGONOMICS EFFORT   20

3-1   The Beginning   20
3-2   Types of Applications   22
3-3   Opportunities for Improvement   25
3-4   Organization   27
3-5   Resources   30
3-6   Selling the Concept to Management   34
3-7   Summary   38
      Acknowledgment   38
      Questions   38

4   IMPLEMENTING AN INDUSTRIAL ERGONOMICS EFFORT   39

4-1   Introduction   39
4-2   Broad-Based Support   40
4-3   User Needs   41
4-4   Creating Interest   44
4-5   Making It Work   46
4-6   A Lasting Change   52
4-7   Summary   52
      Questions   52

5   PROBLEM IDENTIFICATION   54

5-1   Introduction   54
5-2   Traditional Problem Identification   56
5-3   Advanced Problem Identification   64
5-4   Using Problem Identification to Prevent Problems   69
5-5   The Use of Technology   72
5-6   A Philosophy of Problem Identification   72

Contents ix

5–7 Summary 74
Acknowledgment 74
Questions 75

Part 3 Classic Ergonomic Problems **76**

6 THE PROBLEM JOB **76**

6–1 Introduction 76
6–2 The Problem Job Model 78
6–3 A Few Examples 94
6–4 Summary 99
Acknowledgments 99
Questions 100

7 THE WORKPLACE AND THE WORK SPACE **101**

7–1 Introduction 101
7–2 Workplace Design 104
7–3 Work Space Design 116
7–4 Design Processes and Concerns 121
7–5 Anthropometry 131
7–6 Summary 138
Questions 138

8 THE HEAVY JOBS **140**

8–1 Introduction 140
8–2 NIOSH Lifting Guidelines 143
8–3 Job Severity Index 150
8–4 Other Manual Materials Handling Guidelines 164
8–5 Engineering Solutions 169
8–6 Administrative Solutions 176
8–7 Managing Materials Handling 185
8–8 Cumulative Trauma Disorders 188
8–9 Summary 192
Questions 192

9 THE HARD AND HOT JOBS 194

- 9–1   A Fair Day's Work   194
- 9–2   A Brief Description of Work Physiology   195
- 9–3   The Hard Job   204
- 9–4   The Hot Job   216
- 9–5   Summary   232
        Questions   233

10 THE ENVIRONMENT 235

- 10–1   Introduction   235
- 10–2   Thermal Concerns   237
- 10–3   Lighting   240
- 10–4   Noise   246
- 10–5   Vibration   252
- 10–6   Summary   253
         Questions   253

11 THE HUMAN ERROR 254

- 11–1   Introduction   254
- 11–2   Error-Producing Situations: A Model Plus Discussion   256
- 11–3   Error-Producing Tasks   260
- 11–4   Summary   274
         Questions   274

Part 4   Additional Applications of Ergonomics   276

12 THE PURCHASE OF ERGONOMICALLY DESIGNED PRODUCTS 276

- 12–1   Introduction   276
- 12–2   Difficulties in Buying Ergonomically Designed Equipment   277

Contents   xi

12-3	Purchased Equipment	280
12-4	The Design of the Product	288
12-5	Summary	291
	Questions	291

13 JOB DESIGN   292

13-1	Introduction	292
13-2	Utilization of People	294
13-3	The Job Design Process	298
13-4	Automation and Job Design—A Strong Marriage	301
13-5	The Other Factors in Job Design	307
13-6	Summary	311
	Acknowledgments	311
	Questions	312

14 ADMINISTRATIVE DECISIONS   313

14-1	Introduction	313
14-2	Work Times and Schedules	316
14-3	Work Task Expectations	325
14-4	Work Environment	328
14-5	Pay and Personnel	331
14-6	Summary	333
	Questions	334

15 THE DISABLED EMPLOYEE   335

15-1	Introduction	335
15-2	Overview of Employer Concerns	337
15-3	The Accommodation Process	344
15-4	Predesign for Accommodation	348
15-5	Accommodations and Aids	349
15-6	Summary	358
	Questions	358

Part 5 Keeping It Going   **360**

16 SUSTAINING THE ERGONOMICS EFFORT   **360**

    16-1   Introduction   360
    16-2   The Natural Growth of an Ergonomics Program   361
    16-3   Benefits Greater Than Costs   363
    16-4   Winning Converts to Ergonomics   364
    16-5   Summary   367
              Questions   367

INDEX   **369**

PREFACE

The purpose of this book is to explain the practical use of ergonomics in the industrial setting. For far too long, ergonomics has been perceived to be "interesting, but not practical." This book explains how to use the principles of ergonomics to solve common industrial problems. Examples are extensively used to clarify major points.

The audience of this book includes industrial engineers, safety professionals, industrial hygienists, design engineers, health practitioners, personnel specialists, and line managers. It explains problems from their viewpoint and provides the information that they need to resolve those problems.

In a conversation with Dr. Jerry Ramsey in 1984, he encouraged the intent of this book by saying: "Stick your neck out and generalize from the principles. That is what the practitioner has to do. It's about time that someone provided the rules of thumb that those in industry use to get the job done."

With over ten years of practical experience in the practice and management of industrial ergonomics, I feel qualified to author this text. Included in the formulation of this material are hundreds of conversations with practicing ergonomists and engineers, researchers, and line managers. They have shared their problems and solutions with me and have led me to believe that ergonomics is indeed a practical tool.

I have reviewed thousands of pages of ergonomics information, hundreds of technical reports and presentations, and dozens of books. From that wealth of information, I have gleaned the critical material useful to the practitioner. It is presented as the practitioner needs it, with concise tables and figures, and with a text that explains the problems that a practitioner will face. Countless times, I have been surprised when in a conversation with peers, I find them sharing an incident from the real world similar to one already included in the text.

I especially want to acknowledge the initial encouragement from Dr. Leo A. (Tony) Smith and Dr. Thaddeus M. (Ted) Glen to author this text. Their early support was critical for the start of this project in 1982.

I also want to acknowledge the help provided by those who reviewed my early drafts and provided comments for improvements:

Dr. M. M. Ayoub, Texas Technological University
Mr. Harvey Foushee, AT&T Technologies, Inc.
Dr. Joel Greenstein, Clemson University
Dr. John Hungerford, University of Tennessee
Mr. Hart Kaudewitz, Tennessee Eastman Company
Dr. David Kiser, Eastman Kodak Company
Dr. Suzanne Rodgers, Consultant
Dr. Roger Rupp, Tennessee Valley Authority
Dr. George Smith, Ohio State University
Dr. Tony Smith, Auburn University
Mr. Terry Wales, Monsanto Corporation

I am indebted to my colleagues from Tennessee Eastman Company, Bob Terrell and Hart Kaudewitz, who shared and taught me their skills in the field of ergonomics. I am also indebted to my supervisors, Nick Grabar and Bill Burgess, who faithfully supported my initial efforts and largely created my vision of what industrial ergonomics is all about.

The ergonomics professionals within Eastman Kodak Company also deserve my greatest appreciation. To those I've worked closest with—Harry Davis, Sue Rodgers, Terry Faulkner, Richard Pugsley, Stan Caplan, Paul Chapney, Brian Crist, Richard Little, Dave Kiser, Wally Nielsen, and Richard Lucas—thank you.

I especially want to thank my family, Janie and Lucas, for their patience and support during the past two years. They have unquestioningly shared my dreams for this book.

David C. Alexander

Part 1
INDUSTRIAL ERGONOMICS

Chapter 1

INDUSTRIAL ERGONOMICS

Part 1, which consists of Chapters 1 and 2, discusses the field of industrial ergonomics and provides a framework for the application of ergonomics. Certain philosophies appropriate for the application of ergonomics in industry are presented.

This chapter begins with a definition of industrial ergonomics. That definition is expanded to cover a philosophy of applications within the industrial setting. The chapter concludes with the intended use of the book.

1-1 GOALS

During the vice-presidential debates associated with the 1976 presidential campaign, the Republican candidate, Robert Dole, was asked why he wanted to be vice-president. His answer: "It's an inside job with no heavy lifting." In Senator Dole's case, it was merely campaign humor. However, his answer describes some very real concerns shared by many people on their jobs every day. Exposure to the environment and the wear and tear of highly physical jobs are only two of the more obvious industrial ergonomics problems.

In a recent ranking, ergonomics-related problems captured a number of the top places on the most pressing occupational disease and injury categories. Musculoskeletal injuries, traumatic injuries and deaths, cardiovascular diseases, and hearing losses were cited among the top 10 categories, as noted in Table 1-1.

The field of industrial ergonomics is devoted to the alleviation of the

TABLE 1-1 Ten Leading Work-Related Diseases and Injuries in the United States

1. Occupational lung disease
2. Musculoskeletal injuries[1]
3. Occupational cancers
4. Amputations, fractures, eye loss, lacerations, and traumatic deaths[1]
5. Cardiovascular diseases[1]
6. Disorders of reproduction
7. Neurotoxic disorders
8. Noise-induced loss of hearing[1]
9. Dermatologic conditions
10. Psychologic disorders[1]

[1] Ergonomics related.

Source: Centers for Disease Control, "Leading Work-Related Diseases and Injuries—United States," *Morbidity and Mortality Weekly Report,* 32, January 21, 1983, pp. 24–26, 32.

rigors of the workplace and to the improvement of the person's performance on the job. This goal is shared by many people—workers on the line, management in industry, and safety and health professionals. To achieve this goal, changes are made in the jobs that people do, in their workplaces and work spaces, in their work methods, and in the equipment and tools they use. These changes, however, cost money and should not be made without the promise of reduced injuries, reduced costs, and/or improved efficiency.

The challenge, then, is twofold. First, a means of predicting the benefits of a change must be established. Second, recommendations that result in workable changes must be developed and implemented.

1-2 DEFINITION OF INDUSTRIAL ERGONOMICS

Industrial ergonomics is the application of those sciences relating human performance (physiology, psychology, and industrial engineering) to the improvement of the work system, consisting of the person, the job, the tools and equipment, the workplace and work space, and the immediate environment.

There are several important concepts to be noted here. First, there is the concept of human performance. Ergonomics is centered on the person, not on the equipment or facilities. If people are not involved with the system, ergonomics problems simply do not exist.

The second concept is that of the system centered around the person. The equipment itself may be important, but only to the extent that the person must operate, service, install, and/or repair that equipment. The same holds true for the environment. It is important only to the extent that a person is in that environment. Again, with no person, the environment cannot pose a concern from the industrial ergonomics standpoint.

The third concept is that of the work system. Although there are other situations where the concept of ergonomics is important, within the definition of industrial ergonomics, the boundaries will include work duties and will exclude recreational, military, and aerospace applications.

The fourth concept is that of improvement. The use of industrial ergonomics is the improvement of the work system that surrounds the person. Improvement is not just an abstract term, but rather a generally measurable concept. Therefore, improvements in the work system should be measurable in a quantitative or qualitative sense.

This definition of industrial ergonomics will serve as a guide for the rest of this book. The direction will be toward ways to improve the performance of human beings in the work system.

One special note in the discussion of terminology is the use of the word *ergonomics*. Ergonomics has its origins in the United Kingdom when "ergo" (work) was joined with "nomics" (the study of). There is disagreement on the proper terminology to use to describe this field of study. Some prefer the words *human factors* or *human factors engineering*, feeling that the consideration of the human factor best describes this area. Still others choose the terms *bioengineering, biomechanics, human engineering,* or *engineering psychology* to describe their work. Although the term *ergonomics* is more widely used in Europe, *human factors* was the common terminology in the United States as recently as ten years ago. Now, however, *ergonomics* is being used to describe many efforts that arise from industry, but work within the transportation and defense communities continues to be labelled *human factors*. Adding the modifier to form *industrial ergonomics* is intended to distinguish the practice of ergonomics in the plant and office environments from product design applications of ergonomics. While the debate goes on, this author chooses to use the term *industrial ergonomics* as a cover phrase for the activities described in this text.

1-3 ERGONOMICS IN THE INDUSTRIAL AND SERVICE SECTORS

Ergonomics is of importance when it can provide answers to problems. The problems may be well known and of long-standing duration, they may have been recently identified, or they may yet be undiscovered. When the practice of ergonomics can aid in the resolution of these problems, it has value to the organization and to the people in that organization.

Traditionally, there are two methods to use for measuring improvements in the industrial setting. One method is tangible, and it is often thought to be the measure of more importance to the organization. On the tangible scale, dollars are used to equalize comparisons. When choices can be put into dollars, it often becomes easier to make a decision concerning these choices. On a tangible basis, then, it is important to be able to quantify choices—to be able to show that the resolution of a problem will result in fewer dollars being spent by the organization.

The other method is intangible. It is often thought that if choices cannot be stated quantitatively, they are of less importance. Yet in many situations, the intangible scale is thought to represent more heavily those choices of importance to the worker. For example, increased safety might be cited as a reason for making a change. This, if true, would clearly be a substantial benefit to the worker. Yet some would say that since it cannot be quantitatively assessed, it has little value. This, however, is a fallacy and the organization that realizes it as such will soon benefit. One must learn to distinguish between value and the ability to measure. In some cases, the value is there but it is just not worth much to develop the quantitative measures necessary to prove it.

The important point here is that the objectives of the organization and the objectives of the workers are not in opposition. It is a major benefit to the organization to have safe, healthy, and productive workers. The costs to the organization for injuries and illnesses are very real and substantial. Ergonomics, by helping provide improvements to this work system, can clearly be seen as the ally of both the organization and the workers.

1-4 THE PHILOSOPHY OF APPLICATIONS

The world of industrial ergonomics is the real world. It is worth the time to outline some of the constraints of the real world and to show how these can be turned into a philosophy regarding the application of industrial ergonomics. This philosophy will carry through the book and will serve as a guide for many of the ideas discussed.

1-4-1 Workable Solutions

Within the industrial and service sectors, solutions to problems are needed that can be implemented and that will work. It is of little value in the real world merely to be aware of a problem. Results are obtained from the implementation of solutions to these problems. For example, an area may have had a higher-than-normal injury rate for the past few years. Only when something is done to control that injury rate is there value to the organization and to the workers. Here the ultimate objective is the improvement itself, not the use of a particular tool or technique in the solution of that problem. Therefore, industrial ergonomics can be of value only to the extent that it can provide solutions to problems—solutions that can be implemented, that work, and that result in improvements.

1-4-2 High Tech, Low Tech, and No Tech

In the real world of industrial ergonomics, the solutions may or may not be technically sophisticated. One certainly does not have to make detailed analyses of each situation to realize that improvements are needed. For example, an operation or job may have an injury rate three or four times the plant average. With this high of a rate, a detailed analysis to show statistical significance may just slow down the solution. It may be more important to just get on with the business of making the improvement to the job. The solutions to problems need not be highly technical either. For example, a detailed analysis to place safety shower handles at just the right height (to allow shorter people to reach the handle and to allow taller people the clearance to avoid hitting their heads) may be unnecessary if one places the shower head high enough to avoid the tallest people and uses a rope to activate the shower, thus meeting the needs of the shorter-statured people.

1-4-3 On Time, All the Time

A common saying, "time is money," holds true in the industrial and service sectors. The timeliness of an answer can often be of critical importance. When an important new design is being prepared, many design criteria must be pulled together. It is rarely feasible to delay the total design while seeking up-to-the-minute technical information, regardless of the discipline. When faced with the choice to delay an important and money-saving project while waiting for more information, or to proceed using data on hand, the tendency is to make do with the best information already available. Since projects are designed to increase sales or reduce costs, the delay of these projects can be expensive. Generally, many people are willing to trade a little accuracy for a timely answer.

1-4-4 Multiple Criteria

Within the industrial and service sectors, there are many design criteria other than ergonomics. Although the consideration of ergonomics is important, one must realistically understand that a designer or manager is constantly in the business of trading off one factor against the others. (One definition of management is the allocation of scarce resources.) It seems as though the constraints of space, time, and money are always cropping up. To effectively design for people, one must understand the importance of the other criteria. Failure to do so will only result in frustration and poor design. Only by realizing that there are many other criteria, and that these other criteria are more easily understood and more widely practiced, is one really ready to practice industrial ergonomics design. At that point one can begin to develop the perspective to see the whole picture and to understand what is important in each design. With this understanding, one can then propose sound designs that won't get shot down or develop the sales pitch that gains ergonomic criteria the proper importance. The real world does not revolve around ergonomics—it never has and it never will. The realization of that fact is, however, the best way to face that uphill battle.

1-4-5 Best Solution Not Always Chosen

The correct ergonomic solution may not be chosen for any of a number of reasons. Even if one develops the best ergonomic solution to a problem, that solution may not be chosen. An example of this can easily be found by turning to the automotive industry. Maintainability of automobiles is decidedly easier when one has access to the part needing service. However, for years the industry has chosen to pack the engine compartment with equipment. The lack of accessibility has become legend, yet changes are very slow to come. This is a case where improvement is recognized but is not implemented due to other, seemingly more important criteria.

1-4-6 Recommended Actions

It is best to give firm recommendations for the solutions to problems. As the specialist, your advice is sought because you have an understanding of the information that is needed. By failing to give a firm recommendation, you are abdicating that judgment to someone else; and that someone else will rarely have the knowledge to put all the information in the proper perspective. If a decision is to be made, the person with the most and the best information should be the one making the recommendation. Therefore, make a firm recommendation, with the understanding that if that is not acceptable, additional discussions will take place and further recommendations will be developed.

1-5 THE USE OF THIS BOOK

The purpose of this book is to promote the application of ergonomics in industry. Ergonomics will be practiced in industry when it can provide practical solutions to real problems. Through the promotion of practical applications, it will become increasingly easier to gain acceptance for the discipline.

This book is also dedicated to the development and use of sound approaches to the practice of industrial ergonomics. The approach that one uses may have a strong correlation with the success of that application. By the promotion of strong fundamentals, and the use of proven techniques, applications will become sounder.

It is important to learn the proper use of ergonomic data. This book will show how that is done. The inappropriate use of information can lead one to an incorrect solution to a problem. Even the most common data, data about human body dimensions, are very easy to misuse and misinterpret. In addition to promoting the proper use of ergonomic data, this book provides sources of ergonomic data. Reference material is noted throughout the book.

There are many professionals who need to understand the practice and management of industrial ergonomics as part of their primary job duties. Such professionals include industrial engineers, industrial hygienists, occupational nurses, safety professionals, design engineers, development and improvement engineers, and quality professionals. Figure 1-1 illustrates the concept of designing a reference for the practice and management of industrial ergonomics, rather than designing another book for the full-time ergonomics professional.

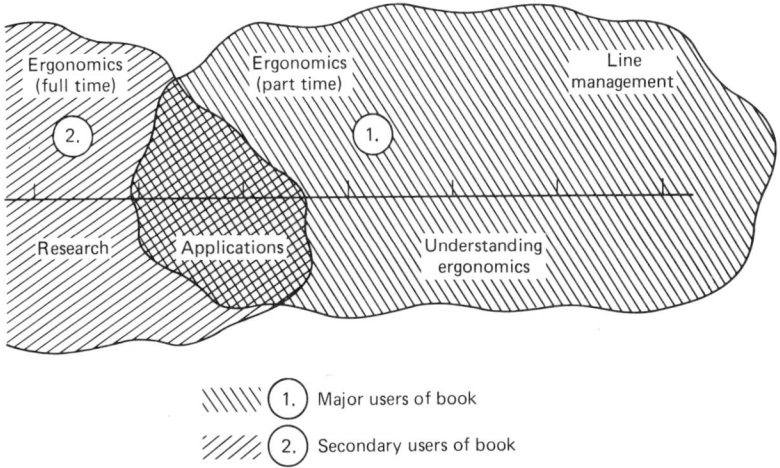

Figure 1-1 Primary users of industrial ergonomics information.

1-6 SUMMARY

The practice of industrial ergonomics is a relatively new phenomenon. For it to become more widespread, it is important to have practical applications. The practice of industrial ergonomics is not impossible, but must mold to the form already in practice in industry. With an understanding of how design and improvement are carried out in this setting, it becomes easier to have applications of this discipline. Proper information and the proper use of that information are very important. Obtaining that information and learning the approaches that make the applications successful is the goal of this book.

QUESTIONS

1. Why does the preference for "an inside job with no heavy lifting" reflect ergonomics concerns?
2. List the major occupational health and safety problems related to poor ergonomics design.
3. What are the major goals of industrial ergonomics?
4. Define *industrial ergonomics*. What sciences are utilized in applying industrial ergonomics?
5. List five key concepts in the application of industrial ergonomics in the "real world."
6. List five disciplines, professions, or areas of industry with a need for information from the field of industrial ergonomics.

Chapter 2

THE APPLICATION OF ERGONOMICS IN INDUSTRY

This chapter provides a framework for the application of ergonomics in industry by discussing when ergonomics should be used. Prevention of problems and correction of problems are both fertile areas. The fields of safety and health, productivity, cost, and human comfort are all suited to the practice of ergonomics. In this chapter, six general types of problems that ergonomics can deal with effectively are discussed. The depth of knowledge required for different questions is also discussed. Finally, the role of the industrial ergonomist is presented.

2-1 INTRODUCTION

The question is often asked, "When do you use ergonomics?" The correct answer is, of course, "Whenever people are present." Ergonomics is the science of people at work. Therefore, ergonomics is important and useful in those work situations when people are present, such as when people operate production systems and when people repair and service equipment.

Theoretically, ergonomics, like most technology, becomes cheaper with increasing applications, as shown in Figure 2-1. The first "ergonomics study" is likely to be an expensive venture. Not only must the problem be solved but the technology must be learned and understood and the data must be located and researched. Typically, as the number of applications increases, the commonality of approaches and the repeated use of data will allow these costs to drop. This chapter is intended to provide an overview of the types of problems

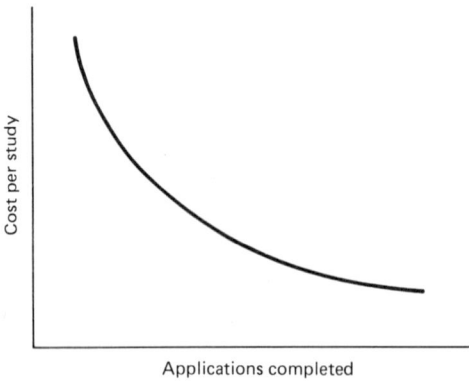

Figure 2-1 Ergonomics costs per application versus the dollars per application.

that one may encounter and the details of questions asked about those problems. This should both broaden the scope of potential applications and shorten the learning process, thus allowing the cost-effectiveness of industrial ergonomics to be reached that much sooner.

2-2 PREVENTION VERSUS CORRECTION

Historically, the practice of ergonomics has been one of correcting existing problems. The very foundations of ergonomics are rooted in this area, first with the elimination of airplane crashes in the military during World War II, and later with the improvement of safety and health conditions in industry.

Once problems became obvious, it was important to correct them. The use of ergonomics personnel for these duties was not often or seriously questioned. With the improvements made to operations as the result of these investigations, the concept of the prevention of the problems received increasing attention.

Unfortunately, it is sometimes difficult to fix something if it isn't broken, or more appropriately, if no one knows that it is broken. That seems to be the case with a number of areas that deal with safety and health. Money is spent on the analysis and correction of problems (when the cow is already out of the pasture), but often that same money is not available to prevent those problems initially (to close the gate). Certainly, there are limitations on financial resources, and this does prevent unlimited study.

With the science becoming sophisticated enough to predict many operating problems, there is increasing emphasis on the prevention of ergonomic problems. Many organizations now have specialists in this field, with the primary duty to prevent system-induced operational errors. Anticipating and preventing problems is the best mode to be in, and although the trend is slow, it is moving in that direction.

2-3 GENERAL AREAS FOR IMPROVEMENTS

Ergonomics, as it is practiced in industry, can have an impact in three separate areas:

1. Employee safety and health concerns
2. Cost- or productivity-related fields
3. The comfort of people

Depending on the location of the ergonomics function within each organization, one of the three areas may be emphasized over the others, but all the areas are familiar to those who practice ergonomics.

Employee safety and health concerns are those where the answer will affect the safety and/or health of those performing the tasks involved. In these situations, the primary desire is always to correct the situation, with cost being a secondary concern.

Cost- and productivity-related areas are those where the application of ergonomics can help minimize the operating costs and/or increase productivity, by ensuring the effectiveness of the personnel on the job. The ergonomist helps to weigh the costs of the various alternatives and to choose the most effective one. The criteria to balance are the avoidance of health and safety concerns, yet minimizing costs and/or maximizing productivity.

Comfort-related ergonomics problems involve the comfort of the operator. Since there are no health- and/or safety-related concerns, the cost of any improvements will receive the most scrutiny. These costs will be weighed against any benefits, such as morale improvements, that are expected to occur. An illustration will demonstrate these three areas of concern and how the ergonomist should deal with each.

Heat stress is common in many industrial situations. Once it comes to the attention of the ergonomist as an operational problem, the first question that must be asked is: Is it a health hazard? If it is a health concern, the situation must be corrected or avoided, whatever the costs involved. (It should never be policy to compensate someone for working in an area involving safety and/or health hazards rather than correcting the hazard. The hazard must be dealt with in every way possible.) There may be decisions about which of several actions to take to improve the situation, but there must not be a decision about whether to take any action. Some of the solutions to this health problem are to cool the area, to increase air flow, to slow the pace of the task, to add workers, and to increase the recovery periods.

If a similar situation becomes apparent, but in which health problems are not anticipated when the work pace is slowed or extra recovery breaks are provided, the situation is one of productivity and/or cost. In this situation, the ergonomist will be concerned about the productivity coupled with cost control. (The balance of the criteria will shift, yet health problems must still be avoided.) The solution may be to change nothing in the current operation, or to change only those factors that can be cost justified. For example, if the costs of extra cooling (or any other improvements) can be offset by increased productivity, extra cooling should be implemented. The decision is primarily financial and can be measured economically.

A third situation may also occur. This situation may be similar to the others, except that little or no increase in productivity is expected from the improvements. In this case, there are no health concerns. The decision to implement the improvements is one of examining the costs relative to intangible benefits that are expected. Since this situation improves comfort factors, the decision is primarily judgmental rather than quantitative.

In summary, depending on the situation, sometimes action is definitely required. In other cases, the decision can be based on economic criteria. In the remaining cases, the decision is based on intangible benefits.

2-4 TYPES OF PROBLEMS

The categorization of industrial ergonomics problems can often be an aid to their solution. The different types of problems have different solution methodologies. By classifying a problem into one of six basic groups, the beginning of a solution is more easily undertaken.

Since ergonomics is the study of people, it is convenient to think of industrial ergonomics problems by the type of body system that is affected. Different types of ergonomic problems affect different body systems, so that is a useful categorization for ergonomic concerns. For example, there is a significant difference between a problem dealing with back injuries from overloading and a problem that results in operating errors from too much data. The six types of problems are discussed below.

2-4-1 Physical Size: Anthropometric

Anthropometry is the science dealing with the measurement of human body dimensions. Measures of stature, body weight, arm reach, and eye height are just a few of the body dimensions that are available through the science of anthropometry. Anthropometric problems are those that deal with a physical conflict between the person and some aspect of the work area. Problems with the workplace and work space are the most common anthropometric problems. These problems are characterized as problems involving fit, where someone is too large or too small to fit the equipment. These problems can be resolved with the use of data about expected populations and the work area that will accommodate them. Additional details are given in Chapter 7.

Figure 2-2 Anthropometric problem.

2-4-2 Endurance: Cardiovascular

Endurance problems are characterized by the stress that they place on the cardiovascular system. This stress may come from a heavy job requiring extensive physical effort and the resulting need to get oxygen and energy to the muscles. Similarly, heat stress can result in the need for extra blood flow to cool parts of the body, causing increased circulatory load with an accompanying elevation of the heart rate. In both of these cases, the concern is from the elevated heart rate and increased blood pressure, which can result in a heart attack or stroke. Endurance problems can result in the establishment of work/recovery cycles or of standards for a fair day's work. Designing tasks and jobs to conform to a predetermined expenditure of energy (or kilocalorie expenditure) is a suitable way to resolve endurance problems. Additional details are presented in Chapters 6 and 9.

Figure 2-3 Endurance problem.

2-4-3 Strength: Biomechanical

Strength problems are often characterized by the need for large muscular effort. Typically, strength problems manifest themselves through injuries on the job. Although most of the work in this area is in the correction of existing problems (those that have resulted in injuries), there is increasing effort to design new jobs so that there are few stressful lifting situations present. The identification of such problems before an injury occurs is a good preventive measure, although it is difficult to predict all injury-prone situations with a high degree of certainty. Strength problems can be analyzed through biomechanical techniques or through the psychophysical methodologies used by Stover Snook. When a strength problem is identified, the use of strength information can help to assess the severity of the problem. The same information can help to predict the success and suitability of alternative solutions. Additional details are given in Chapters 6 and 8.

Figure 2-4 Strength problem.

2-4-4 Manipulative: Kinesology

Manipulative problems are characterized as an inability to perform the fine motions required on the job or a difficulty in performing tasks at the required speed. These problems usually show up in assembly tasks or tasks requiring fine control of dials and instruments. Assembly errors, alignment problems, and dropped pieces are the outcome of manipulative problems. Additional details are presented in Chapters 6 and 11.

2-4-5 Environmental: External

Environmental problems are those which involve the surroundings of the worker. Heat stress, cold stress, lighting, noise, and vibration are the typical environmental problems that one will experience in the practice of industrial

Chap. 2   The Application of Ergonomics in Industry   15

Figure 2-5   Manipulative problem.

ergonomics. When there are problems between the person and the environment, usually some changes must be made to the environment, the work regimen must be altered, or physical training such as heat acclimatization must be introduced. Additional details are given in Chapter 10.

Figure 2-6   Environmental problem.

2-4-6 Cognitive: Thought

Cognitive problems typically show up as operating errors of some type. The limits of short-term memory and the associated difficulty with long and complicated strings of numbers are common cognitive difficulties. Similarly, perceptual problems associated with vision and hearing can result in errors. When situations occur that violate the human mind's ability to function properly, these errors can be predicted. These problems should not be confused with motivational problems, since system-induced operating errors will occur even with highly motivated workers. Additional details are presented in Chapters 10 and 11.

Figure 2-7 Cognitive problem.

2-5 DEPTH OF KNOWLEDGE

Within the application of ergonomics to industry, there are different questions that can be asked that require varying depths of knowledge of the topic. It is important to understand this fact and to use its strength in your particular ergonomics program.

Some questions do require a detailed analysis and a thorough understanding of the field of ergonomics. Many other questions, however, can be answered simply and routinely. This is the point to exploit, so that your time can be used effectively.

1. There are questions that merely require an answer. For example, one question might be: How wide does an aisle need to be to allow clearance for two people to pass each other without touching?

These questions are self-contained, usually have quantitative answers, and occur frequently. Since answers must be provided for designers, it is possible to provide the answers without direct contact, to save consultation time. A published reference book may suffice, the answers can be built into engineering design standards, or in-house design guidelines can be provided.

These questions are the easiest to answer and offer little room for incorrect interpretation, so they are excellent candidates for design reference books. They can be put in this form with almost no trouble.

2. There are questions that require the judgment of prior experience and

a sensitivity for the field of industrial ergonomics in making suitable recommendations. One typical question is: How much should a person lift?

These questions usually require additional information before an answer can be provided. The quality and accuracy of the additional information will determine the effectiveness of the answer. (Usually, with these types of questions, the same facts will produce the same answer. There is some judgment, but not too much.) The answer is normally situation specific and will have only the most general transferability to other situations.

Often, some experience in the field of ergonomics is required to answer these questions. If it is necessary to have the answers to these questions available in reference books, then extra care will have to be provided to ensure that the information is provided in table form with the options clearly outlined. Otherwise, the designer must be capable of and fully prepared to read and understand a narrative that explains the possible choices in detail, including the pros and cons of the choices and how to make the choices.

3. There are questions that require extensive experience to provide the detailed analysis needed before the answer can be formulated. An example of this type of question, "Is there a heat stress problem, and if so, what should be done to economically correct it?", illustrates the complexity of these questions.

These questions usually require a detailed understanding of the practice of industrial ergonomics plus an understanding of the situation to be analyzed. These answers are almost always situation specific and have little transferability to other situations. To answer the question, there is a need for additional information about the task and about the people and their capabilities. The answer will depend on the quality of information available, the analysis performed, and the experience of the ergonomist.

It is uncommon for a novice to be able to answer this third type of question in detail. Reference books are generally unsuitable as an aid in answering these questions unless the designer is quite dedicated and has ample time to read, study, and analyze the situation. This amount of time is not often available to the initial designer in industry, so if in-house ergonomics expertise is not readily available, a consultant should be retained.

2-6 THE ROLE OF THE ERGONOMIST

The industrial ergonomist can play several roles. It is useful to have an understanding of these roles and to know the ones that you may be expected to perform or that you can ask of others.

An ergonomist may be a designer and actually specify the design of products, equipment, facilities, or jobs. They usually have latitude in their work and the measure of their work is the completion of designs.

The ergonomist, working as a consultant, provides answers and analyses to other designers and to managers. For example, the consultant may be asked to provide information on aisle widths necessary for two people to pass. The consultant provides information but has little control over its use. A designer may choose to design the aisles to the specified width or may use a wider or narrower aisle; the choice rests with the designer and is balanced among other important design criteria. The consultant may also provide answers to questions asked by line managers. Again, the answer is only a recommendation and may or may not be used by the manager.

In the role of consultant, it is much more difficult to have influence, since there is no requirement for the designer or manager to use the information. In addition to developing the answers requested, the ergonomist is also faced with the added burden of presenting them so that they are utilized.

Although there may be requirements to use "ergonomics in design," that is done more fully when the designer clearly sees the need to design for the human being. The sales aspect of consulting is important and should not be overlooked.

The part-time ergonomist is a third role, yet it includes parts of the designer and consultant roles. The part-time ergonomist has the ability to include the practice of ergonomics into his or her everyday job. This role clearly affects the application of ergonomics and should be encouraged for as many people as possible. The occupational health nurse who asks questions about a lifting injury to help diagnose the problem, the industrial hygienist who gets involved in identifying ergonomics problems while in the work area, and the industrial engineer who includes ergonomics in job designs and layouts are examples of the part-time ergonomist. This is a powerful role and should be promoted.

2-7 SUMMARY

This chapter discusses the application of ergonomics to industry by discussing the types of questions and answers required in that setting. This understanding is important in building a framework to support the ergonomics effort and to ensure its success. The information in this chapter, covering prevention versus correction of problems, areas of application, general types of problems, and depth of knowledge, can be utilized in building the framework for Part 2, resulting in a complete program being planned and developed.

QUESTIONS

1. When do you use ergonomics?
2. Why do the solutions to ergonomics problems become cheaper with increasing applications?

3. Why is prevention of ergonomics problems more difficult than correction of known problems?
4. List three general levels of criticality for ergonomics problems.
5. Illustrate the three general levels of criticality for ergonomics problems with a heat stress problem. Discuss solutions required.
6. List and illustrate the six different types of ergonomics problems based on body systems affected.
7. Discuss the three increasing levels of difficulty for ergonomics questions that are presented to the ergonomist.

Part 2
STARTING THE ERGONOMICS EFFORT

Chapter 3

PLANNING AN INDUSTRIAL ERGONOMICS EFFORT

Part 2, which consists of Chapters 3 through 5, covers the initiation and start-up of an industrial ergonomics effort. It begins with planning the effort and all the things to consider. The implement plan is developed and implemented, and problems are identified. The problem identification block is an appropriate lead for Part 3, which deals with the solution to problems.

This chapter details the plans and decisions that one must make when starting an industrial ergonomics effort. That process begins with the types of application that will be pursued. This leads to the benefits of ergonomics, the organization needed, and the resources used. Finally, selling the completed concept to management is discussed.

3-1 THE BEGINNING

An engineer wrote me last year saying that he was interested in developing an ergonomics program and wanted to know how to begin. "Should I begin with a few projects, or should I develop a proposal for top management?" he asked. "What areas should we work on? How much time will it take? What resources do I need?" These are excellent questions, and they provide the framework for planning an industrial ergonomics effort.

Planning is important if the ergonomics effort is to function as effectively as possible. Rarely, however, does someone from management come to you and say: "Spend the next month planning our new industrial ergonomics effort." Often, the planning is the result of those with a background in safety and health realizing the need for the focused expertise that this discipline can provide. Therefore, planning is not so much a formal sitting down and writing out plans as it is a collaborative series of discussions over a much longer period.

Frequently, the impetus to start an effort of this type will come from one of the following sources:

1. An article in a trade magazine or technical journal covers some useful aspects of ergonomics and gets distributed within the organization.
2. A competitor successfully uses some aspect of ergonomics and obtains a marketing advantage because of it.
3. A trade association provides information about ergonomics and develops some curiosity and interest at the upper management levels.
4. An accident or injury results from failure to consider the proper use of ergonomics, and interest develops overnight.
5. A conversation with colleagues turns toward a new way of better utilizing the person in the production process.
6. A recent graduate begins to talk about some applications of ergonomics that seem to have a real benefit in the operation of the business.

There are many ways in which the seeds for the industrial ergonomics effort can be planted. Some are fortuitous, some are the normal evolution of business, and some are the natural spreading of technical knowledge about any new discipline. Some efforts, however, begin as the direct result of efforts to promote the discipline. A concentrated effort by a few people can result in the knowledge and understanding that cause the ergonomics effort to happen. You can plant the seeds that result in applications of ergonomics in your organization.

Regardless of how the opportunity to work in the area of ergonomics presents itself, there are some important factors to be considered. These include:

- Types of application
- Types of studies or projects
- Placement within the organization
- Types of resources that may be needed

There is a logical way of thinking through these interlocking factors and the associated decisions. The heart of any effort is the types of applications that are pursued. This, in turn, determines many of the other factors that need to be considered. The major studies depend on the applications. Similarly, the placement within the organization should be a result of the studies anticipated. The resources needed also depend on the applications and the studies expected.

3-2 TYPES OF APPLICATIONS

There are three basic ways of framing the work that is done. The three general types of applications are (1) in-plant studies, (2) design of facilities and equipment, and (3) product design.

3-2-1 In-Plant Studies

These studies usually center on improvements to existing operations in manufacturing, office functions, or other areas. In an address to the Human Factors Society, Richard Lucas makes the point quite well:

> There are also some Human Factors departments in industry with "process approaches." Here, the objective is to incorporate past experience and experimental evidence . . . into the jobs of people involved in the manufacturing act itself—to "build a mousetrap better.". . .
>
> Process oriented recommendations, if implemented, should make the job a better one. They will make the company a better place to work. They may open up jobs to more employees, making a company more flexible and maybe, just maybe, make it more efficient and productive. Fine, but do these justify spending dollars on less-than-tangible benefits? The answer is yes.
>
> The aim is to protect human assets as we would any other valuable part in our production process.[1]

Typical in-plant studies involve:

1. Analysis of accidents
2. Work methods improvements of operations
3. Analysis of inspection tasks
4. Analysis of potential heat stress areas
5. Investigation of hot, heavy jobs
6. Investigation of operational errors
7. Job design

This is a traditional area of work for ergonomists. In Europe, work of this type has developed at a more rapid pace than in the United States. It is not uncommon to find applications well developed in Europe before the need even becomes widely known and understood in the United States.

---

[1]Richard L. Lucas, "Human Factors in Industry—Getting the Point Across," *Proceedings of the Human Factors Society, 1984* (Santa Monica, Calif.: Human Factors Society, 1984), pp. 731–732. Copyright 1984 by The Human Factors Society, Inc. and reproduced by permission.

3-2-2 Design of Equipment and Facilities

This area refers to the design of in-house equipment and limited-use items. Often, this is done by in-house engineering design staffs. High-volume product design is discussed next.

The design of equipment and facilities is often improved by the use of knowledge from the field of ergonomics. Much of the traditional emphasis in the United States has been in this area, particularly in the military and aerospace fields.

Many years ago, tools and equipment developed slowly through generations of workers, allowing evolution to refine the designs to the point where the equipment was suitable for all people. Usually, by the time the tool or equipment had widespread usage, it was refined enough to "fit the user." With today's emphasis on the immediate utilization of new technology, there is no longer time for this trial-and-error process to take its full course. In many situations, from manned space flight to military operations, testing takes place literally under fire. By realizing the capabilities and limitations of the human being during the design process, many mistakes can be avoided. Typical equipment and facilities design projects include:

1. Design of equipment controls and displays
2. Layout of equipment for ease and speed of operation
3. Design of equipment so that parts to be serviced and maintained are accessible to the human being
4. Layout of facilities so that there is adequate room for one or more human beings

Initially, these applications will show the benefits and costs of design improvements, but that will eventually change. Beevis and Slade point out, and very appropriately, that "as ergonomists become more fully involved in the design of new tasks and equipment, rather than in the redesign of existing ones, there will be less opportunity to make such before and after cost comparisons."[2]

There are a number of texts that provide information to the designer of facilities and equipment. In fact, the traditional human factors engineering design book is oriented quite heavily in the direction of equipment, facilities, and tool design.

3-2-3 Product Design

Product design is the third application area for industrial ergonomics. Product design is similar to the design of in-house equipment. Both are con-

---

[2]D. Beevis and I. M. Slade, "Ergonomics—Costs and Benefits," *Applied Ergonomics*, 1970, pp. 1.2.79–84. Published by Butterworth Scientific Limited., Guildford, Surrey, UK.

cerned with the design of equipment for a wide variety of users. Product design, however, has some special characteristics. With consumer products, the manufacturer has little control over the end use of the product. The producer cannot control the skill level of the user, the placement of equipment, the raw materials used, or subsequent modifications to the product. Under these conditions, ergonomics is even more important.

Product design is a relatively new application area for the practice of ergonomics. However, it is in this field that many people first become acquainted with ergonomics. The ease of use of products is readily becoming associated with ergonomics, all the way from automobiles to CB radios and lawn mowers. It is now common to find advertisers touting the ergonomics aspects of their designs. For years now, Volvo and Mercedes have listed ergonomic designs as an important part of both the comfort and safety of their automobiles. Other automakers are also finding it worthwhile to include this aspect of design in their advertising. The computer industry, looking for a marketing aspect the consumer can understand, has widely promoted the value of ergonomic design in their hardware (keyboards, tables, etc.) and in their software ("user friendly").

Weiman explains the use of ergonomics in product design: "Twenty years ago, when electronic instruments were simple to use, there was little interest in applying human factor principles to products. Now, however, as products become more complex because of increased capabilities, there is a real need for human factor specialists to play a key role in the design and final use of those products."[3]

The ergonomist can help design initial prototypes, and test and evaluate mock-ups of initial designs. Analysis of accidents and misoperations can result in further improvements. Design of labels and instructions can easily be done by someone trained in this area. The uses of human factors in product design and development are also expanded to include providing human factors guidelines for initial design, product evaluations on existing instruments, and prototype evaluations on future instruments.

3-2-4 Summary

Based on these descriptions of three general areas of applications, it is important to define the possible areas of applications within your organization. Some operations will involve two or even all three of these areas. The design of your ergonomics effort will result from this decision, and the work

---

[3]Novia Weiman, "Integrating Human Factors into Product Development within Industry," *Proceedings of the Human Factors Society, 26th Annual Meeting 1982* (Santa Monica, Calif.: Human Factors Society, 1982), pp. 731-732. Copyright 1982 by The Human Factors Society, Inc. and reproduced by permission.

developing the ergonomics effort will be directed toward the applications that are defined.

3-3 OPPORTUNITIES FOR IMPROVEMENT

In addition to defining those areas where the applications of industrial ergonomics are possible, it is also important to look at the opportunities for improvement in each area. Those opportunities are your bread and butter. They determine the attention and resources that top management will be able to provide to this effort. If the opportunities are large, it will make a tremendous difference in the resources, such as staffing, provided in this area.

An opportunity for improvement can be defined as any situation where improvements can be obtained. This definition is intentionally inclusive. It includes problem solving in the traditional sense, since solving the problem will obviously result in an improvement to the organization. But it also includes those areas where no apparent problems exist, but where improvements can result in benefits to the organization, its equipment, and its products. A surge of back injuries is an obvious problem that should be corrected; however, a low rate of injuries over a longer period of time is just as expensive but not as readily apparent. Both types of improvements abound; thus it is important to broaden the thinking when determining the ways that ergonomics can benefit your organization. Descriptions of some organizational benefits are provided. They are by no means all-inclusive, but merely serve to show the kinds of opportunities that may be expected.

3-3-1 Production-Cost-Savings

Some of the more common opportunities for improvements lie in the actual production area. Following are a few example areas:

1. With better work methods, productivity improves.
2. By improving accessibility, the time to repair equipment is reduced.
3. Improved inspection results in fewer "double checks" and/or fewer shipping errors.
4. Better control systems result in fewer operational errors.
5. A better work environment results in the need for fewer work breaks.

3-3-2 Accident and Medical Costs

The costs associated with accidents and personal injuries are quite large. In many cases, particularly with injuries and lost time, these costs are not fully charged to the task or process where the injury took place. This may stem from a somewhat erroneous belief that accidents are a random occurrence or depend more on the people involved than on the production process.

A more suitable way of fully allocating the injury costs is to pass them back to the process involved. In this way it becomes apparent where the injuries are taking place (these really do increase the cost of products and services) so that improvements can be made. Allocating costs to the appropriate area ensures that production management is an advocate for improved operations. Solving these problems will then result in lower costs of operation. Hiding the costs prevents detection and eventual solution of these problems.

Some of the costs in this area are as follows:

1. The cost of the injury, including medical treatment, lost work hours, and any makeup wages paid to supplement disability income
2. The costs of long-term disability paid to a worker no longer able to work
3. The costs of workers no longer able to work at the 100% level on a job, such as not meeting the pace or not being able to perform all the tasks
4. The cost of damaged equipment
5. The cost of damaged or lost products
6. The time to get production operating again after an accident or injury
7. The time for review meetings, accident investigations, and reports

The area of accidents and personal injury is a very expensive one to both the individual and the organization. By portraying the true costs, the mechanism is put into place to begin the corrective process.

3-3-3 Design and Development Engineering Time and Costs

One area that may be overlooked in the search to justify the practice of industrial ergonomics is the cost of redesigning equipment, jobs, and facilities. These costs are large, because of (1) the design personnel involved, (2) the cost of transition to the new design, (3) the losses involved in using the old design, and (4) the costs associated with implementing the original design, especially when it has to be torn out and redone.

A better way is to design it right the first time and avoid the trial-and-error method. For any design that requires a human being for operation, service, or maintenance, the "specifications" of the human user should be incorporated into the original design. Only with this knowledge, like any other information that the designer needs, can the proper design be executed. That is the reason we need the designer to begin with: to get it right the first time.

A few brief examples of the information that can help a designer follow:

1. The proper clearances needed for maintenance
2. The proper height of workplaces

Chap. 3   Planning an Industrial Ergonomics Effort   27

3. The work pace used in production operations
4. The need for cooling workers
5. How people react to various information coding systems
6. How much people should lift

It is not easy to think of all the ways that time is lost in the design process, but the area of redesign for lack of proper knowledge about the human operator is one that is more common than it needs to be. As such, it is an opportunity for improvement that must be explored. Keep track of these costs as you redesign projects, for they can be the basis of projecting savings associated with involvement in new design work, even after there is an understanding that ergonomics is required in the design of new processed and equipment.

Schulz and Pine discuss this extensively. They point out some of the special challenges that redesign efforts require.

> Design alternatives are more limited, and those that remain are likely to be more costly to implement as retrofits than to include in a new design.
> Stricter criteria for adoption of alternative designs. Unless the system fails utterly, the original design will have already demonstrated some degree of operability, giving it a strong apparent advantage over untried alternatives.
> Potential for negative transfer of training. Any change affecting human-machine interaction carries the risk that "old" human responses may recur which are inappropriate to the new design.
> All of these factors act to limit the extent of changes which can be made, creating a risk that the enhancement program ends up with an uncoordinated collection of quick fixes—"band aids."[4]

3-4 ORGANIZATION

Ergonomics is a multidisciplined science. As such, there is no one "right place" for an industrial ergonomics effort. It is important to realize the various groups that will interface with and support an ergonomics effort. In the end, one of these groups will house the effort and the others will have strong ties with the effort. Experience has shown that the following groups or functions are closely allied with the practice of industrial ergonomics:

1. Industrial engineering
2. Medical

---

[4]Kenneth A. Schulz and Steven M. Pine, "A Human Engineering Approach to the Enhancement of Operational Systems," *Proceedings of the Human Factors Society 26th Annual Meeting, 1982* (Santa Monica, Calif.: Human Factors Society, 1982), pp. 727-730. Copyright 1982 by The Human Factors Society, Inc. and reproduced by permission.

3. Industrial hygiene
4. Safety
5. Design engineering
6. Materials handling engineering
7. Product engineering

Each of these groups will occasionally have the need for the expertise that ergonomics can provide.

Earlier, three types of applications—in-plant studies, design of facilities and equipment, and product design—were discussed. The foundation of the whole program begins with this need. The type of applications planned will now help to determine the placement of the ergonomics effort in the organization. If there are a lot of "in-plant studies," the effort may naturally fall into the industrial engineering area. If cost reduction is heavily emphasized, industrial engineering will also be a suitable home.

If identification of in-plant injury-producing situations is of primary importance, the medical department will probably be the proper location for such an effort. Should the effort emphasize the improvement of occupational health (heat stress, environmental contaminants, chronic injuries) or occupational safety (traumatic injuries, misoperations, biomechanics), the effort should probably be located, respectively, in the areas of industrial hygiene and safety.

With an emphasis on the design of in-plant equipment, facilities, or layouts, the effort may be more at home in the design engineering department. Materials handling engineering may emphasize the entire spectrum of handling material, from both the equipment and the human standpoints. With this broad scope, ergonomics is certainly an appropriate part of the materials handling job. If the emphasis is on the design of new products, the location of an ergonomics effort in the product design department is certainly in order.

Another criterion for the location of the ergonomics effort is well stated by Singleton. "There is one consideration which should override all others in deciding the right place for the ergonomist. The head of the department or some other senior person should understand what the ergonomist can do, and how he does it, and should be prepared to support him against the men of action who will inevitably want an immediate answer to every man–machine problem."[5]

The placement of the ergonomics effort should be made to maximize its effectiveness. No other reason is quite as important. Thus, by defining the appropriate areas of need for the expertise that ergonomics can provide and by determining the strong areas of support, it becomes easier to place the

[5]W. T. Singleton, "Part 1, A First Introduction to Ergonomics, Chapter 1, The Industrial Use of Ergonomics," *Applied Ergonomics*, 1, December 1969, pp. 26-32.

effort in the proper place within the organization. In some large organizations, where many people are involved full time in the practice of ergonomics, two or more groups may be formed. For example, both Eastman Kodak Company and General Motors Corporation have chosen to separate their product design ergonomists and those ergonomists who lead "in-plant studies." Only with larger organizations is this level of specialization worthwhile.

In no case, however, should the group be limited to practicing in only one of the many functional areas. The location of the effort should ensure that the other areas will have access to the expertise, information, and guidance that ergonomics can provide. It may be appropriate to establish methods and procedures that will allow the group to practice in other areas. Another system that works well is to establish local "experts" so that the more common ergonomics questions arising in a particular area can be answered by personnel actually working within that area. Experience has shown that the easier it is to consult a resource, the more likely that a problem will actually be recognized and/or brought up to be solved. No one seems to enjoy continually going to others for help, especially if those persons are not convenient or readily accessible.

Fortunately, many of the individual requests for ergonomics design involve the same information. The Pareto principle tells us that 80% of the requests should involve about 20% of the knowledge. For that reason, the use of "tiers of knowledge" is very useful (see Figure 3-1). This requires an ergonomist to develop on-the-job specialists in the various areas of application. This specialist serves as an intermediate source of information for designers with questions. It works like this: A designer needs information about clearances for maintenance tasks. The designer first consults reference books, guidelines, or design standards for the information. If the information is not readily available in that form, the designer contacts the "local expert" or "ergonomics specialist" for assistance. In most cases this will resolve the problem, but if not, the ergonomist is contacted for assistance. This method has several advantages. It frees the ergonomist for the more technically difficult problems. It builds a core of knowledge throughout the organization. It is an efficient use of the organization's resources. It also works very well—more designs do begin to accommodate the human user.

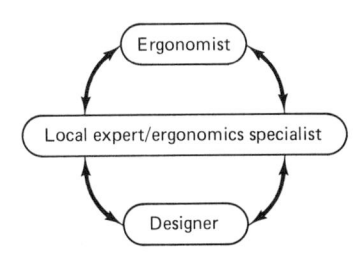

Figure 3-1 The "tiers of knowledge" concept.

3-5 RESOURCES

The resources for the ergonomics effort are the people, their equipment, and their information. The personnel staffing the industrial ergonomics effort are a very important resource. They need specialized background experiences or training, as well as ongoing training to stay current in the field.

There is a need for some equipment and instrumentation. Ergonomics problems often require detailed data collection to determine the best solution.

The third resource is information. It is not possible for one person to retain and recall all the information needed for the full range of industrial ergonomics activities. Most active industrial ergonomics groups keep a full range of books, studies, proceedings, and papers on topics relevant to their jobs and their applications. This information and the sources from which it comes are critical in keeping the effort up to date and accurate.

3-5-1 People

The people are the most important resource. They perform the studies and also utilize the other resources. Singleton lists the options open to the firm that decides to undertake an effort in ergonomics. "If a firm decides to make use of ergonomics, what is the next step? Should it set up a separate department and embark on a research programme? Few firms are in a position to do this, and in fact only in exceptional circumstances would it be desirable. There are several ways open to a firm: it can take ergonomics problems to its research association (trade association); or it can call in a consultant; or it can either recruit an ergonomist, or send a member of staff on a training course in ergonomics."[6]

It is important that any new personnel have the appropriate background or that the training be available. The "types of applications" come into play here. If these situations involve a large degree of in-plant industrial applications, someone with a background in industrial engineering may be more appropriate than someone with a background in the biological sciences. However, if the emphasis is on the prevention of back injuries and reduction of cardiac problems, someone with a background in the biological sciences will certainly understand the problems better. For equipment design, a background in traditional engineering disciplines with supplemental work in ergonomics might be desirable.

Although it is possible to cross-train someone, it is also expensive and time consuming. A good analysis of needs will point to the proper background of the person needed. There are an increasing number of opportunities for continued professional training in this area. The Institute of Industrial Engineers, the American Industrial Hygiene Association, and the Human Factors

[6]Ibid.

Society all have an emphasis in industrial ergonomics. In addition to journals and newsletters, there are annual conferences and programs, and on a less frequent basis, seminars or colloquiums. In addition, several universities offer programs and training in this area. Some that frequently offer opportunities for continuing education (and/or which have several faculty members involved in industrial ergonomics efforts) are the University of Michigan, Virginia Polytechnic Institute and State University, Texas Technological University, North Carolina State University, Auburn University, and Purdue University.

3-5-2 Equipment

The equipment used in industrial ergonomics need not be extensive. Since some quantitative data are needed, however, it is necessary to be able to make measurements. The areas of applications also dictate the specific equipment needs. If the problems are of the anthropometric or strength type, a tape measure and a push-pull gauge will be required. This will allow one to determine the spaces available at the work site or the actual forces needed on the job. Then, by comparing the capabilities of the population (see Part 3) with the requirements at the work site, it is easy to determine whether there will be problems or not. Figure 3-2 shows the tape measure and push-pull gauge used in industry.

If the questions are more complex, such as determining the physiological requirements to do various jobs, more complex equipment will be required. Even so, the equipment need not cost a lot. The worst thing to do is to purchase a full laboratory when it will not be used. If equipment is needed, it can often be borrowed or rented from a nearby hospital, medical facility, or university. In addition to the use of the equipment, often trained personnel can be provided, resulting in less start-up time than might otherwise be necessary.

The basic physiological equipment is used to determine heart rate and oxygen consumption during the work and rest periods. A Max Planck respirator (Figure 3-3) is a self-contained "backpack" unit used to measure the volume (and to sample a designated percentage of the volume) of air expired by the person being studied. Several heart rate recorders and transmitters are available which will allow a complete recording of the person's heart rate. Although these are commonly used by cardiac specialists to examine heart wave patterns for an extended period of time, they do an excellent job of determining heart rates on the job. Figure 3-4 shows several heart rate transmitters.

Heat stress is also a common problem for the ergonomist. Temperature measuring and recording instruments are needed for heat stress work. It is desirable to be able to record temperatures for periods of time, so recorders should be utilized if possible. The instruments must measure ambient temperature, radiant temperature, wet-bulb temperature (to determine relative hu-

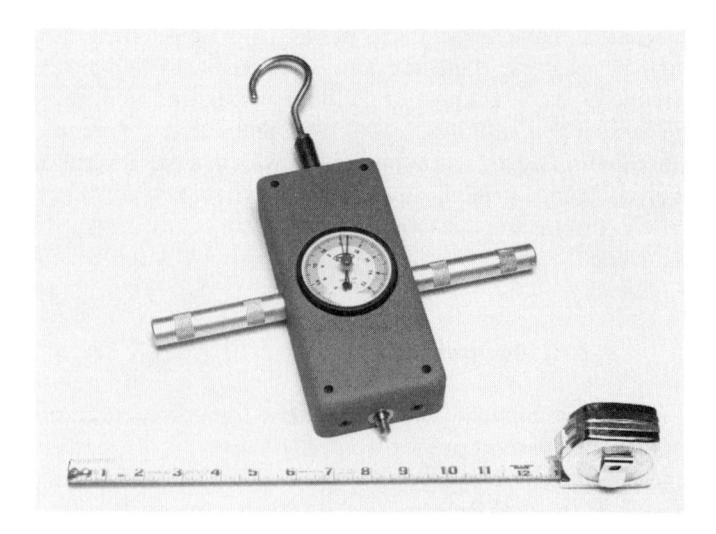

Figure 3-2 Tape measure and push-pull gauge. *Source:* Eastman Chemicals Division of Eastman Kodak Company.

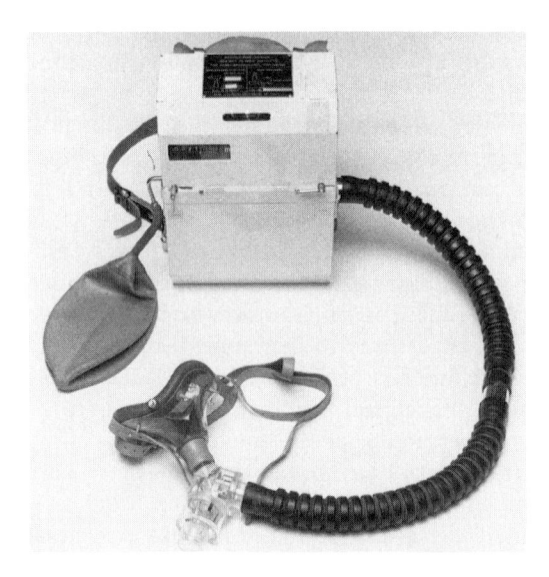

Figure 3-3 Max Planck respirator. *Source:* Eastman Chemicals Division of Eastman Kodak Company.

midity), and air velocity. Photographs of two units are shown in Figures 3-5 and 3-6. The first unit contains chart recorders to provide temperature recording for an entire 24-hour day. Although it is somewhat large to carry about, it is self-contained and can be set up in almost any location. The second unit is smaller and more mobile. It is used to obtain many measurements in several areas during a short period.

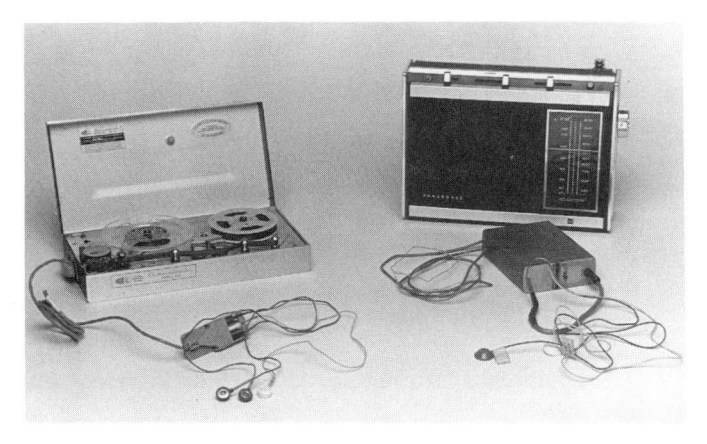

Chap. 3   Planning an Industrial Ergonomics Effort   33

Figure 3-4   Heart rate transmitters. *Source:* Eastman Chemicals Division of Eastman Kodak Company.

Figure 3-5   Large temperature recording device. *Source:* Eastman Chemicals Division of Eastman Kodak Company.

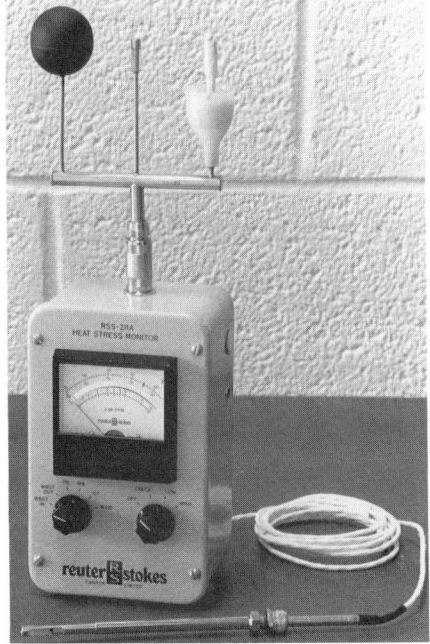

Figure 3-6   Small temperature recording device. *Source:* Eastman Chemicals Division of Eastman Kodak Company.

Observation of people at work is enhanced through the use of slow-motion cameras and/or video cameras. These are excellent tools to observe the work methods/techniques utilized by excellent performers and those who are experiencing problems on the job. The speed of the operation will often hide critical details of the operation. Occasionally, there may be a need for specialized equipment for specific measurements. Such equipment as the electromyogram (EMG), the electroencephalogram (EEG), and eye-scan cameras are not often used in industry.

3-5-3 Information

The reference information collected will reflect the special needs of the user. A reference file of articles (cross-referenced by topic, title, and author) will help to retrieve information when it is needed. The reference file is easily built by reading the following professional journals:

Ergonomics
Applied Ergonomics
Industrial Engineering
Human Factors Journal
American Industrial Hygiene Association Journal
Journal of the American Medical Association
National Safety Council Journal

3-6 SELLING THE CONCEPT TO MANAGEMENT

The final part of planning the industrial ergonomics effort is to sell a proposal to management. This involves two distinct parts: planning the proposal and making the presentation.

3-6-1 Planning

Planning the proposal involves putting the information from this chapter into a cohesive plan (and one that can be implemented). An important part of the proposal is tangible proof that industrial ergonomics can be beneficial to the organization. A very strong selling point can be made if ergonomics in some form is already in use and is benefiting the organization. Empirical evidence can do more good than the most persuasive argument. In many organizations, there has been substantial work of an ergonomics nature. It was done by industrial engineering and labeled a methods study, or it was done by safety and called an injury reduction study, or it dealt with shipping errors, or with heat stress, or with any of the other aspects of ergonomics. Look for these examples of prior work to use in the presentation.

If there is no previous or ongoing work, this is the time to try to get some experience, if you can. Perhaps before the presentation, it is possible to bootleg a study or two that just prove your point. Of course, if they do not prove your point, it is back to the drawing board to figure out just where it went wrong. Better to have it occur now, with just a few people knowing what happened, than during your big trial period when everyone is watching.

Smith and Smith have extensively explored the justification of an ergonomics program in industry (see Figure 3-7). They state:

> Three basic approaches to the problem of justification of ergonomics activities in production facilities have been identified. These are:
> 1. Justification on the basis of increased productivity
> 2. Justification on the basis of reduced non-productive time and reduced overhead expenses
> 3. Justification on the basis of social (or legal) responsibility
>
> The first justification approach is the one utilized traditionally by engineers. . . . The justification will be on the basis of a benefit/cost analysis.
>
> Justification on the basis of reduced non-productive time and reduced overhead costs typically is made on the basis of reductions in absenteeism, labor turnover and time lost due to the occurrence of events that interfere with the routine operation of the manufacturing facility. Time saved by prevention of an occupational accident or a task-induced medical case fits into this category. Included would be time spent going to, coming from, waiting for and receiving medical attention. Time spent filling out accident reports and investigating accidents or near accidents, productivity lost when less qualified workers must tem-

Figure 3-7 Justification of an ergonomics program. *Source:* Leo A. Smith and James L. Smith, "How Can an IE Justify a Human Factors Activities Program to Management?" *Industrial Engineering,* February 1982, p. 40. Reprinted by permission. Copyright © Institute of Industrial Engineers, 25 Technology Park/Atlanta, Norcross, Ga. 30092.

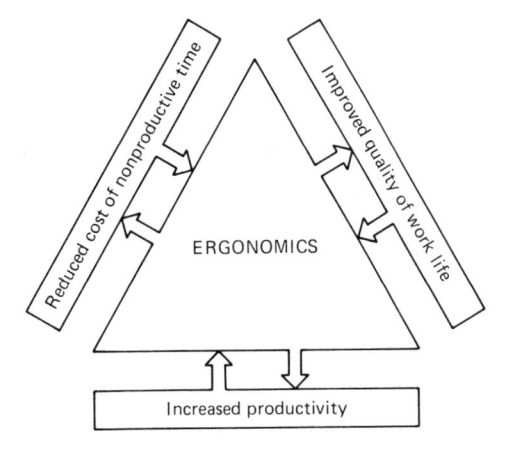

porarily replace those absent and workers' compensation costs associated with the incident are also contributors to excessive non-productive and overhead costs.

The last justification approach is concerned with improvements in the quality of working life of the employees and the general social responsibility of the firm. Justification arguments in this category range from purely altruistic appeals to the social consciousness of the firm to the scare tactic of emphasizing the firm's potential legal liability. The potential penalties associated with violations of occupational safety standards and equal employment regulations and the potential of product liability suits imply that the company *must* be concerned.

Economic justification is not completely absent from this third approach, since reduced exposure to the possibility of fines and product liability judgments implies potentially significant financial savings. Nevertheless, the thrust of the argument here is not economic, but rather social responsibility or avoidance of a situation that could potentially be misinterpreted as the firm's lack of concern for its employees.

Production management is more sensitive to this argument than might be initially expected.[7]

Beevis and Slade also discuss the reasons for establishing an ergonomics effort.

Historically . . . the emphasis was therefore primarily on improving the performance of given man–machine combinations, rather than producing improvements in efficiency measured in terms of value added per man hour. This attitude is still prevalent today, coupled in some quarters with the idea that ergonomics is some form of welfare service to be provided for the employee by improving his comfort, health or safety. Indeed, although financial savings may be shown to accrue from applying ergonomics to job or equipment redesign, they have seldom been the reasons for establishing an ergonomics service or department within an organization.[8]

Although there are other approaches to the justification of ergonomics, an approach at General Motors indicates that cost reduction is important, and that the "intangible costs" can and should be quantified. Blum states:

Initial approaches for introducing HFE principles to the plant workplace which were somewhat less than successful included emphasis on social responsibilities, legislation changes and increasing numbers of women entering the manufacturing workforce. The present, very successful, approach involves quantification of costs which can be favorably impacted by an emphasis on HFE principles at the

---

[7]Leo A. Smith and James L. Smith, "How Can an IE Justify a Human Factors Activities Program to Management?" *Industrial Engineering*, February 1982, pp. 39–43. Reprinted by permission. Copyright © Institute of Industrial Engineers, 25 Technology Park/Atlanta, Norcross, Ga. 30092.

[8]Beevis and Slade, "Ergonomics—Costs and Benefits."

plant workplace. . . . Health and safety factor costs will become a "per piece" product cost. . . .[9]

The possible justifications of the ergonomics effort must be fully explored. The final plan should reflect the full variety of ways that ergonomics can benefit the organization. Planning the proposal can be fun and it should be interesting. Involve others or have them review your material and your outline.

3-6-2 The Presentation

The actual presentation is where your labors begin to pay off. If you have done a thorough job analyzing the organization's needs and how ergonomics can meet those needs, things should go smoothly. Some things that can help with your presentation are covered below.

Your audience is probably unfamiliar with ergonomics, so the burden of defining ergonomics in a favorable light rests with you. Examples of projects and, particularly, successful results will help acquaint others with ergonomics faster than any other methods. Photos or slides can demonstrate many points that are otherwise difficult to make.

By having firm convictions of the importance of ergonomics, your proposal will have an offensive slant, not a defensive one. The end result of any ergonomics work should result in a benefit to the organization as well as to the employees. This should come across clearly in any presentation.

It is up to you to demonstrate the need for ergonomics. The work previously done in this area is invaluable to you in making a satisfactory presentation to management. Management is interested in solutions to problems. By showing potential problem areas, together with typical solutions, the credibility of your proposal will be firmly established.

Tailor your proposal to the various concerns at each level of management. Lower-level management may be concerned about an accident that occurred last week; middle management may be concerned about meeting safety goals despite a sudden rash of accidents over the last three months; and upper management may be concerned about the spiraling cost of long-term disability payments. To be effective with your proposal, you must address the appropriate concern at each level. Questions on the costs/benefits of your proposal and the practicality of services will inevitably arise. Preparation of this information will help make your presentation much smoother.

---

[9]Max A. Blum, "Human Factors Engineering: Getting Started at the General Motors Workplace," *22nd Proceedings of the Human Factors Society Annual Meeting, 1978* (Santa Monica, Calif.: Human Factors Society, 1978), p. 183. Copyright 1978 by The Human Factors Society, Inc. and reproduced by permission.

3-7 SUMMARY

This chapter provides a broad range of considerations for the person who is planning an industrial ergonomics effort. One must begin with an understanding of the areas of application. From that, the types of problems and the opportunities for improvement can be pursued. The benefits of ergonomics can then be identified and the ergonomics effort placed in the organization. The resources needed can be planned and the concept presented to management.

ACKNOWLEDGMENT

This chapter utilizes and builds on material from David C. Alexander, C. G. Givens, and H. R. Kaudewitz, "Human Factors Engineering in Industry—How Do You Begin?" *Proceedings of the Human Factors Society, 25th Annual Meeting, 1981* (Santa Monica, Calif.: Human Factors Society, 1981), pp. 359–363; and from the presentation and paper by David C. Alexander, "Human Factors Engineering in Industry—How Do You Begin?" at the 1979 American Institute of Industrial Engineers Spring Conference.

QUESTIONS

1. What are some ways that an industrial ergonomics effort might get started? Which can you initiate?
2. List the major types of applications for ergonomics in industrial organizations.
3. List the ways that the application of industrial ergonomics can save your organization money.
4. Ergonomics is a multidisciplined science. List the disciplines that should contribute to ergonomics in an industrial organization.
5. What specific steps are involved in selling the "ergonomics concept" to management?

Chapter 4

IMPLEMENTING AN INDUSTRIAL ERGONOMICS EFFORT

This chapter covers the steps involved in actually implementing an ergonomics program. The steps begin with developing support and interest for the program through training others in the practice of ergonomics.

4-1 INTRODUCTION

An engineer was planning to ship more material on each railcar leaving the plant. A thorough analysis was done to determine loading patterns and shipping volumes. It was discovered that by having the labor force stack the material in each car to a height of 8 feet rather than the present 7 feet, approximately $200 per month in freight costs could be saved. This was a very thorough analysis, yet the engineer had failed to consider the impact that this higher lift would have on the people doing the job. A more detailed investigation of the ergonomic aspects revealed that the injury rate was already considered too high and was well over the plant average. Any additional injuries would have greatly offset the projected cost savings. Fortunately, the change was never implemented. Even though the ergonomics program was in place, the engineer felt that "ergonomics couldn't really answer a question like this."

There are many ways to solve operational problems, and an ergonomics approach is only one of them. To ensure that ergonomics is fully implemented, its value as a practical tool for solving problems must be recognized by many people. As long as the existing methods are perceived as adequate, any new programs such as ergonomics will fail to prosper.

As one begins to implement an industrial ergonomics program, there is a transition from selling the concept to management to making it work on the job. The ergonomist's task now becomes how to do the best job within the constraints at hand. Usually, in the beginning, the job looks very big and the resources equally small. There are several critical aspects important to the implementation of an ergonomics program. These include (1) a broad-based support for the program, (2) an understanding of the needs of all the users, (3) a way to create interest in ergonomics, (4) a method to make it work with limited resources, and (5) a way of creating a lasting change. The following sections discuss these areas.

4-2 BROAD-BASED SUPPORT

That there is a need to develop broad-based support for a beginning program is not particularly startling news. All new efforts must have strong support to survive in a competitive business environment. This support must not be overlooked, nor can it be taken for granted. Some organizations use multilevel committee structures to develop and enhance support, while others rely on the ability of the program to sustain itself. This last approach is the harshest, particularly in the beginning. Yet if the program can survive this approach, it will have demonstrated a strength that will probably last.

Overall support will come when the effort is seen as being more of a help than a hindrance in meeting the organization's goals. Equally important, individual support will come when the program does something good for each person. For the person on the line, that means correcting an unsafe condition or removing an arduous task. For the manager, it means improving performance parameters such as safety, quality, or productivity. For the engineer, it means helping to solve a thorny design problem.

The correction of a problem involves much more than just determining a solution. It is often said that 90% of solving a problem is simply to recognize that there is a problem that can be solved. Within the field of ergonomics, especially as it is practiced in industry, this is more often the case than not. With this in mind, there is a two-step process that can help to build the support that is needed. First, the users must have a favorable introduction to the program, and especially what it can do for them. With an emphasis on the identification of problems and opportunities, the potential users can gain a clear understanding of the ways that ergonomics can benefit them. Second, the users must have a favorable experience when they use ergonomics initially. The use of ergonomics technology must help them correct the problem, and the technology must solve the problem better than other methods.

To accomplish the foregoing steps successfully, the needs of the user must be carefully analyzed and planned. An understanding of the needs of the various user groups should help to shape the awareness campaign. In Chapter 3, the need to tailor proposals to the various levels in the organization was emphasized, in order to achieve the maximum benefits for any proposals made. The same holds true with any of the information that you are planning to dispense relative to industrial ergonomics. For example, safety personnel are interested in injury reduction, while engineering personnel will need information about human sizes and capabilities for design. Managers may be interested in cost control and quality improvement.

4-3 USER NEEDS

Within the industrial setting, there are several distinct user groups of ergonomic information. Each group has unique needs for ergonomic information, and each group should be examined carefully to determine that need. To implement an ergonomic program effectively, the correct information must get to each group.

The different groups normally found in the industrial setting are (1) operators/first-line supervisors, (2) line supervisors, (3) managers, (4) designers and engineers, and (5) ergonomics specialists. Within this order of the groups, the depth of knowledge of ergonomics needed will generally increase. However, since a major portion of the solution to a problem is its recognition, there is a strong foundation needed in the area of problem identification. Consequently, problem identification is the minimum level of understanding that is necessary for each group. What follows is the thought process used in analyzing the informational and awareness needs of each of the various user groups.

4-3-1 Operators/First-Line Supervisors

The person who is the most knowledgeable of any job is the person who performs that job. This experience is required for full understanding and knowledge of the details of each job. With this knowledge of all the details, then, this group is in the best position to identify problems within the existing jobs.

The main reason to create awareness among the operators and first-line supervisors is so that they may recognize the problems that currently exist. Although a designer, engineer, or manager may be a very astute observer of operations, the operators are there all the time. Developing their capability to recognize a problem will ensure that they will find any that do exist. A strong knowledge of problem identification techniques will work quite well at this level of the organization. This step is important if your goal is safe and efficient operations. It may initially result in more identified problems to correct, however. One international union has demonstrated a strong understanding of this principle by developing a manual that covers many aspects of ergonomic problems at the workplace: *Strains & Sprains: A Worker's Guide to Job Design.*[1]

4-3-2 Line Supervisors

Further information is needed in the line operating areas. This is necessary to support the efforts of the operators and first-line supervisors in problem identification. As they bring up situations, they need a receptive audience by their supervisors, who must listen to their concerns and evaluate these situations fairly.

The target level of knowledge for those in line supervision is to be able to recognize ergonomics problems, just like the operators and first-line supervisors. They must also be able to recognize the benefit of allocating resources in the solution of these problems and to anticipate problems in new designs. To ensure that the resources are used properly, line supervisors must have a strong appreciation for the results of long-term ergonomic problems and for the benefits to be gained by the organization with successful resolution of these problems.

4-3-3 Managers

Management is the group that can see overall trends more clearly, since managers are generally removed from the details of daily operations. They are still in the line organization, however, and need to have an appreciation for the levels of knowledge discussed previously. Since they allocate resources, just like the supervisory group, their support is just as imporant.

They can help in two additional ways. First, from their perspective, they can often spot trends that others are not yet able to see. Second, they can ensure that ergonomics is applied widely within the organization. Often, all that is necessary to get a new program off the ground is for the management level to inquire regularly whether ergonomics was used in the new design.

An appeal to the management level should be made based on the long-term interests of the whole organization, with specifics that detail the correction of problems that are affecting the organization today. There is a desire for management (1) to allocate resources to investigate and correct anticipated

---

[1]Dan MacLeod, *Strains & Sprains: A Worker's Guide to Job Design* (Detroit, Mich.: International Union, United Automobile, Aerospace and Agricultural Implement Workers of America, UAW, 1982), Publication 460.

problems, and (2) to support the allocation of resources by others to correct the problems that are vexing operations today.

4-3-4 Designers and Engineers

Up to this point, the emphasis has been on understanding the benefits of ergonomics and on the identification of ergonomics problems. The designer/engineer is the first person with a strong need for design skills and all the associated information that entails. So the design and engineering group needs both a thorough understanding of the points above and of problem solution as well.

They are the people who will struggle with the various solutions to the problems that are identified; and they will be the ones who recommend the appropriate solutions to problems. Designers and engineers will be balancing all the design criteria. This group must have a strong appreciation for ergonomics so that important principles are not violated or forgotten. Often they will use ergonomics information on a design without your knowledge—provided, of course, that they see a need to do so. This makes it almost impossible to assess the "level" of ergonomics actually practiced.

A special case of the designer is the design manager. The needs of the design manager are explained by Rogers and Armstrong: "The design manager must conduct a cost-effective program under definite schedule constraints. He and his designers, therefore, may tend to resist changes which increase cost or delay schedules with no apparent immediate benefit. Manager and designer resistance . . . may be reduced by providing convincing evidence that human factors can be cost effective within normal schedule constraints."[2]

Every attempt should be made to meet the needs of the designers, since often, the actual application of industrial ergonomics will rest completely in their hands.

4-3-5 Ergonomics Specialists

The people who need the most complete information and training in the field of ergonomics are the ergonomics specialists. They must be able to do everything that has been mentioned previously and, in addition, they must be able to teach others to do these things. The strength of the ergonomics program will rest on this person or these people.

A summary of the various user needs is provided in Table 4-1.

---

[2]Jon G. Rogers and Richard Armstrong, "Use of Human Engineering Standards in Design," *Human Factors*, 19(1), 1977, pp. 15-23. Copyright 1977 by The Human Factors Society, Inc. and reproduced by permission.

TABLE 4-1 User Group versus Information Needs

	Target Group				
Level of Knowledge	*Operators/ Foremen*	*Line Supervisors*	*Managers*	*Designers and Engineers*	*Ergonomics Specialist*
Problem identification	×	×	×	×	×
Benefits of ergonomics		×	×	×	×
Trends of problems			×	×	×
Design solutions				×	×
Wide knowledge					×
Ability to teach ergonomic skills					×

4-4 CREATING INTEREST

The step of creating interest may be by far the most important in this whole process, simply because it is the most visible. It is the time when those important first impressions are made.

The whole concept of the awareness training is (1) to create the need for the application of ergonomics, and (2) to provide a means of increasing the skill of the users. The next section, "Making It Work," deals more with the methods of increasing the skills of the users, although skill development is often touched on in some of the initial awareness presentations.

The more common methods of broad-scale selling of ideas in industry are live presentations, published media, and "canned" presentations such as video presentations and slide-tape shows. The medium will affect the presentation and will determine the need to provide a high level of details. For example, with live presentations, there is the possibility for discussion and clarification of critical points.

Some of the methods of selling human factors are noted by Lucas:

> Process oriented human factors is very dependent on the ability of the Human Factors specialist to sell his point. He must convince engineering and management that his duty is to assure smooth operations after a facility or a piece of equipment is in operation. He is trying to make their jobs easier in the long run.
>
> The quickest route to this goal is the education of engineering and line

management in the principles and the purpose of Human Factors. Seminars and newsletters have been used. The most successful approach, however, has been the actual applied situations themselves that have served to develop understanding."[3]

Regardless of the medium, however, the general outline of the oral presentations should remain the same. Briefly, an effective presentation (1) will show the benefits to be derived from the application of ergonomics, (2) will define ergonomics in terms that are meaningful to the audience and, (3) will tell the audience where to seek further information, both to assist in the identification of problems and in the solution to problems.

Many people find the concept of ergonomics completely new. However, one dilemma that often shows up in the industrial setting is that some people have already been involved in the practice of ergonomics, although they may never have heard of the word. It is common to find that designers have had a concern about people and how to design for them. Many designers have become quite adept at designing facilities and equipment for people without any formal training. Yet even these people can benefit from an understanding of ergonomics as a formal discipline. Now they know that even more design information is available and where to get it. At the same time, this orientation provides them with a legitimate way of designing for the human being and of transferring these skills to others.

The awareness presentations should take on a different flavor with each of the different user groups. If there is a need for the group to understand only the identification of problems, there is little need to go into a long explanation of design principles or of where to go for more detailed design data. Similarly, when the group is involved with the management of operations, the time that they can spend on design is somewhat limited. It is probably not necessary to provide the detailed design data since they will not have the time to apply it.

Even within the group of users that will be doing the actual designing, it may be important to selectively provide design data that are needed rather than overloading users with details that simply show how much information is available. For instance, if you are dealing with classical facilities designers, they probably do not need much information on how to design "user-friendly" computer software. Similarly, software designers do not have a great need to know the accessibility requirements for maintenance of the company's process equipment. Therefore, it is important to target your audience and to show them exactly what the proper ergonomics information can do for them.

The order of presentation for the material to the potential users is also

---

[3]Richard L. Lucas, "Human Factors in Industry—Getting the Point Across," *Proceedings of the Human Factors Society 18th Annual Meeting, 1974* (Santa Monica, Calif.: Human Factors Society, 1974), pp. 404-405. Copyright 1974 by The Human Factors Society, Inc. and reproduced by permission.

important. Designers and engineers are usually free to pick up information as they need it. However, when presentations are made within the line organizations, it is best to start from the top down. Nothing is more frustrating than to build someone's enthusiasm only to have the person's supervisors kill the ideas because they were not familiar with the importance of the technology.

One part of each presentation must include where to get additional information. This is critical. It is easy to forget that there are two aspects to the use of ergonomics data in the industrial setting: the potential users of the information must be both willing and able to use the information. Although the presentations up to this point have concentrated on increasing the desire of the user to want to use the information and design with the human being in mind, that desire will fade very quickly if the user cannot find the information needed for design. The next section covers some methods of providing design information and design practice to potential users.

Another way of creating interest is through the use of committees. Blum describes the use of corporate human factors engineering (HFE) committees and of plant HFE committees as an effective part of the implementation of that human factors engineering effort.[4]

4-5 MAKING IT WORK

To make the ergonomic program successful, problems must be solved; and problems can be solved only when the needed information is readily available. A key to the success of the ergonomics program, then, is to provide design information to the potential users.

If the program is to be well regarded, it is important to provide timely answers to the questions that are asked. The designer has certain needs during the process of design. These needs, as covered below, should be met for the designer to do an effective job.

1. The designer must be concerned with how the equipment and the system work and must also be concerned with how the human operator "fits" into the system.
2. The designer must consider how the equipment will be used and how the operator will interface with it. The economics, productivity, and quality of the entire system are important, not just optimizing one part of it.
3. The designer should know how to solve most of the problems that arise in the design, including problems that involve use of the human being.

[4]Max A. Blum, "Human Factors Engineering: Getting Started at the General Motors Workplace," *Proceedings of the Human Factors Society, 1978* (Santa Monica, Calif.: Human Factors Society, 1978), p. 183.

4. In solving a problem, a designer must be able to gather the appropriate information about the design parameters, such as the capabilities and/or limitations of people.
5. The designer must know how to access and understand information about people contained in design references.

From the designer's viewpoint, nothing is worse than asking someone what seems to be a simple question, only to have to wait a long time for an answer. So, by generating interest in ergonomics, it becomes important to provide a way to get correct and timely answers to those with questions. Since the ergonomics consultant is normally the person asked for the answer, it is very easy to become overloaded with routine and mundane inquiries. Three practical ways to increase your coverage without increasing your staffing complete this section.

4-5-1 Tiers of Knowledge and Area Specialists

This concept was discussed in Chapter 3. It involves the development of "local on-the-job specialists" who can provide answers to many of the questions that arise in a particular area. For example, an area specialist in the materials handling area would become familiar with lifting guidelines and with solutions to typical problems. Materials handling designers with questions in that area would ask the area specialist for assistance. Only if the area specialist were not able to answer the question would the ergonomist be called in. With the benefit of this experience, in the future the area specialist should be able to address similar problems.

This system gradually builds the skills of the area specialist and allows more complex problems to be dealt with at that level. The concept also allows faster answers to requests and should provide more workable solutions, since the area specialist is able to bridge the gap and communicate between the designers and the ergonomist. The area specialist can put the problem in terms that the ergonomist can more easily understand; for example: "This is a musculoskeletal problem that involves lifting 45 pounds from the floor to 6 feet at a frequency of five lifts per minute. . . ." The area specialist is also in an excellent position to assess the practicality of the solution and to ask for additional design options if needed.

4-5-2 Widespread Training

The ergonomist may need to provide training for the designers/engineers. Training seminars usually last from 4 hours to 3 days when done in-house. The seminar needs to focus on:

- Problem recognition and identification
- Problem formulation
- Information gathering
    Information from the job/task
    Textbook data on human capabilities
- Solution development
- Justification of the solution

These elements are discussed further in Chapter 6.

Case studies are useful in showing "before" and "after" situations. They aid greatly in developing an ability to recognize the "cues" that identify a problem.

Formulating a problem is important. This includes the scope and magnitude of the problem: How many people are involved and how serious is the situation? Is the problem new or an existing one? What human systems are affected—the musculoskeletal, the cardiovascular, or the anthropometric?

Information gathering on the job allows the job/task duties to be fully detailed. Information on the load, lift frequency, temperature, and other task or environmental requirements can be specified. Gathering textbook data on human capabilities will then allow hazardous situations to be pinpointed. Solution development provides some insight into the classic types of solutions used, including equipment and job redesign and mechanization.

Justification of the solution requires an understanding of the tangible and intangible costs associated with ergonomic problems. Since many solutions are not justified on cost/benefit analysis alone, this is a good place to discuss safety and health implications and the benefits that they bring to the organization. A seminar should provide an opportunity to solve some case problems or to discuss and solve some "real-world" problems brought in by the class participants.

The major limitation of widespread training, if it is done with short seminars, is a superficial knowledge of ergonomics that seems to encourage the participant to recognize only the simpler problems and then to develop "band-aid" solutions. Complex problems may never be identified and detailed solutions may be too time consuming to pursue. The strength of widespread training, however, is that it gets many people to think more about the human being during the design process.

4-5-3 Design References

The field of ergonomics is replete with information and data. The government has been funding research in this area for years and there are volumes of data readily available within the public sector. At the same time many private researchers have also been busy and have generally been very open in providing their information to the ergonomics community at large.

Thus the quantity of information is not the problem with providing information to users. The problem is with locating the appropriate information in a form that the designer can use quickly and easily. This problem was addressed in some detail by Meister and Farr. They identified two major weaknesses in getting designers to utilize design reference information. The first weakness was that the designers did not regularly consult either design references or ergonomic personnel. The second weakness was that information was not in a form that was readily usable by the designer. In a scaled ranking the designers preferred information in the following formats:[5]

1. Graphic (88%)
2. Pictorial (85%)
3. Tabular (33%)
4. Verbal (9%)

The degree to which the information is used, then, depends largely on the format used to present the information. This is not surprising since, for years, the ergonomics community has told other groups that "design" is critical if the user is expected to perform properly.

The ergonomist has several choices in providing the user with information. These include:

- The use of standard references in the field
- The development of a user-specific manual
- The development of engineering standards

Depending on the particular situation, it may be appropriate to mix several alternatives to provide information to the potential users.

Standard references have the advantage of ready availability, but they frequently provide far more information than each designer needs. The overload of information may discourage use of these books. It is unusual to find one reference that suits your needs completely, so often, two or more references must be provided to each designer. This compounds the problem just mentioned.

The development of user-specific manuals offsets these problems, but they involve up-front development time to put together, and later, time to update. They can be tailored to your specific needs and can reference all the current information available.

---

[5]David E. Meister and Donald E. Farr, "The Utilization of Human Factors Information by Designers," *Human Factors*, 9(1) 1967, pp. 71–87. Copyright 1967 by The Human Factors Society, Inc. and reproduced by permission.

Design standards are usually requirements of design. Each specific standard must be followed unless there is an approved explanation of the variance. Whereas the other references are usually guidelines or recommendations, design standards are much more rigid. With the flexibility needed to apply ergonomics data, there usually are few items of information that can be covered with design standards. These usually are in the environmental area, where light and noise levels are specified and where temperature and ventilation standards may be used. In some cases anthropometric-based standards are used for aisle widths, doorways, and so on. Detailed designs and layouts almost always require judgment to apply successfully.

The decision on what design references to use will probably be based on the dual criteria of cost and usability. The cost of providing the information is an important criterion but may be minor when placed against the more important criterion of usability. Regardless of the cost, if users do not find the information they need, or find the information so cumbersome or awkward to use that they quit using the reference material, all the efforts are wasted. Ease of use should never be compromised in an effort to reduce the costs of this information.

In addition to the type of material used, the presentation of the material will also affect the quality of the final design. Use by the user can be enhanced by ensuring that the information is in a form that is easily processed. For illustration, two forms of information are shown in Figures 4-1 and 4-2. One

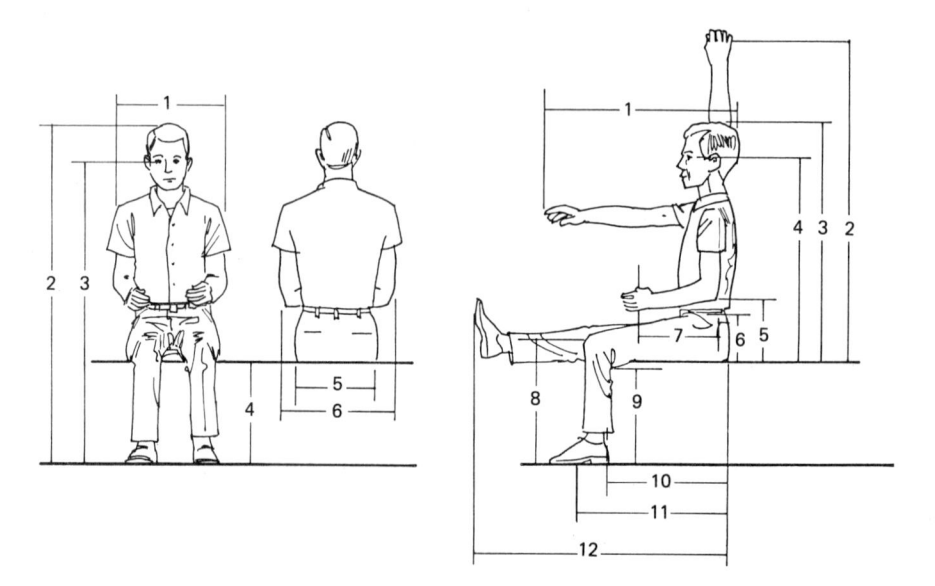

Figure 4-1 Raw anthropometric data. *Source:* Adapted from Wesley E. Woodson and Donald W. Conover, *Human Engineering Guide for Equipment Designers,* 2nd rev. ed. (Berkeley: University of California Press, 1964), pp. 5-16.

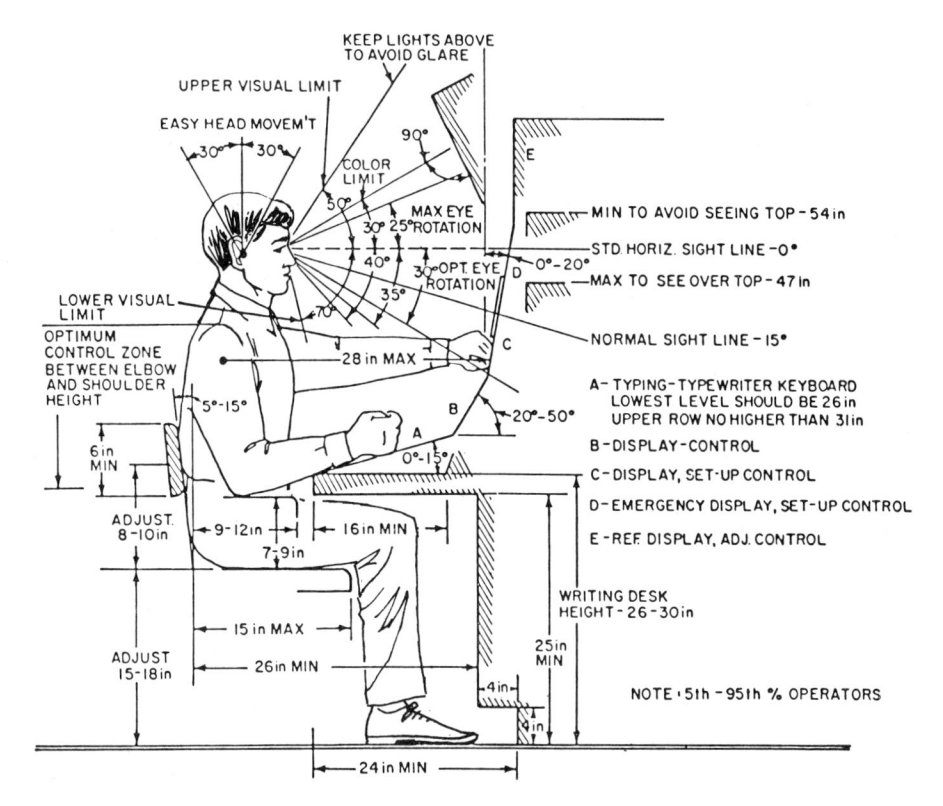

Chap. 4 Implementing an Industrial Ergonomics Effort 51

Figure 4-2 Applied anthropometric data. *Source:* H. P. VanCott and R. G. Kinkade, eds., *Human Engineering Guide to Equipment Design* (Washington, D.C.: U.S. Government Printing Office, 1972), p. 393.

provides raw anthropometric data to the user, while the other provides design guidelines in a finished form. The user can more easily assimilate and use the data in a finished format.

Following are some valuable resources to use when selecting or building a general design reference.

CHAFFIN, DON B., and LEO GREENBERG, *Workers and Their Tools*. Ann Arbor, Mich.: University of Michigan Press, 1977. (143 pages.)

GRANDJEAN, E., *Fitting the Task to the Man: An Ergonomics Approach*. London: Taylor & Francis Ltd., 1980. (379 pages.)

George C. Marshall Space Flight Center, *NASA Standard Human Engineering Design Criteria*. Huntsville, Ala.: Marshall Space Flight Center, 1966. (418 pages.)

MCCORMICK, E. J., and M. S. SANDERS, *Human Factors in Engineering and Design*. New York: McGraw-Hill Book Company, 1982. (512 pages.)

Military Standard 1472 B, *Human Engineering Design Criteria for Military Systems,*

Equipment, and Facilities. Washington, D.C.: U.S. Department of Defense. (239 pages.)

MURRELL, K. F. H., *Human Performance in Industry*. New York: Reinhold Publishing Corporation, 1965. (496 pages.)

VANCOTT, H. P., and R. G. KINKADE, eds., *Human Engineering Guide to Equipment Design*. Washington, D.C.: U.S. Government Printing Office, 1972. (752 pages.)

WOODSON, W. E., and D. W. CONOVER, *Human Engineering Guide for Equipment Designers,* 2nd rev. ed. Berkeley, Calif.: University of California Press, 1965. (474 pages.)

4-6 A LASTING CHANGE

Experts in the field of organizational development talk about how to achieve lasting change within an organization. When they speak of the most enduring type of change, they mention cultural change. This occurs when there is a common desire, usually unstated and usually widespread, to perform certain activities. This widespread desire to apply ergonomics is the ultimate acceptance of the ergonomics program.

This is no easy job. What you want is to go from the situation of having someone say: "Why do we have to redesign this? I'm sure that people can use it like it is. It will only cost us more to redo it" to the situation of having people ask: "Why wasn't this designed so that people can easily operate and maintain it?" The change is from having to sell the implementation of ergonomics every day, to having people in the organization freely support and enforce its use. It then becomes difficult for any designer to ignore the human being in the design process. The designer then has the uphill battle instead of the ergonomist. This cultural change is difficult to achieve, but it will have a far more long lasting impact on your organization.

4-7 SUMMARY

Implementing an ergonomics effort is no simple task. It involves building an extensive base of support to build interest in the program. Information is necessary for others to become involved in the design process; and for ergonomics to last, a cultural change to consider the human being fully is critical.

QUESTIONS

1. What is the critical transition that the ergonomist must make as the ergonomics program is beginning to be implemented?
2. What are two steps required to develop a satisfied ergonomics customer and another supporter for the ergonomics program?

Chap. 4 Implementing an Industrial Ergonomics Effort 53

3. List the different user groups for ergonomics information and the types of information each requires.
4. What are the requirements necessary to ensure that designers design for people?
5. What elements must be included in ergonomics training for designers?
6. What are the major weaknesses in getting designers to utilize ergonomics data? How should information be presented to make it the most useful?
7. What are the pluses and minuses of developing a design reference manual specifically for your organization?

Chapter 5

PROBLEM IDENTIFICATION

The intent of this chapter is to help with the identification of problems in the real-world work setting. This chapter begins with the classical method of identifying problems, the use of checklists. Following that are methods that involve the client and methods that rely primarily on the analysis of data. This leads to methods to prevent problems rather than simply detecting those that now exist. Finally, there are ways to anticipate problems that may occur from the introduction of new technology. Problem identification is the key to problem solution and will build the foundation necessary for the solution methodologies discussed in Part 3.

5-1 INTRODUCTION

After shipping errors were reduced at a loading dock, the supervisor pointed with pride to his solution. A system of double and triple checks had been implemented for each load before the truck left the dock. When asked if the initial cause of the errors could have been the eight-digit product code used on the bill of lading, the stated reply was "There's no problem now," yet what went unsaid was "Don't bother me, I can solve my own problems." The value of the ergonomic solution was lost because this was perceived to be either a motivational problem or an unavoidable problem. By failing to recognize that it was a design-induced ergonomic problem, the appropriate solution was not implemented. As this example shows, a broad-based level of understanding and an ability to recognize ergonomic problems is essential for each program to succeed.

The crucial hurdle to preventing and/or solving many industrial problems is simply their identification. Many serious problems are overlooked because they are not recognized as problems. Many examples like the one above abound in the real world.

This book is about solving real-world ergonomic problems. But the first part of problem solving is the identification of those problems. That is why this chapter on problem identification is so important and why it precedes the chapters dealing with the solutions to problems.

The identification of problems is as much an art as it is a science. This chapter provides a wide range of ideas on the identification of problems. The identification of problems can be interactive with the client. In some cases, however, it may be preferable to identify potential problems on your own, prior to the involvement of the client. As you might expect, the techniques are situationally dependent.

Another dimension in the problem identification arena is the detection of potential problems rather than the identification of existing problems. This is an interesting situation since you may find yourself with the added burden of "selling the problem" in addition to developing and selling the solution to the problem.

The ultimate answer to the question of problem identification is simply to have people identify their own problems, without your intervention. By teaching others to identify problems (and eventually, how to correct those problems) you will have the best method of eliminating problems and of providing a safe and efficient operation.

The use of technological change can have both a positive and a negative impact on your work. It may allow correction of long-standing problems. At the same time, technology can cause problems that would not have appeared otherwise. The use of robots to solve lifting problems and the use of computer workstations that create glare and headaches are two examples that represent both sides of this coin. The addition of technology in problem identification must not be overlooked.

Finally, the ergonomist must be able to evaluate the various problems and choose those with the largest positive benefit to work on. It is common to find one's self with more projects than time. There are some methods of evaluating the projects to determine the best approach to use.

The identification of problems is an appropriate conclusion to Part 2 since this will ultimately determine the need for an ergonomics program. If there are few or no problems, the ergonomics effort is not necessary. If there are problems, however, continuing the implementation of the ergonomics effort is entirely appropriate.

5-2 TRADITIONAL PROBLEM IDENTIFICATION

The traditional method of identifying problems is to look until you find something that is wrong. This hit-or-miss approach is simply not very effective. Unless one has had extensive experience, a checklist of some type can be valuable.

There a number of checklists available. Primary differences are that some of the checklists are more extensive than others and that some are designed for specific industries or special uses. Shorter checklists are valuable if one is making a cursory inspection or explaining ergonomics to others. It is possible to use one checklist as an example of the utility of designing a customized checklist to fit a particular need. Often, special checklists have been designed by trade associations or user groups. It is worthwhile to check with appropriate associations to see what they may have to offer in this area.

5-2-1 Checklists/Questions

There are many checklists designed to highlight ergonomics problems. A few examples are provided in Figures 5-1 through 5-4. The value of these checklists is to provide a complete list of potential problem areas so that no problems are overlooked. The checklist is a valuable tool for the less experienced ergonomist since it can direct the person to look for problems that have not yet been encountered. The checklist is also valuable for the experienced practitioner since it can serve as a reminder to prevent overlooking any areas.

The checklist is helpful in explaining to others the types of problems an ergonomist can help to solve. It is very useful during interactive introductory discussions as a way of explaining what will be expected during a walk-through inspection.

There are a number of different checklists, with a variety of uses. It is important to use this variety and flexibility to your advantage rather than letting it be a liability. Find the checklist that meets your specific needs rather than using the first one found. Figure 5-1 presents a checklist suited to evaluate the general workplace, task, and work environment.

The checklist shown in Figure 5-2, is designed specifically to solicit information from the personnel department or the line manager. It is very useful to use during interviews since it presents initial questions to determine problem areas and then uses detailed questions to diagnose the causes of these problems.

The checklist shown in Figure 5-3 begins with symptoms (cues) that one may actually see in the work area and then leads to possible causes and solutions for that situation. It is effective when shared with the client in addition to being useful for the ergonomist.

One additional checklist, Figure 5-4, specifically addresses back prob-

Human Factors Engineering Checklist

Developed by Leo A. Smith and James L. Smith

Purpose:
The purpose of this checklist is to assist the responsible person in remembering and systematically considering all of the important human factors engineering points when creating an efficient and safe working environment. A positive response to any of the questions posed in the checklist indicates a need for careful human factors engineering evaluation of the work situation under consideration.

General indicators of the need for human factors engineering evaluation:

1. Is absenteeism on this task too high?
2. Is turnover on this task too high?
3. Is production efficiency on this task too low?
4. Do employees complain frequently about this task?
5. Is personnel assignment on this task limited by age, sex, or body size?
6. Is the training time for this task too long?
7. Is product quality too low?
8. Have there been too many accidents?
9. Have there been too many visits to medical?
10. Does this task result in too much material waste?
11. Is there excessive equipment damage on this job?
12. Does the worker make frequent mistakes?
13. Is the operator frequently away from his workplace?
14. Is this workplace utilized on more than one shift per day?

Indicators of the need for workplace redesign:

1. Does the work surface appear to be too high or too low for many operators?
2. Do workers frequently sit on the front edge of their chair?
3. Must the worker assume an unnatural or stretched position in order to see dials, gauges, or parts of the work unit and in order to reach controls, materials, or parts of the work unit?
4. Is the worker required to operate foot pedals while standing?
5. Does the operation of foot pedals or knee switches prevent the worker from assuming a natural, comfortable posture?
6. Are foot pedals too small to allow foot-position changes?
7. Is a footrest necessary?
8. Do workers frequently attempt to modify their work chair by adding cushions or pads?
9. Are workers required to hold up their arms or hands without the assistance of armrests?
10. Are dials and equipment controls difficult to operate or poorly labeled?

(Continued)

11. Is the design and layout of equipment a hindrance to cleaning and maintenance activities?
12. Does the workplace appear unnecessarily cluttered?
13. Is the worker required to use a nonadjustable chair?
14. Can the worker be relieved of static holding by providing clamps or support for the work units?

Indicators of the need for task redesign:

1. Is the worker required to lift and carry too much weight?
2. Is the worker required to push or pull carts, boxes, rolls of material, etc., that involve large breakaway forces to get started?
3. Is the worker required to push or pull carts and hand trucks up or down ramps and inclines?
4. Does the task require the worker to apply pushing, pulling, lifting, or lowering forces while the body is bent, twisted, or stretched out?
5. Is the work pace rapid and not under the operator's control?
6. Does the worker's heart rate exceed 120 beats per minute during task performance?
7. Do workers complain that their fatigue allowances are insufficient?
8. Does the task require that one motion pattern be repetitively performed at a high frequency?
9. Does the task require the frequent use or manipulation of hand tools?
10. Does the task require both hands and both feet to continually operate controls or manipulate the work unit?
11. Is the worker required to maintain the same posture, either sitting or standing, all the time?
12. Is the worker required to mentally keep track of a changing work situation particularly as it concerns the status of several machines?
13. Is the rate at which the worker must process information likely to exceed his or her capability?
14. Does the worker have insufficient time to sense and respond to information signals that occur simultaneously from different machines?

Indicators of the need for special consideration of the working environment:

1. Does process noise interfere with the reception of speech or auditory signals?
2. Is process noise of an irritating nature so that it interferes with the worker's attention to his or her task?
3. Is process noise loud enough to cause hearing loss?
4. Do the work tasks contain significant visual components, thus necessitating careful attention to lighting?
5. Does the worker's eye have to move periodically from dark to light areas?
6. Are there any direct or reflected glare sources in the work area?
7. Do lights shine on moving machinery in such a manner as to produce stroboscopic effects or distracting flashes?
8. Does task background coloration interfere with the color codes or knobs, handles, or displays?

Chap. 5 Problem Identification 59

9. Is the air temperature uncomfortably hot or cold?
10. Is the relative humidity uncomfortably high?
11. Are radiant heat sources located near an operator's workstation?
12. Is the worker exposed to rapid thermal or visual environmental changes?
13. Do hand tools or process equipment vibrate the worker's hands, arms, or whole body?
14. Does process dust settle on displays, making them difficult to see?

Indicators of the need for restricted employee selection:

1. Is process equipment so designed that only very tall persons can effectively reach them?
2. Is process equipment so designed that very tall persons must bend too much while performing their task?
3. Is process equipment so close together that large persons cannot easily and safely attend it?
4. Is the task thermal environment such that only individuals in good physical condition who are acclimatized to work in the heat can safely perform the task?
5. Does the task require normal color vision?
6. Does the task require heavy physical exertion so that only persons in good physical condition can perform it?
7. Are any aspects of the task unnecessarily awkward or difficult to perform if the worker becomes pregnant?
8. Is the personal space under the work table so restricted that only small persons can properly sit at the table and work?
9. Does the task require normal hearing ability?
10. Does the task require rapid decision making and manual response to machine signals?
11. Does the task require unusual eye–hand coordination?

Indicators of the need for more effective supervisory control:

1. Do workers appear uninformed concerning proper work postures?
2. Are chairs properly adjusted?
3. Do employees continue to work during scheduled breaks?
4. Do employees appear uninformed concerning proper lifting, carrying, lowering, pushing and pulling techniques?
5. Do employees often fail to wear their personal protective equipment?
6. Do employees appear uninformed concerning the effects of noise and heat stress on their health?
7. Do employees fail to drink sufficient liquid when they work in high-heat areas?

Figure 5-1 Ergonomics checklist. *Source:* Leo A. Smith and James L. Smith, "Human Factors Engineering Checklist," personal communication, 1983.

Questions for Identifying Human Factors Engineering Needs and Opportunities

Developed by H. R. Kaudewitz

Are there jobs in the work area:

1. That require a long training period or are very difficult to learn?
   a. Is the job mentally very demanding?
   b. Is a high level of skill and/or experience required?
2. With a high error rate or a high error consequence?
   (Process Control)
   a. Are the errors from inattention?
   b. Are the errors from insufficient training or job knowledge?
   c. Is sufficient process control information available?
   (Inspection)
   d. Is a high degree of concentration required?
   e. Is lighting sufficient for the task?
   f. Are the errors from insufficient training or job knowledge?
   (Packaging, Labeling)
   g. Is the product code and/or labeling information clear and easy to understand?
   h. Are the errors from insufficient training or job knowledge?
3. That require a high level of motivation, alertness, or concentration?
   a. Is automation a possibility, or is human control required?
   b. Is the consequence for an error serious?
4. That are repetitive or boring?
   a. Is the job machine paced?
   b. Does the job allow social interaction while working?
   c. Can the job be automated?
5. That have a high accident and/or injury rate (especially hand and musculoskeletal injuries)?
   a. Are the injuries similar in nature?
   b. Are there tasks which require significant strength?
   c. Do job tasks require awkward or uncomfortable body positions or postures?
   d. Is repeated lifting or lowering required?
   e. Are materials handled or lifted below the knees, above the chest, or out away from the body?
6. That have a high absence, turnover, and/or job transfer request rate?
   a. Is there frequent leaving of the work area for nonwork reasons?
   b. Is there a high rate of medical visits?
   c. Are real medical problems reported or just minor complaints?
   d. Is the situation a general one or does it occur mainly when certain types of work activity are scheduled?
   e. Is the problem age-or sex-related?
   f. Is the work difficult, tiring, or boring?
   g. Is the work area hot?
   h. Are worker complaints general or related to specific problems?

7. That employees refuse promotion to?
   a. Are refusals age-or sex-related?
   b. Is the job difficult, or does it require above-average strength?
   c. Is employee perception of the job accurate?
8. That workers complain a lot about?
   a. Is the work area very hot?
   b. Is the work area dirty or dusty?
   c. Is the work considered "too hard"?
9. That involve a lot of manual materials handling?
   a. Is the manual materials handling the major part of the job or a small part of it?
   b. Is mechanization possible?
   c. What is the frequency of handling?
   d. What is the length of handling activity?
   e. What size, shape, and weight of material is involved?
   f. What lifting heights are involved?
   g. Can the material be held in close to the body?
   h. Is there a high accident, injury, or medical visit rate for this job?
   i. Are mechanical assists required or needed?
10. That require awkward postures or body positions?
    a. Is the worker stooped over much of the time?
    b. Does the work require body limbs to be held in the same position for long periods of time?
    c. What is the length of time and frequency of the activity?
    d. Are heavy tools used overhead?
11. That females and/or older workers find difficult?
    a. Is the work hot or dirty?
    b. Is the job physically demanding?
    c. Is physical strength required?
    d. Does the job appear to require an employee with certain physical characteristics?
12. Where environmental conditions like heat or noise affect production or worker productivity?
    a. Do workers complain about job conditions in general, or are there specific complaints?
    b. Is the problem one of comfort, or do work conditions present a health concern?

Figure 5-2 Ergonomics checklist. *Source:* Hart R. Kaudewitz, "Questions for Identifying Human Factors Engineering Needs and Opportunities," personal communication. Courtesy of Tennessee Eastman Company, Division of Eastman Kodak Company, 1983.

Cues That May Indicate Possible Human Factors Problems

Cues	Possible Causes and Solutions
High vision demands (color, fine discrimination, glare, contrast)	Change lighting; specialized lighting; vision testing; use of optical scanners
Twisting, rotating, or unstable body postures while lifting	Change workplace layout; automate; change weight of lifts
Lack of maintenance accessibility	Plan for maintenance and service work; provide access to equipment; make accessibility part of the equipment design
High data load	Change information rates; use better coding systems; look at and balance workload requirements; automate
Use of CRTs	Remove glare; workplace design to provide adjustability and to remove constrained postures; work pace and schedule; task variability
Make-shift workplaces	Provide well-designed workplace; create adjustability for a variety of workers
Interference with verbal communication	Lower noise level; change mode of communication to visual or tactile; reduce need to communicate
Stretching to reach	Lower the items being reached for; raise person; selection of people; lower shelves; workplace redesign
Constrained or rigid postures	Change job to provide/require other postures; avoid overuse of equipment that requires a fixed point of contact between the human and the equipment, such as eye scopes and CRT screens
Workplace not adjustable	Provide adjustability for workplace (chair, table, footrest); vary task requirements
Excessive weight handled	Reduce weight; select people; use handling aids or equipment; automate
Can't work double shifts	Reduce workload if job is excessively difficult; provide suitable work/recovery regimen
High job turndown rate, requests for transfers	Redesign for excessive workload or changing work population; reputation of job
Multiple shift operation (concerns for shift work system, use of equipment by different people, communications, etc)	Provide for adjustability; check operating methods for consistency; training; suitable shift schedule

Chap. 5 Problem Identification   63

Cues	Possible Causes and Solutions
High error rate	Improve coding systems; change sensory input methods; change pace; training
Population stereotypes incorrect for control	Change to correct stereotype; color code wrong stereotypes; purchase ergonomically designed equipment
Legibility of labels and signs	Size of letters; type styles; color of letters and background; viewing distance from label or sign
Sore hands and wrists	Change tool design; change task requirements; reduce forces required; reduce multiple motions
Static muscle loading	Provide rest for load; use alignment pins; use quick connects; reduce weight; slide material instead of lifting
Different manufacturers' equipment causes lack of conformity	Place nonstandard machines in different areas; color code different machines; purchase standard equipment; use ergonomic criteria in purchasing decisions
High injuries	Reduce task overload; employee selection; workplace design
Long training time	Poor methods; incorrect stereotypes; redesign of complex tasks

Figure 5-3 Ergonomics checklist.

Questions on Back Pain

Developed by Dan MacLeod

1. Does your back hurt when you work or by the end of the day?
2. Have other people at work had back problems?
3. Does your job entail:
    Lifting from high spots?
    Lifting from low spots?
    Pulling or pushing loads?
    Leaning over excessively?
    Twisting, reaching motions?
    Working with a bent neck or spine?
    Standing still too long?
4. Is the height of your workbench or the equipment you are operating correct for the work you are doing? The chair or stool, if you have one?
5. Are there ways to change your job to reduce the pressures on your back?

Figure 5-4 Ergonomics checklist. *Source:* Dan MacLeod, *Strains & Sprains—A Worker's Guide to Job Design* (Detroit: International Union, United Automobile, Aerospace, and Agricultural Implement Workers of America, UAW, 1982), p. 30.

lems. It is easy to use and understand and illustrates the value of checklists designed to focus on a specific problem or task.

5-2-2 Effectively Using Checklists

Checklists are effective to begin the problem identification process during the inspection of a work site. The checklist can help to identify items to look for or it can serve as a good memory jogger for those more experienced ergonomists.

Checklists are also a very good tool to share with operations personnel, such as managers, supervisors, and operators, to aid in the identification of ergonomic problems in their work areas. As mentioned in Chapter 4, the United Auto Workers have used this approach with their membership through an internally developed brochure.

The use of checklists can be enhanced by using a tiered approach. For example, some people may be turned off by a very long checklist with a seemingly endless list of questions. In these cases, it is useful to start with a short, general checklist, and later go to one that is more comprehensive. The simpler checklist can serve to develop the interest that is needed to get a foot in the door, yet not overwhelm one so much as to turn them off to the concept. At this point, it may be appropriate to begin a comparison of the effectiveness of the different methods of identifying ergonomic problems, with Table 5-1.

5-3 ADVANCED PROBLEM IDENTIFICATION

Although checklists remain the most common of the methods to identify ergonomic problems, there are other methods with which the ergonomist must be familiar. These methods are broken up into two major areas: on-site and off-site activities. The on-site activities include interactive discussions, walk-through inspections, and on-site studies. The off-site methods include the use of data bases and professional resources.

TABLE 5-1 Effectiveness of Various Strategies in the Identification and Solution of Ergonomics Problems

Method	Identifies Problems
Checklists	Effective
Sharing checklists	Effective

5-3-1 On-Site Problem Identification

An effective method of identifying problems is to work directly with the client. This allows the two factors needed in the identification of problems to come into direct contact: a strong knowledge of ergonomics and its applications, and a strong knowledge of the operations, including the jobs and equipment. Several methods of working with the clients are discussed below. These can be used individually or integrated together. A comprehensive format that integrates these techniques is outlined later. This format allows an efficient and thorough exchange of information. Of course, there are other ways to solicit this information, and the practicing ergonomist will develop different methodologies to suit the client's needs and the client's understanding of ergonomics.

Interactive discussions. Often, a one-on-one discussion of safety and health and operational problems with line management will reveal some ergonomics problems. This discussion may be difficult to initiate and lead for the inexperienced ergonomist, but the format is effective for those with more experience. During an open discussion of the operation, potential problems usually begin to surface. Since the ergonomist is usually seeking information from the client and has the most experience with applications of this type, the ergonomist usually leads the discussion.

Physical inspections. One method that has proven its utility is the physical inspection. These inspections let the ergonomist physically inspect the work area looking for possible ergonomic problems. The basis for this type of problem identification process is that the ergonomist can probably see and identify problems more easily than when they are described or talked about. The inspection is more time consuming than some other methods, but it has the advantages of allowing the ergonomist to view the actual work area and to see work in progress. This will help to identify those problems that are observable during the inspection. The limitation of the physical inspection is that all the work activities may not be performed during the inspection period.

Normally, the inspection will be preceded by an orientation and discussion of the types of products manufactured, the processes used, the jobs performed, and any work-related problems that have occurred. This information will alert the ergonomist to look for specific problems.

Walk-through inspections. The walk-through inspection is an important part of the industrial ergonomist's work. These are similar to physical inspections, except that they are usually done collaboratively with operating personnel. The inspection allows the ergonomist to see firsthand what the job site is like, what the job duties are, and how the tasks are carried out.

In addition to providing a good inspection of the job and work site, the inspection allows the ergonomist to demonstrate to the client the nature of applied ergonomics. For example, it may be easy to talk about accessibility on the work site but showing actual examples in the client's shop prove the importance of accessibility. This also allows the client to move from a base of knowledge to a base of understanding. This understanding is important because it allows the client to begin to identify operating problems as they occur. The physical inspection also allows the ergonomist to make more detailed contact with the operators and supervisors who are most involved with the jobs.

In summary, then, this type of inspection serves several useful purposes:

- It is used to show by example how to evaluate jobs.
- It demonstrates how one can go from theory to applied ergonomics.
- It helps to uncover poor jobs or equipment, depending on what is being done on that day.
- It allows discussion with operators and foremen.
- It allows one to sell ergonomics.

On-site studies. Frequently, on-site studies are performed by the ergonomist. These studies aid the ergonomist in looking for specific problems in the operation. The on-site studies are most useful when there is a symptom, such as an excessive number of low back injuries, without an obvious task that should be corrected to solve the problem. The ergonomist, by working alongside those on the job, can frequently begin to pinpoint the tasks that contribute to these types of injuries.

These studies are fairly common in industry. An example can help to clarify how this can work. In one instance, a group of laundry employees began experiencing a sudden increase in skin rashes on their forearms. There were no obvious solutions since this particular operation had been done for years with no symptoms of this type. The workers sorted laundry from tubs by lifting handfuls of laundry, briefly holding them up to drain, and then sorting them in bins for drying and, subsequently, for folding. Since the cleaning materials were known to be slightly allergenic, gloves were required to protect the hands of those doing the job. Armed with these facts, the ergonomist had little reason to suspect any particular cause for the increasing number of skin rashes. In the observation of the job, however, one problem was noticed that seemed to contribute to the symptoms. When working with larger pieces of laundry, the people had to hold their arms higher while draining the laundry. This allowed the water to run down their gloves and drip on their forearms. The frequency of this contact was responsible for the increasing number of rashes that were occurring. When looking for an explanation as to why this had not shown up earlier as a serious problem, the ergonomist found that the work population had changed recently and now contained a much larger number of immigrants from southeast Asia. These people, who generally were of shorter stature than the traditional employees, had to reach proportionately higher to allow the laundry to drain. This higher reach then allowed the laundry water to drip down the gloves and onto the arms rather than just dripping from the gloves back into the sinks. The problem, once confirmed, was relatively easy to solve. The wooden platforms that stood by the sinks were raised slightly so that the dripping was once again contained by the gloves. This example, of course, would have been very difficult to solve without the direct observation provided by an on-site study.

During on-site studies, it is common for the ergonomist to find additional problems that were not apparent earlier. Although this is not the purpose of the study, it can provide fortuitous results on occasion.

Priority setting. Few organizations have the resources to correct every problem that is uncovered. A priority system can help establish priorities to correct the problems in the most appropriate order. An example of the type of system that might be used is as follows:

A	Fix immediately; an urgent safety or health danger
B	Fix soon; within a week; safety concern, but not life threatening
C	Fix when equipment is shut down next; safety concern, but no serious accident probable
D	Redesign and fix if cost/benefit ratio is acceptable
E	Redesign next time the equipment is built

This type of stratification will enable the client to use limited resources to correct the most urgent problems first, based on the experience and judgment of the ergonomist.

5-3-2 Off-Site Problem Identification

It is not uncommon for the ergonomist to work off-site in the identification of operational problems. There are many good tools that lend themselves to the identification of problems that are not available at the job site. Here the ergonomist plays the role of an analyst rather than of a detective. As a detective, the ergonomist is looking for clues at the job site that may reveal operational problems or the causes of operational problems. As an analyst, the ergonomist is looking at groups of data to find a trend inconsistent with other data. This trend is then explored further at the job site or with management in the area to determine if it is significant.

Data bases. In most organizations, health, safety, attendance, and other related information is logged and retained. These data can help the ergonomist in the search for operational problems. Some of the more useful data include:

- Accident and/or injury data
- Lost time and/or absenteeism data
- long-term disability data and costs

These data can usually be analyzed by each operating section or department, by job, by historical time (to see if trends are increasing or decreasing), by gender, by age, by length of employment time, and/or by other attributes. Within these categories, useful trends often begin to appear. Safety data are a good example of the way that data can be categorized and then analyzed. For example, when the percentage of back injuries in one department greatly exceeds the average of the work force, the data can often provide more insight about problems. How are the injuries occurring, what are the causes, and what trends are there: by job, by gender, by time of day? This extra information can make a trip to the operating area more productive. Relating these data to operating conditions causes the identification of the real problems to occur.

The data can also reveal the magnitude of the problem and provide some sense of urgency toward its solution. For example, if the percentage of injuries is unusually high in an area, that is important and should be investigated very soon. Other indicators of severity include high rates of injuries or high costs associated with the injuries in the area. It is better to look at more than one index of severity if that is possible.

Professional contacts. Although many organizations prefer not to exchange information on operating procedures or costs, they usually do not find it objectionable to discuss safety- and health-related matters, within the context of appropriate trade associations or professional and technical organizations. For example, the Chemical Manufacturers' Association has had a number of committees and task groups investigate safety and health problems common to chemical manufacturing, including a task group on ergonomics. These trade groups are a good source for aid in identifying the types of ergonomic problems common to your industry. Often a request to such an organization can provide information about what others in your industry have found to be serious problems. This may then lead you to investigate those areas of concern first.

It is also useful to develop a network of contacts within the field of ergonomics. These contacts are useful in identifying those areas where ergonomics can provide solutions to problems. Their experience can direct you toward the more serious problems that they have faced, thus allowing you the benefit of their experience. These contacts can be made through technical societies and ergonomic-related subgroups. The short courses mentioned in the sections on the startup of an ergonomics effort can also be useful in providing information about the more serious problems that occur in actual practice.

Intuition. One resource that the experienced ergonomist can use effectively is intuition. Often, the ergonomist can sense that something is wrong before the details of the situation even become clear. This intuition, although difficult to explain or demonstrate, is a tool that should not be overlooked by the experienced practitioner.

5-3-3 Pulling It All Together

There are many ways of identifying problems. One method of pulling these ways together to effectively analyze operations and identify problems is shown in Figure 5-5. This outline is an overview of how the process might be used in a real-world setting. It is comprehensive and will probably be modified to suit your particular needs, after you gain familiarity with its use.

It is useful to explore the effectiveness of the various strategies discussed so far. Their use can then be planned, depending on the needs that have been identified for the operating area (see Table 5-2).

5-4 USING PROBLEM IDENTIFICATION TO PREVENT PROBLEMS

It is a short step from problem identification to problem prevention, yet many people fail to make the step. The same technology and information that allows us to detect ergonomic problems can also be used as design input in the prevention of problems. The key to the identification and prevention of problems is being able to predict what will happen to people given a known set of conditions. Our ability to predict problems leads to our ability to prevent those problems to start with. When we know that certain things result in ergonomic problems, we should then take action to ensure that those conditions are not permitted in new designs.

If it is acceptable to use a checklist to critique a finished installation, it is acceptable to use that same checklist to critique the design plans. The ergonomist may find it useful to assume this role temporarily, until this critique has proven its usefulness.

Ultimately, it is essential to have the designer assume the responsibility for the critique of each design. It is difficult to mandate effective critiques, so it is better to take a slower approach to implement this concept. One might start with physical inspections of the finished installation with the designer. As problems are noted, it becomes easier to sell the idea of a critique prior to construction. It is also easier to sell the idea of the self-critique during this process. That can then lead to the development of specialized critique check-

Comprehensive Method of Identifying Ergonomics Problems

Step 1: Introduction (led by the ergonomist)

1. Brief definition of ergonomics
2. Some examples of applied ergonomics

Step 2: Introduction to the operating area (led by the client)

1. Purpose of operation
2. Equipment and operation
3. Jobs
4. Potential and/or suspected problems

Step 3: Detailed interview (led by the ergonomist)

1. Checklists
   a. Human problems
   b. Equipment interfaces
   c. Diagnostic questions
2. Specific questions
   a. Accidents/injuries
   b. Promotion problems
   c. Training problems
   d. Acceptance of jobs

Step 4: On-site work (ergonomist and client participate)

1. Physical inspections
2. Walk-through inspections
3. Detailed job studies

Step 5: Investigation of additional data sources (ergonomist and others participate)

1. Lost time
2. Accident/injury reports
3. Medical and disability records

Step 6: Closing discussion (ergonomist and client)

1. Possible problems
   a. How identified
   b. Severity
   c. Priority
2. Possible solutions
   a. Description
   b. Implementation
3. Where to go from here

Figure 5-5

Chap. 5   Problem Identification   71

TABLE 5-2 Effectiveness of Various Strategies in the Identification and Solution of Ergonomics Problems

Method	Identifies Problems	Develops Independence of Client	Identifies Trends
Checklists	Effective	Good	No
Sharing checklists	Effective	Effective	No
Discussions	Effective	Effective	Good
Physical inspections	Effective	No	No
Walk-throughs	Effective	Effective	Good
In-area studies	Effective	No	No
Data bases	Effective	No	Effective
Professional contacts	Good	No	Effective

lists for use by design personnel. From there it is a short step to making the checklists a fundamental part of the design manuals and of the design process. Table 5-3 presents a summary of the various strategies for the identification and solution of ergonomics problems.

TABLE 5-3 Effectiveness of Various Strategies in the Identification and Solution of Ergonomics Problems

Method	Identifies Problems	Develops Independence of Client	Identifies Trends	Anticipates Problems	Justifies Solution
Checklists	Effective	Good	No	Some	No
Sharing checklists	Effective	Effective	No	Some	No
Discussions	Effective	Effective	Good	Good	Good
Physical inspections	Effective	No	No	Some	No
Walk-throughs	Effective	Effective	Good	Some	Good
In-area studies	Effective	No	No	Some	Effective
Data bases	Effective	No	Effective	Some	Good
Professional contacts	Good	No	Effective	Some	No

5-5 THE USE OF TECHNOLOGY

For a long time now, ergonomists have been telling operations personnel: "This task is too tough for people to do, find another way to do this task." If the answer came back, "Okay, how should it be done?", the ergonomist could be at a loss for workable ideas to propose.

Fortunately, with the introduction of more and more automation, there is more often an answer to this dilemma. Often there is a mechanical or electrical system that can replace the human being in those critical areas. Robots can now perform repetitive and complex maneuvers, bar-code scanners can find and relay information with a high degree of accuracy, and computer information systems can store and retrieve vast amounts of information.

Technology has some very strong points for the practitioner of ergonomics. Now, solutions to some problems are easy to visualize and easy to sell. The ergonomist must keep up with technology, since it can offer the answers to problems that have been difficult to deal with in the past.

However, technological change can have both a positive and negative impact on your work. It may allow the correction of long-standing problems, yet at the same time, technology may cause problems that would not have appeared otherwise. The introduction of the personal computer workstation continues to be the source of many ergonomic problems—headaches, eyestrain, and back problems have all been attributed to the use of these workstations.

The introduction of technology, then, must be monitored, to ensure that it is compatible with the human user. Problem identification methodologies must include ways of monitoring the introduction of technology and assessing its impact on the ergonomics of your organization.

5-6 A PHILOSOPHY OF PROBLEM IDENTIFICATION

A question often asked is: "When should one expect ergonomic problems?" The most appropriate answer is: "Whenever there are people at work."

With this much larger overview (that there is a potential for ergonomic problems whenever people are present), the identification of problems takes on a totally different form. The ergonomist's job now becomes one of confirming problems and of selecting appropriate problems to work on. This philosophy is not for the beginner, but it can serve to guide the experienced practitioner toward a broader view of problem identification.

The concern with problem identification for most people is that they only want to identify the most significant problems or the problems that result in the largest losses. Unfortunately, it just does not work that way. The best way to identify the most significant problems is first to identify all potential problems and then look for the problems that provide the largest reward. The key, then, is not in problem identification, per se, but in one's ability to separate the important and significant problems from those of lesser significance.

One of the problems with this method is that all too often the identification of a problem means a lengthy investigation, a report, justification, and finally the implementation of changes. The best way to handle these situations is to resolve to investigate only those situations that really require it and to make changes only when there is a strong assurance that the changes will result in some improvements.

One useful concept is a three-part stratification of the problem seriousness. There are situations (category III) where the situation is obviously bad and a solution is necessary. In these cases, no study is necessary to realize that improvements are required, so just spend your time developing solutions.

There are other situations (category I) where things are obviously not a problem and where improvements are very costly. In these cases it is probably not worthwhile to make any changes, nor is it worth any time spent working to study the situation in the first place.

Another category (category II) falls in between these two extremes. This last category of concerns is the most difficult to deal with. In these situations, the concern is neither black nor white and some study is required to clarify the situation and, as necessary, to formulate a solution (see Table 5-4).

The key to this philosophy is reasonable judgment about the severity of the problem and the possibility of a solution. In category III, when everyone agrees that there is a problem, it is best to get on with the solution to the problem as soon as possible and not waste time and effort trying to prove that it is a problem. At the other end, where reasonable people agree that there is not a problem, it is best not to spend your limited resources on another study. The difficulties of problem identification come when the suspected problem falls into that gray area of category II. At that time it becomes important to gather just enough additional information about the situation to allow a decision to be made. Is there a need for a solution or not? This works best as an iterative process where successive amounts of information are gathered and the situation is evaluated again. This will allow the decision to be made (Is action required or is everything okay as it is?) at a minimum cost. One sig-

TABLE 5-4 Stratifying Problem Difficulty

Category I	Category II[1]	Category III
Situation okay	?	Clearly a problem

[1] ?, Unsure of whether situation is a problem or not; additional data required to determine whether a problem exists and the magnitude of that problem. The data will successively narrow the gray area shown as category II.

nificant waste of the ergonomic resource is trying to justify that a problem needs a solution, only to find out that everyone knew that all along.

An illustration of this point will help. When one is involved with lifting, there are some weights that are difficult for almost everyone (category III). Almost everyone will agree that lifting 100-lb bags of cement from the ground to a truck bed is a task that should be mechanized. There is no need for a thorough biomechanical analysis to show the high stresses placed on the body and the increased risk of injury.

On the other hand, lifting reams of paper in an office is such a light task that almost everyone is capable of it (category I). There is little need to study the effects of this task on the human being, especially if one has other project work to do.

There are many tasks that fall in between these two cases (category II tasks). To determine whether or not changes are necessary, additional information should be gathered. Gather the most critical information first, to see if this will clarify the situation. How many lifts are required, how often, for what duration, at what pace? What is the capability of the personnel in the area? Then, with this detail, is the task clearly okay or are changes clearly needed? If it is not clear, gather a little more information and check again. This iterative process will provide the answer for the least cost of time and effort.

The research report *Safe Practices for Manual Lifting*, sponsored by NIOSH, uses a similar format. Using limited information about each lifting task, the tasks are placed in one of three categories: acceptable, unacceptable, or an administrative control area. This is a useful strategy, and one that quickly allows the ergonomist to determine whether or not there is a serious problem.

5-7 SUMMARY

The identification of problems is an appropriate conclusion to Part 2 since it determines the need for an ergonomics effort. If there are few or no problems, the ergonomics effort is not necessary. If there are problems, however, continuing the implementation of the ergonomics effort is entirely appropriate.

Finally, the identification of problems is an appropriate introduction to Part 3, where the solution to problems begins. With an understanding of the problems in your industry and your organization, the proper focus on problem areas and solutions can be made in your progression through this book.

ACKNOWLEDGMENT

This chapter utilizes and builds on material from David C. Alexander and Leo A. Smith, "Evaluating Industrial Jobs, Determining Problem Areas and Making Modifications," *Industrial Engineering*, 1982, pp. 44-50.

QUESTIONS

1. What are the advantages of using checklists in the identification of ergonomic problems?
2. When should one expect ergonomic problems?
3. List the methods available to the ergonomist to identify ergonomic problems.
4. What two types of knowledge are required before an ergonomic problem can be identified? Which is more critical?
5. Briefly describe the types of on-site problem identification activities. Include the advantages and disadvantages of each.
6. Describe a typical priority system for correcting ergonomic problems. Why are these valuable?
7. When utilizing data bases to identify possible ergonomic problems, what types of information are most useful? What are some ways to analyze this information to help pinpoint problems?
8. Describe briefly the concepts of the three-part stratification of problem seriousness. What are its advantages?

Part 3
CLASSIC ERGONOMIC PROBLEMS

Chapter 6

THE PROBLEM JOB

Part 3, which consists of Chapters 6 through 11, is designed to cover the traditional aspects of industrial ergonomics. The elements of workplace design, lifting, heat and other environment concerns, and human error are covered. The part begins with a model for analyzing a problem job which is useful in utilizing the material in the remaining parts.

There are two ways to avoid ergonomics problems: (1) design it correctly the first time, or (2) detect operational problems, diagnose the cause, and then determine a solution. For those in industry, the second route is often more common. This chapter provides an overview or a road map of that second process—detection, diagnosis, and solution development. The process begins, for the ergonomist, with a problem job. From there, it must be determined to be an ergonomics problem and one that is worthwhile to work on. Further clarification of the problem is done to pinpoint specific problems and then redesign efforts or administrative controls are developed to correct the problem. A flowchart of the model is provided and examples are used to illustrate parts of the model and the whole model. This process will be used time and again in the solution of the classic industrial ergonomics problems discussed in Chapters 7 through 11.

6-1 INTRODUCTION

"Once I learn of a problem, how do I approach it?" asked my colleague. I had worked with the engineer to start an ergonomics program, and now that potential problems were being identified, he needed a way to approach the problems effectively. "I have a model of that in my mind," I replied. "Let me put it on paper and send it to you." The model allowed the engineer to evaluate problems as if he had many years of experience in the field.

As industrial ergonomics is more widely practiced, there is an increasing need to provide some systematic approaches to its practice. This will allow the effective transfer of knowledge and skills to more personnel. One of the most common symptoms that an ergonomist receives is that "This job is too tough" or that "People can't do this task." Although this statement is not very descriptive, it generally is the starting point for another project for the ergonomist.

This chapter is intended to provide a conceptual model for dealing with a project that begins with that description and ends with the successful resolution to that problem. Unfortunately, nonspecific problem descriptions are more often the rule than the exception. The ergonomist's job often begins with a need to define the problem and with an investigation to determine how big the problem is. Successful resolution of the problem is another point that it may be necessary to define—it means a resolution that effectively deals with the problem from the human standpoint and one that the organization will implement. Solutions written but not implemented are not successful resolutions to problems.

Dealing with the "problem job" is one of the best ways to improve the physical aspects of the quality of work life. The resolution of these problems prolongs working careers, prevents injuries (both traumatic and chronic), and allows people with a wider range of physical capabilities to perform each and every job. Initially, these sound like benefits only for the person on the job, but they are also clear and strong advantages to the employing organization. The costs of long-term disabilities, of occupational injuries and illnesses, and of special selection of employees for the more demanding jobs is very costly for any organization. Thus the alleviation of the problem job will benefit everyone involved.

It is one of the best uses of ergonomics, simply because of the tangible improvements that can be realized. In this setting ergonomics serves both the worker and the organization, and for their mutual benefit.

The *problem job model* provides a way to transfer practical skills on the practice of industrial ergonomics. This model is the result of many discussions on how to approach the analysis and resolution of problem situations at the workplace.

The problem job model follows a logical sequence of questions, to aid in the diagnosis and evaluation of the given situation. It is heavily based on industrial engineering techniques. One fundamental skill is the ability to carefully analyze jobs and tasks. It is a central theme in the analysis of the problem job. Since many users will already be well versed in the use of task analysis, there is only a brief description of this technique in Section 6-2-7.

The model is a conceptual approach to problems with the symptom of "This is a tough job." Seldom will the ergonomist be required to use the whole model in any single analysis, yet it is important to understand the complete model. This understanding helps the ergonomist to ask the right questions. It also allows an effective and efficient approach to resolving any problem situation that is identified.

6-2 THE PROBLEM JOB MODEL

In this section we develop the problem job model and explain and illustrate the individual steps.

6-2-1 Identification of a Problem Job

The model begins with the identification of a "problem job." There are several ways that the identification can be made and/or come to your attention. There may be the complaint of a tough job from an employee. A supervisor may call to see if there are any ways of reducing the effort required for several jobs in their area. An in-house survey may reveal several jobs that are physically demanding and in need of improvement. A safety investigation following an injury or accident may result in a request to analyze a particular job. There may be lower-than-expected performance on a job, or the error rate may be too high. Or you may simply observe a job or task that appears difficult during a tour of the operations area.

The point is that a potential problem area is identified that needs to be clarified and possibly corrected. At this time, it is much better to identify a situation that later proves to be acceptable than to pass over a situation that may later prove to be serious. There are means to screen out those areas where there are not problems, but there are no sure means to recall those that are erroneously culled out at this point. Figure 6-1 shows step 1 of the problem job model.

6-2-2 Major Task Analysis

The second step is one of clarifying the information that is presented in the first step. Depending on the method in which the initial request is received, it may require additional clarification or detailing. If the request is made by an outside party, it will probably be a little vague. The major task analysis clarifies the problem and determines whether it is appropriate for analysis with this model or not. The tough job model is designed for anthropometric problems, lifting problems, endurance problems, and even human error situations.

If a task/job is designated as tough because it is monotonous or boring, it is not appropriate for this model. The major task analysis should screen out

Figure 6-1 Problem job model, step 1.

```
┌─────────────────────────┐
│  PROBLEM IDENTIFICATION │
└─────────────────────────┘
```

these types of concerns, and as needed, one can deal with them in another manner. These situations may be dealt with more appropriately through an analysis of the motivational aspects of the job. It is not the intention of this book to deal in detail with the motivational aspects of the design of jobs.

The major task analysis could identify the problem as one that is physically strenuous. The task may be performed only occasionally, or it may be performed routinely throughout each work shift. Some physically strenuous problems that can be further analyzed by this model include handling 50-lb bags of material, pushing a cart loaded with material, using a work tool, such as a heavy impact wrench, or pulling a welding rig over uneven terrain. These tasks usually involve physical forces and/or endurance.

Those tasks involving force are usually easier to analyze, simply because the activities are easier to quantify. Tasks requiring endurance can be quantified, although those techniques usually require specialized equipment or training. Typically, the physically difficult tasks that the ergonomist deals with are those involving force and not endurance.

The major task analysis might also identify an anthropometric problem. This occurred with the wide-scale employment of females in industry in the 1970s—there were now workers of shorter stature and lower reach. This type of situation could be analyzed with the problem job model, resulting in the identification of those situations that must be changed to accommodate the new workers.

Another type of problem might be one involving human error, such as inspection or assembly problems. The problem job model should focus on situations where the person is trying to do a good job but is still producing errors. Motivational problems, although important, are not appropriate for this model, since it is designed to pinpoint and resolve problems with the requirements of the task/job and the person's capacity.

The clarification of the meaning of "problem job" for the particular situation is the objective of this part of the analysis. If the situation continues to be appropriate for the problem job analysis, one should continue with the analysis. The major task or tasks that cause the problem should be identified and some general information about the tasks should be available. For example, if the concern is one of lifting, the weights and frequency of the lifts should be known, as well as the general position of the lifter and the type of load (bags, boxes, drums, etc.) handled. At this step, it is not necessary or appropriate to have excessively detailed information about the whole job or task, such as the exact placement of the load or the position of the hands. That is one of the special aspects of the problem job model; it requires detailed data (which is time consuming and therefore expensive to collect) only as it is absolutely required to answer the next question in the model.

If the problem job and its symptoms are not appropriate for this model, now is the time to stop the analysis and end this particular study. Figure 6-2 shows step 2 of the problem job model.

6-2-3 Is This the Right Problem to Solve? Does This Problem Need Solving?

The third step in the analysis is to question whether the problem needs to be solved. Situations change, and before you invest a lot of time and energy into the problem, it is wise to ask a few questions. If the problem involves lifting bags of material, could they be palletized? Will the jobs be here next year, or is some equipment due to be installed that will make the job obsolete? If it is a heat problem, is air conditioning to be automatically installed before next summer?

Some problems are only part of larger problems that should be resolved all at once. For example, the problem of handling heavy bags is something that could occur at many workstations. That same problem could be addressed through the purchasing organization by requesting the purchase only of bags weighing 50 pounds or less. Or it could be addressed through a workplace design group by providing them with guidelines on how to handle heavy bags. The least efficient method of handling the bags is to do one on-the-job study after another. It is important to look for trends and to try to generalize studies whenever possible.

Sometimes, the symptom that you see now is a clue to a more serious problem that is coming later. For example, some early problems with VDT workstations might be resolved through selection of personnel or rotation during the workday. That is a "problem" that really does not need solving. However, it points to the need for a long-term solution that will be effective when

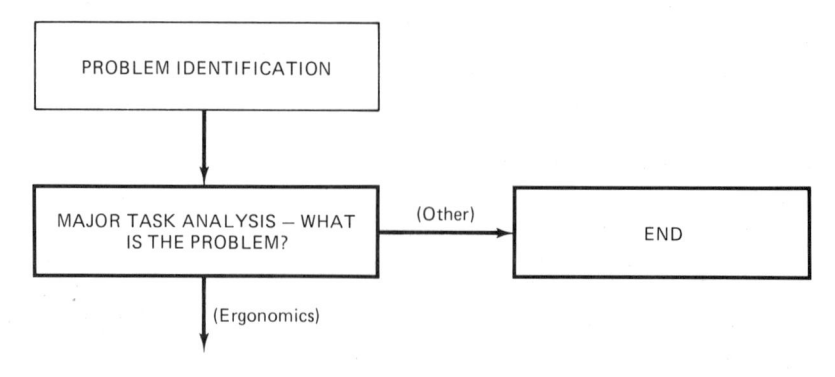

Figure 6-2 Problem job model, step 2.

Chap. 6   The Problem Job   81

more people are required to spend longer amounts of time at VDT workstations.

In these circumstances it is best to restart the process and restate the problem in the new terms. If it is not an important problem, this is the time to stop the process. Figure 6-3 shows the third step in the process.

6-2-4 Is the Problem Real or Perceived?

The next step of the model is the decision as to whether the problem identified is a real problem, or simply one that is perceived to be a problem that warrants attention. Obviously, this is an extremely tough question to answer for many tasks. If the problem is real and can reasonably result in occupational injury or illness, it should be pursued and corrected. On the other hand, if the task/job is just perceived to be a problem and there is no significant risk of injury/illness or problem with lost productivity or quality, the ergonomist probably should address another problem.

Why all the fuss over why a problem is "real" or "perceived"? For most of us, there are limited resources to use in the analysis and correction of problems. When those resources are directed toward situations that do not really warrant them, other situations, which can result in an injury/illness or increased cost are going unattended. The priority of each particular problem must be addressed, and it is best to assess priorities before too much time has been spent on the problem.

Although this determination is a judgment, there are some means available to help determine whether or not the situation should be pursued. (When

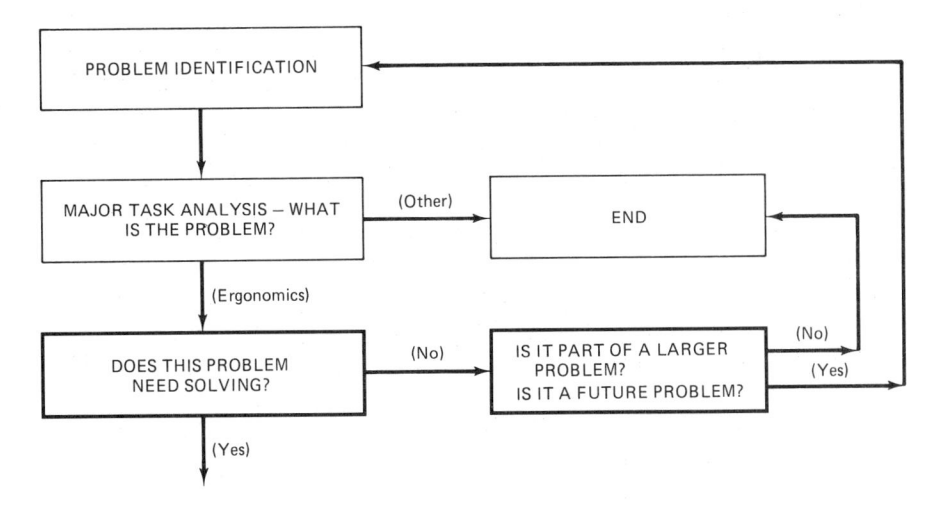

Figure 6-3   Problem job model, step 3.

in doubt, one should continue the analysis by going to the next step in the model.)

Some general questions/information that will clarify the situation follow:

- Are there requests for transfers to other jobs, even jobs that pay less?
- Do people regularly refuse promotion to that job?
- Do the complaints, and so on, come from any identifiable section of the work population (e.g., females, older workers, short-statured individuals)?
- What is the absenteeism rate on the job relative to others in the plant, in the promotional sequence, in the same pay rate, in nearby work areas?

Some questions/information relative to strength and endurance follow:

- The number of injuries on the job; what is the injury rate relative to other jobs and to the rest of the plant?
- Are the injuries all of the same type (e.g., low back, upper arm, etc.)?
- Has the injury rate changed recently, or is there a trend in the data that shows increasing injuries?
- Has there been a change in the work population?
- What do the recent accident investigations reveal? Do they show trends that verify and corroborate the verbal information?
- Is there a history of visits to the medical department that show information similar to the accident/safety data?
- Do the complaints, and so on, come from any identifiable section of the work population (e.g., females, older workers, short-statured individuals)?

Some questions relative to human error situations follow:

- Has the error rate changed lately?
- Do the errors come from an identifiable group (e.g., age related, newer workers, those with glasses)?
- Are the jobs/tasks difficult to learn?

Some anthropometric questions follow:

- Do shorter workers have more problems than taller workers?
- Is there a performance difference associated with heavier people?
- Does a larger size, reach, or stature make a difference?

Chap. 6   The Problem Job   83

Some questions to assess environmental problems follow:

- What are nearby work areas like?
- Is it a combination of the job and the environment, or just the environment? Are there similar jobs in other areas about which workers have no complaints and/or problems?

These and other questions simply help to determine if there is factual support for the problem at hand. Certainly, professional judgment is required. The principle of continuing the analysis until there is clear evidence to stop will help to avoid missing a situation where the practice of ergonomics is important to resolve a problem. Figure 6-4 shows step four of the problem job model.

6-2-5 Perceived Problems

The earlier discussion (on whether the problem identified was real or perceived) was necessary to avoid focusing the limited resources in areas where work was not essential. This is a discussion on how to deal with those situations where the problem seems to be one that is perceived rather than real.

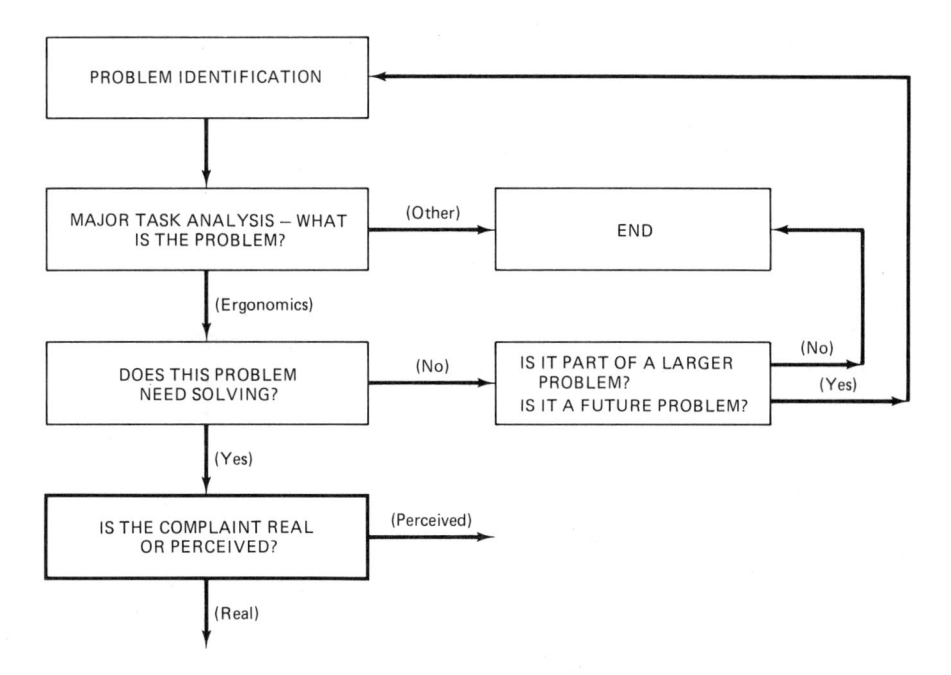

Figure 6-4   Problem job model, step 4.

In many cases, it is appropriate to deal with these situations simply because there are economic benefits that outweigh any costs of implementing changes. For example, a change in methods or equipment may be beneficial financially in addition to reducing work stress, even though there was no overwhelming need from the human standpoint to initiate the change in the first place.

Although the change is not driven by ergonomics concerns, it still may be appropriate to pursue. Even major changes in equipment may be appropriate. If the tasks are very labor intensive, it could be beneficial to automate the task and let the labor savings pay for the necessary equipment. This is a particularly desirable alternative if it is feasible. These events do occur, and with surprising regularity. Several robot systems were initiated by ergonomics concerns but were later sold to management as improvement projects with justifiable economic returns.

In other cases, it may be appropriate to deal with the situation simply because it is relatively inexpensive and shows a responsiveness to the concerns of the employee. The "Hawthorne effect" may come into play, with the result being a significant improvement in the work situation when none would have been predicted based on previous experience or on the facts of the situation. Again, this situation is not that unusual in true industrial situations. If the "solution" removes the perception of a problem situation, (even though there was no real problem), it is effective. Often, cooling fans are purchased when the heat stress level does not justify it (no health or real productivity concerns), but for a low cost, the purchase shows management concern.

When the situation does not warrant further attention, and the costs are relatively high, it is fair to point out that situation and to recommend that nothing be done about it. Confronting the situation directly may be the best approach, even though it is somewhat uncomfortable. The ergonomist is there to demonstrate the facts and correct those situations that are needed. By dropping appropriate situations, the credibility of the ergonomist is often enhanced.

The approach to use in dropping a study depends on where the original complaint came from (the problems that you initiate yourself are obviously the easiest to deal with), the credibility that you have with the client, the cost of making any changes at all, and the degree to which there are hard and indisputable facts surrounding the matter.

In some cases, just pointing out that you have investigated the situation and see no apparent problem is sufficient, especially if you have some technical credibility with the client. In other cases, this may not be enough to convince people that there is no hazard. A short presentation may be needed to compare this job or task with others that are physically as difficult but are not perceived to be a problem. There are many cases of jobs being perceived to be difficult ("man's work") only later to have them populated by females when times changed. Many jobs were done by "Rosie the Riveter" during

Chap. 6　The Problem Job　　85

World War II, only to be declared too difficult for females later. Perceptions of difficult jobs will often lead to the refusals of promotions or transfers, and when you find these perceptions, it is best to correct them as soon as possible.

It sometimes may be wise to make changes in a job before the need is fully demonstrated. If the cost of the study is high, as heat stress studies tend to be, it may be more economical to make minor changes, such as buying a fan, than to continue to study that situation. In other cases, an inexpensive dolly, hoist, or even a pry bar or cheater can be added at little cost. Rotation of certain job tasks can often be done at no cost. Alternative job designs may also help by clustering difficult tasks on one job, or, in other cases, by spreading difficult tasks among many jobs.

Some methods of dealing with the perceived problems are:

- Education of the workers and supervisor (the job is not that difficult)
- Minor job changes (the Hawthorne effect)
- Major job changes (for economic or other benefits)
- Training (a low-cost method of changing perceptions)

Figure 6-5 shows this addition to the process.

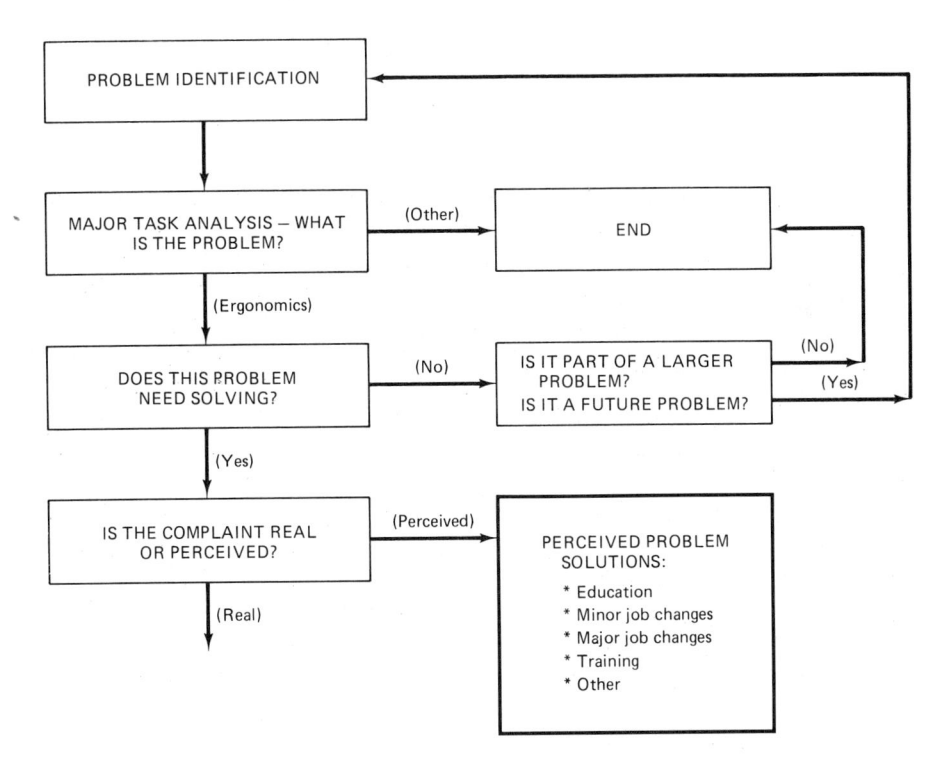

Figure 6-5　Problem job model, step 5.

6-2-6 What Needs to Be Corrected?

Up to this point, the problem has been treated rather generally. If the problem is to be worked on, it needs to be defined very specifically. What is the true problem—not just the symptom—and why is it a problem? The problem needs to be something that can be corrected.

At this point, the symptom of a "problem job" is no longer sufficient. There must be specific symptoms. It is not suitable to say, "Females have problems with this job." The specific jobs and tasks must be identified. At the same time, the ergonomist must develop an idea as to which of the body systems are affected.

After this step, the ergonomist should have a list of specific tasks/jobs, the symptoms, the population affected (if known), and the body system affected. A useful format for these items and some examples are shown in Table 6-1. Figure 6-6 shows the problem job model as it is developed up to this point.

6-2-7 Task Analysis to Pinpoint Problems

If the problem appears to be a "real" one, some type of solution mode is in order. Working in the industrial setting, with limited resources, forces one to make the most effective changes at the least cost. In most situations there will be many alternative solutions; to avoid trial and error it is best to pinpoint the most significant item and correct that item first.

Task analysis is a very useful technique for ergonomic projects. It is a way of breaking a complex job into its components (tasks) and subcomponents (elements) so that they can be studied for improvements. It is rare to find that all tasks and elements of a job are equally demanding. Thus it becomes important to isolate the tasks that do cause problems so that improvement efforts can be focused on those areas.

A task analysis of the firefighter is a good example. Although it is a

TABLE 6-1 Specifying Problem Jobs and Tasks

Job/Task	Symptom	Population Where Problem Seen	Body System Affected
Material handling	High job-refusal rate	Female	Strength, endurance
Inspection	Low detection of errors	Those with glasses	Environmental (vision), cognitive
Keypunching	Headaches, low productivity	All	Anthropometric, environmental (vision)

Chap. 6   The Problem Job   87

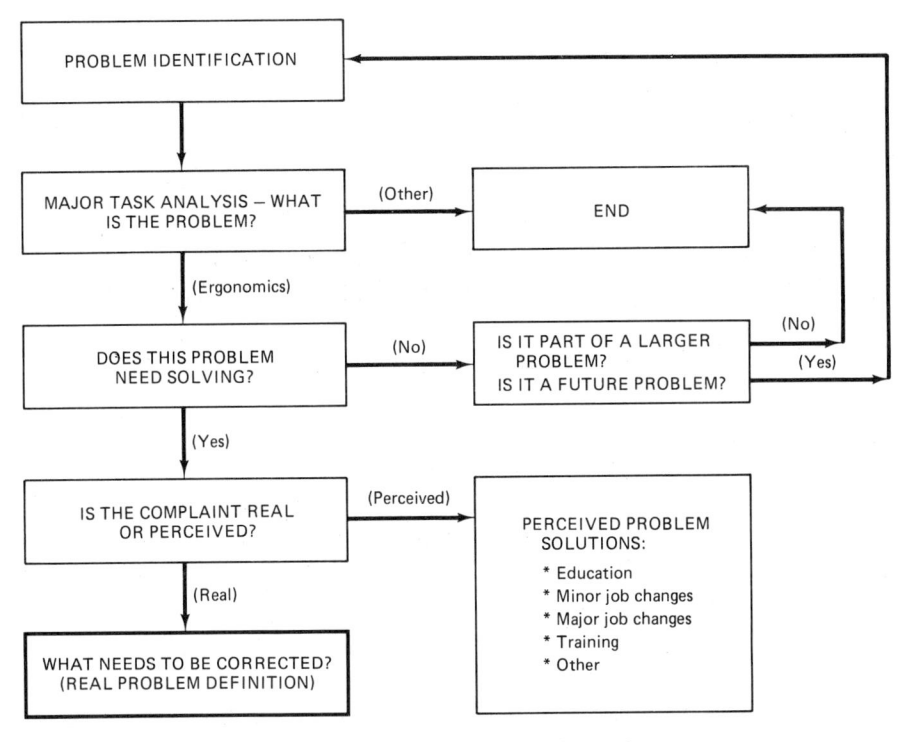

Figure 6-6   Problem job model, step 6.

difficult job, not all tasks are of equal difficulty. Carrying 200 ft of hose 100 yards at a full run is one of the more physically exhausting tasks. Driving a truck requires less physical skill, as do dispatching and fire inspection duties. If the ergonomist were trying to make the job more suitable for females, or trying to reduce lifting injuries, there are tasks where the results would be more beneficial, and these tasks can be pinpointed early.

Task identification is the first step. A task can be identified as a separate unit of work. In Table 6-2, the organization has a function (e.g., shipping and delivery); the job is the part of that function belonging to one person (e.g., local delivery); a task is a separate identifiable part of that job (e.g., to go to the post office); an element is one part of the task (e.g., to get in the truck; to open the door); and the movements (e.g., reach for the door handle) are the lowest level. A focus at the task and element level is the best to pinpoint ergonomic problems.

The main reason for performing a task analysis is to identify and resolve a problem situation. The best way to identify ergonomic problems is to compare what is required by the task with the capabilities of the population that is required (or will be required) to do the task. That comparison can easily

TABLE 6-2 Hierarchy of Industrial Work

Hierarchy of Work	Typical Measurement Units	Example
Function	Days/weeks/months	Shipping and receiving
Job	Hours/days	Local delivery; invoicing; warehousing
Task	Minutes/hours	Go to post office; load truck; unload parcel
Element	Seconds/minutes	Open truck door; get in truck; sign invoice
Movement	Microseconds/seconds	Reach for door handle; pull open door; grasp pencil

pinpoint mismatches that need to be corrected. A useful format is to take a job, list each task, and if necessary, list each element using the format shown in Table 6-3.

When the capacity of the population exceeds the task requirements, there is no problem. The "match" simply indicates that there is a perceptual problem, but not a real one. Population changes can be compared with this method and it is easier to see what effect a change in the capabilities of different populations will have.

When the task exceeds the abilities of the population, there is a problem that needs attention. This situation requires the ergonomist to redesign some aspect of the work environment to correct the problem. The task analysis also helps to find the real cause of the problem, rather than concentrating on the symptom. The real cause needs solution, not simply the addition of another "band-aid." Figure 6-7 shows the additional step added to the model.

One word of caution is appropriate. In some cases, the task will be acceptable for most of the working population, yet there will be a valid com-

TABLE 6-3 Task Analysis

Task/Element	Population	Population Capacity	Problem/Mismatch	Possible Solutions
Grasp knob at 76 in.	Male	95% can reach 76.8 in.	No	—
Grasp knob at 76 in.	All	95% can reach 73 in.	Yes	1. Step 2. Lower knob
Lift 15 lb, 10 lifts/min from 20 to 40 in.	All	19 lb at 12 lifts/min from 20 to 40 in.	No	—
Lift 25 lb, 12 lifts/min from 20 to 40 in.	All	19 lb at 12 lifts/min from 20 to 40 in.	Yes	1. Reduce weight 2. Reduce pace 3. Select people

Chap. 6   The Problem Job

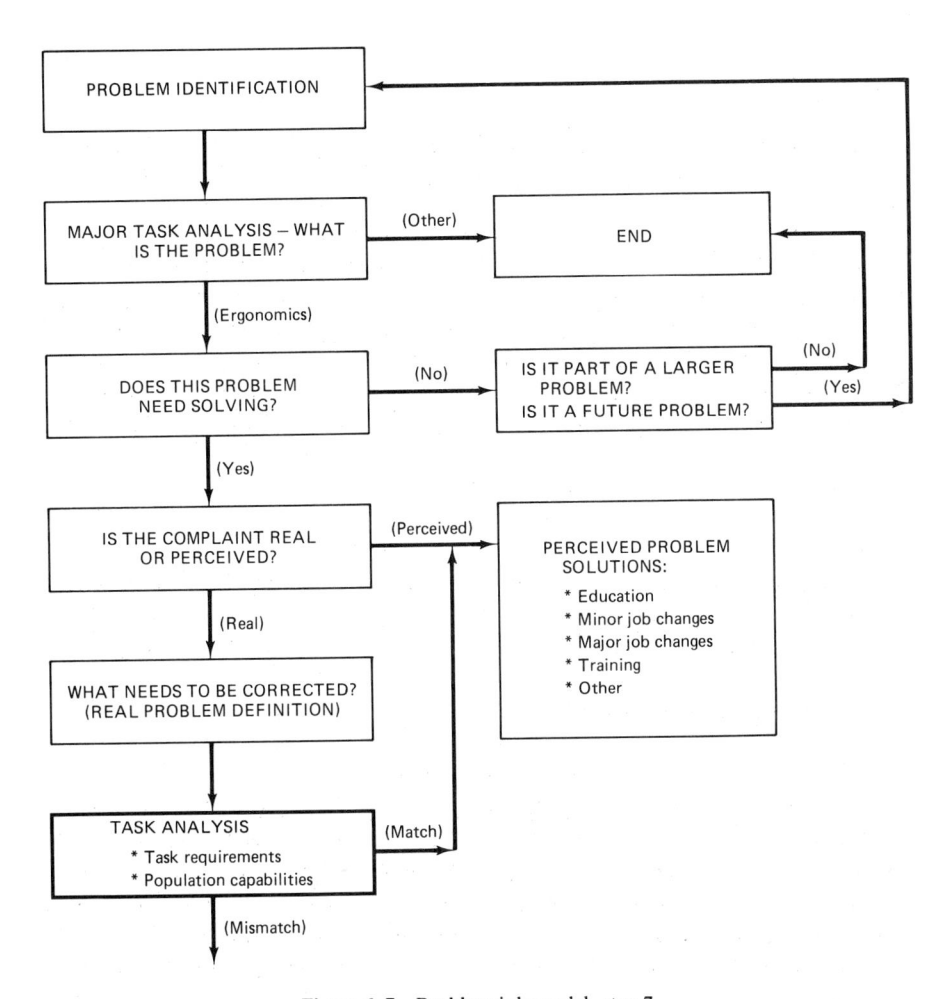

Figure 6-7 Problem job model, step 7.

plaint or operational problem. The answer to this dilemma may be a person with such work-related limitations as a very low strength capacity, a medical problem such as a "weak back," or uncorrectable vision problems. The solution may not be to redesign the job or the equipment; it may be appropriate to have the person transferred to another job which accommodates the physical limitation. Ideally, it is desirable to design jobs so that everyone can do them. Practically, this just is not possible in industry today.

6-2-8 Redesign

When the problem is important, is real, and affects the human being in the system, redesign is in order. There are two approaches, engineering redesign or administrative controls. An engineering design change is a permanent one-time change to fix the problem. It generally requires expenditures, and it usually fixes the problem for good. Administrative changes (job rotation, earplugs, etc.) are usually quicker to implement, cost less initially, and require constant enforcement. The most satisfactory approach is an engineering design solution to the problem, so that approach is evaluated first. Then administrative solutions are examined. Figure 6-8 shows the blocks on redesign and the possible design solutions.

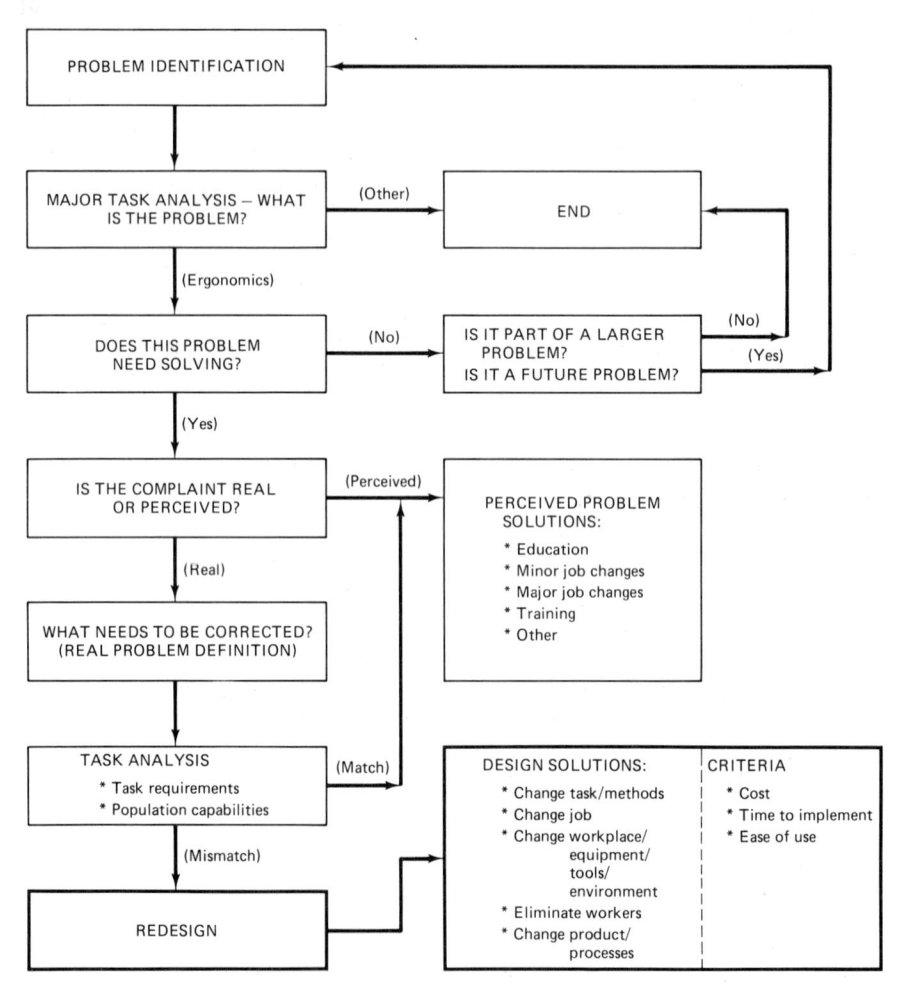

Figure 6-8 Problem job model, step 8.

6-2-9 Design Solutions

The design solutions fall into several major categories. Each is examined in turn. As one begins to examine each major type of solution, there are some criteria that can be used to evaluate the potential solutions generated. The criteria include:

- Cost to implement
- Time to implement
- Ease of use by people

Cost to implement is the total cost involved in implementing each potential solution, including the cost for each design and redesign. These costs can mount up quickly, and engineering costs are expensive for the investigation of detailed design solutions. The time to implement reflects the time it will take to solve the problem, by implementing a workable solution. If the problem is serious/urgent, this criterion is very important. The ease of use by people must be considered because some solutions are more easily utilized by people than other solutions. It is not worthwhile to put in a solution that is more difficult to use than the original problem. Powered hoists that are slow, equipment that does not work, and overly crowded workplaces are not really effective solutions and should not be implemented.

Five general methods of solving redesign problems are discussed briefly below. These descriptions are intended to provide a conceptual framework more than to provide a list of exact solutions. As you examine each alternative relative to your current problem, let your mind wander to encompass as many potential solutions as possible.

Change task/methods. If the task causes problems, change it so that it is easier to do. Break the load into two parts and make two trips. Bring material closer to the operator. Change the sequence of events or the pace of the operations.

Change job. A job is composed of one or more tasks. The job may be too difficult only because of the cumulative effect of the tasks performed. Change the tasks that make up the job so that fatigue is not induced. Data entry on video display units is stressful if done for 8 hours. Change the job by adding other tasks that do not require constant visual attention and that job becomes acceptable. In other situations, periodic rotation during the work shift can allow acceptable performance.

Job design can help in two ways. Distribution of tasks can reduce the work load on any one person. The opposite approach, consolidation of difficult tasks, can also be used effectively to remove some of the more difficult activities from a variety of other jobs. Isolate and consolidate all the difficult elements and use one stronger person who chooses to do that job.

Change workplace/equipment/tools/environment. The task may be difficult, not because of the task or how it is done, but because of the area or place where the task is performed. Change the workplace so that it fits more people; make the equipment easier to operate, service, and maintain; change the tools so that they are more effective or easier to use; or change the environment in which the job is done. Redesign.

Eliminate workers. This is not as harsh as it sounds. Various forms of automation are available to relieve workers of difficult tasks and jobs. Such things as robotics and bar-coding systems have done a lot to relieve workers of undesirable tasks.

Change the product/processes. If the changes are ineffectual in the accommodation of the human being, consider changes to the product or how the product is fabricated. Reduce the "per unit" weight so that everyone can handle it. Or increase the unit weight so that no one will try to handle it. Modify the process so that material is added from a weigh bin instead of being added manually. There are many possibilities if one really asks questions. Changes are not as infrequent as one might expect.

6-2-10 Problem Corrected?

At this point, the question of a successful resolution of the original problem is raised. Can the work population safely and effectively perform the tasks required? If there continue to be problems with people performing the tasks, administrative controls may be appropriate, as shown in Figure 6–9.

6-2-11 Administrative Controls

In those cases where a design solution cannot be found, administrative solutions may be the most suitable alternative. (It should be noted that administrative controls are generally not an alternative to design or redesign solutions. Design solutions should be used if possible because they are more likely to remain in effect.) By definition, the administrative controls must be monitored regularly to be effective; they are not permanent ongoing solutions. Often, these controls must be reinstituted periodically to maintain their effectiveness.

Situations occur where it is impossible to change tasks, methods, or equipment, yet the job is not suitable for everyone. The choices are to modify

Chap. 6   The Problem Job

```
PROBLEM IDENTIFICATION
         │
         ▼
MAJOR TASK ANALYSIS — WHAT   ──(Other)──▶  END
IS THE PROBLEM?
         │
       (Ergonomics)
         │
         ▼                               (No)
DOES THIS PROBLEM  ──(No)──▶  IS IT PART OF A LARGER
NEED SOLVING?                  PROBLEM?
                               IS IT A FUTURE PROBLEM?  (Yes)
         │
       (Yes)
         ▼
IS THE COMPLAINT REAL  ──(Perceived)──▶  PERCEIVED PROBLEM
OR PERCEIVED?                             SOLUTIONS:
         │                                  * Education
       (Real)                               * Minor job changes
         ▼                                  * Major job changes
WHAT NEEDS TO BE CORRECTED?                 * Training
(REAL PROBLEM DEFINITION)                   * Other
         │
         ▼
TASK ANALYSIS   ──(Match)──▶  DESIGN SOLUTIONS:        | CRITERIA
 * Task requirements            * Change task/methods  |  * Cost
 * Population capabilities      * Change job           |  * Time to implement
         │                      * Change workplace/    |  * Ease of use
     (Mismatch)                   equipment/
         │                        tools/
         ▼                        environment
    REDESIGN                    * Eliminate workers
                                * Change product/
                                  processes
         │
         ▼
PROBLEM CORRECTED?  ──(No)──▶  ADMINISTRATIVE CONTROLS:
                                * Rotation
                                * Selection
                                * Training
```

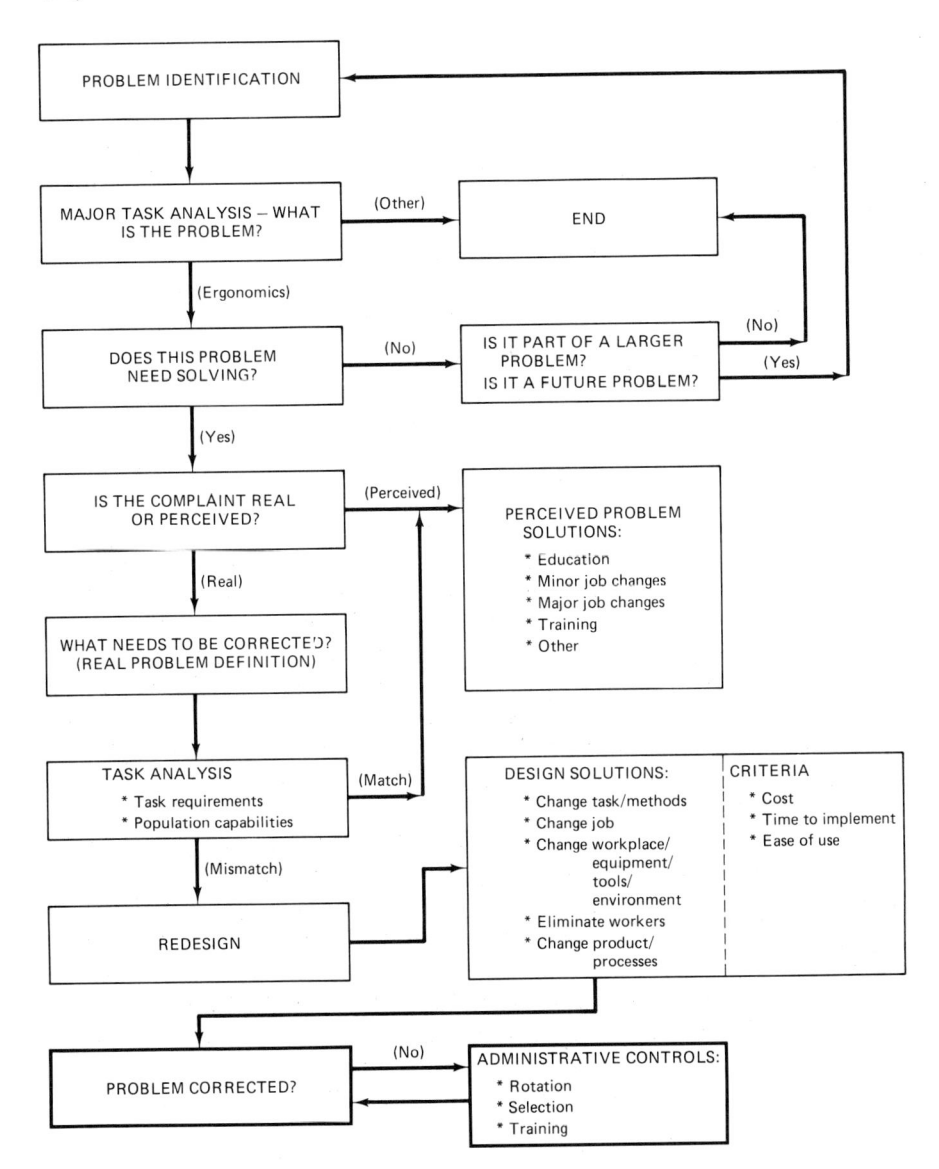

Figure 6-9 Problem job model, step 9.

the skill levels by training, to select people who have skills that are required, or to distribute difficult work tasks by a rotation of workers.

Training. Detailed training on the safe aspects of lifting may allow someone to handle infrequent loads with a wider margin of safety. Similarly, training in "number skills" has been shown to improve one's ability to detect clerical errors.

Selection. The selection of personnel for physically demanding jobs has come under considerable attention lately. There are ways to select personnel, although the validation of the selection tests is rigorous and expensive. The use of forced selection (decision made by management) is open to more questions than the use of natural selection (each person decides whether he or she can do the job).

Selection of people based on strength and endurance characteristics are the more typical situations and the ones that are the most difficult to validate. Some other selection situations that have proven to be less illusive are where the requirements are easier to quantify. For example, seated eye-height requirements for some airline pilots (due to the design of some cockpit equipment years ago and the resulting cost to modify the aircraft) were found to be legally acceptable. Excellent color vision for the jobs of matching colors in the plastics industry (which incidentally screen out a higher percentage of color-blind males) is also a reasonable selection criterion.

Rotation. One administrative control solution that can be used is to distribute a heavy work load by the rotation of workers through the difficult job. This will reduce the effort of each worker, although the job has not been changed fundamentally.

6-2-12 Finish

By this time, the problem should be resolved (see Figure 6-10). If it is not now satisfactory for the work population to perform the task/job, the process should be repeated with some additional people participating in the problem-solving process. One should consider letting those who perform the job participate in this process, for they are the most knowledgeable of the job and how it is performed. Often they may be able to come up with a low-cost solution that is quite effective.

6-3 A FEW EXAMPLES

The examples in this section are intended to illustrate the use of the problem job model. The model was originally designed to help pinpoint operational problems and direct their solution. It is also useful to aid in the analysis of situations where the work population is changing. Finally, the concept of the model, especially task analysis, may be useful to analyze new designs for jobs, equipment, and facilities.

The full model is not always used, but it is important to understand what

Chap. 6   The Problem Job

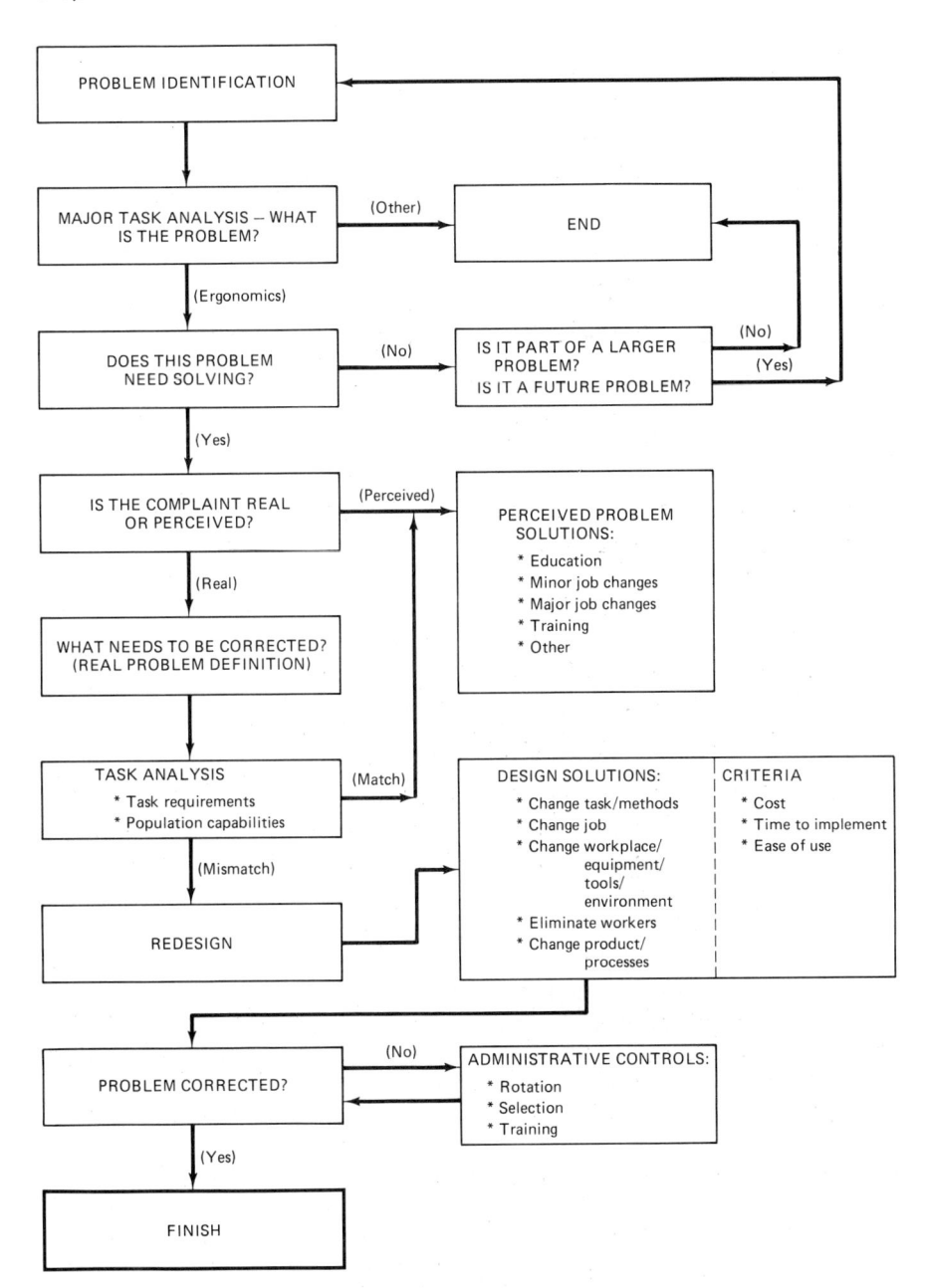

```
┌──────────────────────┐
│ PROBLEM IDENTIFICATION │◄──────────────────────────┐
└──────────┬───────────┘                             │
           │                                         │
           ▼                                         │
┌──────────────────────┐   (Other)  ┌─────────┐     │
│ MAJOR TASK ANALYSIS — WHAT ├──────►│   END   │◄───┤
│    IS THE PROBLEM?   │            └─────────┘     │
└──────────┬───────────┘                  ▲         │
      (Ergonomics)                        │(No)     │
           ▼                               │         │
┌──────────────────────┐  (No)  ┌───────────────────────┐
│ DOES THIS PROBLEM    ├───────►│ IS IT PART OF A LARGER │
│   NEED SOLVING?      │        │       PROBLEM?         │
└──────────┬───────────┘        │ IS IT A FUTURE PROBLEM?│(Yes)
         (Yes)                   └───────────────────────┘
           ▼
┌──────────────────────┐ (Perceived) ┌────────────────────────┐
│ IS THE COMPLAINT REAL├────────────►│ PERCEIVED PROBLEM      │
│    OR PERCEIVED?     │             │   SOLUTIONS:           │
└──────────┬───────────┘             │    * Education         │
         (Real)                       │    * Minor job changes │
           ▼                          │    * Major job changes │
┌──────────────────────┐              │    * Training          │
│ WHAT NEEDS TO BE CORRECTED? │       │    * Other             │
│   (REAL PROBLEM DEFINITION) │       └────────────────────────┘
└──────────┬───────────┘
           ▼
┌──────────────────────┐ (Match)  ┌─────────────────────────┐ ┌──────────────────┐
│   TASK ANALYSIS      ├─────────►│ DESIGN SOLUTIONS:       │ │ CRITERIA         │
│  * Task requirements │          │  * Change task/methods  │ │  * Cost          │
│  * Population capabilities │    │  * Change job           │ │  * Time to implement │
└──────────┬───────────┘          │  * Change workplace/    │ │  * Ease of use   │
       (Mismatch)                 │     equipment/          │ │                  │
           ▼                       │     tools/             │ │                  │
┌──────────────────────┐           │     environment        │ │                  │
│     REDESIGN         ├──────────►│  * Eliminate workers   │ │                  │
└──────────┬───────────┘           │  * Change product/     │ │                  │
           │                       │     processes          │ │                  │
           │                       └─────────────────────────┘ └──────────────────┘
           ▼
┌──────────────────────┐  (No)   ┌───────────────────────────┐
│  PROBLEM CORRECTED?  ├────────►│ ADMINISTRATIVE CONTROLS:  │
└──────────┬───────────┘◄────────┤   * Rotation              │
         (Yes)                    │   * Selection             │
           ▼                      │   * Training              │
┌──────────────────────┐          └───────────────────────────┘
│       FINISH         │
└──────────────────────┘
```

Figure 6-10 Problem job model, step 10.

it is trying to do. Shortcuts may be appropriate, but they should be planned to avoid losing any of the power that the model can provide.

6-3-1 Pinpointing Problems

An example will illustrate the power of the problem job model to pinpoint specific problem tasks. One problem was identified simply as women trying to perform "men's work." A brief discussion proved it to be an ergonomics problem. Women on a new job were experiencing performance problems (they could not always keep up) and had some minor injuries such as muscle pulls. The problem was important, and if left alone would grow in seriousness. The injuries were real and the problem was justified to work on.

The specific problem was a strength-related problem on one task of the whole job. The job consisted of many tasks, but the toughest part of the job was manually moving metal boxes of material. The boxes were large, about 30 in. wide by 24 in. high by 72 in. long. The size was dictated by the production process and substantial changes in the boxes would have been expensive due to box replacement costs and due to production equipment changes. Each box carried about 1500 lb of material when full, plus the box weight of 300 lb. The operator was required only to adjust the position of the box manually, which involved moving it up to 18 in. Specifying the problem task was straightforward:

Job/Task	Symptom	Population Where Problem Seen	Body System Affected
Move box	Injury	Female	Strength, body weight

This analysis was also useful in fully defining the problem. Although the female population seemed to be having all the problems, a review of past injuries in that work section showed that two males had experienced hernias within the last 30 months, both from that same task. The task analysis pinpointed the specific problem:

Task/Element	Population	Population Capacity	Problem/Mismatch	Possible Solutions
Move box; push force is 65 lb.	Female	Push force is 44 lb.	Yes	Powered assists Less weight Use two people Larger casters

Chap. 6   The Problem Job   97

Redesign was the preferred solution and an evaluation of the alternatives showed that changing caster size was the best solution. This solution met the need of the population for a 44-pound push force and was permanent and economical.

6-3-2 A Changing Work Population

In the past it was common for limitations on some jobs to be carried over to other jobs on the promotion sequence. For example, if an entry-level job required someone to be able to lift a certain amount, only people with that strength performed that job and were promoted beyond it. Since the working population had been "screened" by this lifting requirement, all the remaining jobs in the promotion sequence could be designed to utilize that strength.

When the question was asked whether strength was required to do these other jobs, it became apparent that the additional strength was a convenience rather than an absolute requirement. With relatively little effort these jobs could have been designed a lot differently than they had been designed. The jobs would have to be redesigned to accommodate the wider range of population now seeking these jobs, since they were unnecessarily restrictive.

Similar situations occurred over and over again during the 1970s as challenges were made to traditional selection methods. From the ergonomist's viewpoint, it was important to identify those areas where an attribute (such as higher strength) was required and where it was merely a convenience and the job could be redesigned.

A similar problem occurred with stature, an anthropometric measure. One early job in the promotion sequence required a certain reach due to the design of the equipment. The production process dictated the equipment design and it was impossible to avoid this reach to operate the equipment. The reach screened out about 30% of the working population, and since reach and stature are gender biased, that percentage was mostly female. The remainder of the jobs were populated with people who had met this reach criterion at one time and were promoted.

There were planned personnel policy changes that would require the addition of shorter-statured people to those jobs. The ergonomist was asked to determine what job and equipment redesigns were required to accommodate these changes. The initial parts of the model were easy to use. The problem had been identified, and the major task analysis was a change in the current working population. The problem obviously needed solving and was defined to cover all the jobs in that operating department. There were no complaints at this time, but the problem appeared to be a real one. The real problem identification focused on categories of problems rather than on specific jobs. The categories of problems were:

1. Shorter stature
2. Lower arm reach
3. Lower eye height
4. Lower lifting capabilities

The specific jobs and tasks where there were problems were identified further in task analysis. The population capabilities were clearly identified for the fifth percentile female population:

- Stature = 59.0 in.
- Arm reach
    Vertical, standing: 72.9 in.
    Vertical, sitting: 49.1 in. above seat
    Front, pushbutton: 28.6 in.
    Front, grasp 26.6 in.
    Side: 27.0 in. from body centerline
    Side: 54.0 in. arm span
- Eye sight
    Standing: 56.3 in. from floor
    Sitting: 28.1 in. above seat
- Lifting: Maximum weight handled should be decreased above shoulder height (48.5 in.) and again above head height (59.0 in.)

The task requirements were then determined by observation and task analysis and by interviews with production personnel and their supervision. Whenever a reach, a lift, or a visual observation was required, a comparison was done to determine whether there was a mismatch with the task requirement and the capability of the expected population.

This was the heart of the analysis. For those tasks where the new work population could expect problems, design changes were initiated. Space does not permit a detailed explanation of the design changes made, but full use of steps and platforms was made, together with "reach extensions" such as longer tool handles. Administrative controls were not generally appropriate, although some natural selection of jobs by the very shortest people was permitted.

6-3-3 Predesign Analysis

While the problem job model is specifically designed to direct an analysis of problems on existing jobs, it can be useful for anticipating problems in future jobs, especially if they are similar to current jobs.

In one example, a new facility was to be built and questions arose about the need to air-condition the facility. There were equipment costs and operating expenses associated with the installation of air conditioning. Yet management wanted the air conditioning if it was needed.

The task analysis portion of the model was used most heavily. From the literature on heat stress, it was determined what levels of heat stress (combination of work load, temperature, humidity, and airflow) could be experienced without causing operational problems. A similar job was studied to determine work load and a nearby building of similar construction was monitored for temperature and humidity levels. A synthesis of these variables was made to perform a task analysis. The analysis indicated that no health or performance problems should result and that increasing the airflow would have a better effect than lowering the temperature. Substantial money was saved and the operating conditions were improved over the original plans.

6-4 SUMMARY

The problem job model is a useful tool to use in exploring those areas where people have problems performing operational tasks. The model, once fully understood, is also useful in exploring other areas of ergonomic design, such as new designs.

The model is a useful tool to have as the book moves into the traditional areas of industrial ergonomics in Chapters 7 through 11. The model should provide a road map for approaching existing problems. It should also be useful in determining which problems require your time and effort and which problems should be dropped before you get heavily involved.

ACKNOWLEDGMENTS

This chapter utilizes and builds on material from:

1. David C. Alexander and Leo A. Smith, "Evaluating Industrial Jobs, Determining Problem Areas and Making Modifications," *Industrial Engineering,* 1982, pp. 44-50.
2. David C. Alexander, "Job Design You Can Use," a continuing education seminar, sponsored by the Institute of Industrial Engineers, May 1982.
3. David C. Alexander, "Ergonomics and Job Design—Keys to Improved Worker Performance," a continuing education seminar, sponsored by the Institute of Industrial Engineers, May 1984.

QUESTIONS

1. What are the two ways to avoid ergonomic problems? Which is more effective?
2. Why is a "model" for analyzing a problem job useful?
3. What defines the successful resolution of a problem?
4. How are problem jobs initially brought to the ergonomist's attention?
5. What types of "tough jobs" are not appropriate for the problem job model?
6. List the ways to help determine whether a problem is real or just perceived to be a problem.
7. What are some ways to deal with perceived problems? Why should these be dealt with at all?
8. What information is necessary to specify an ergonomics problem?
9. Why is task identification a useful tool for the industrial ergonomist? Give an example of task identification (and the hierarchy of work) for a custodial operation.
10. What determines an ergonomic problem?
11. Explain the differences between an engineering redesign solution and an administrative control solution to a problem. Give an example for a noise problem and for a lifting problem.
12. List five general redesign methods of solving problems.
13. What criterion are useful to evaluate engineering redesign solutions?
14. When redesign solutions are not effective, what general administrative solutions are available?

Chapter 7

THE WORKPLACE AND THE WORK SPACE

This chapter provides the background necessary to design to accommodate the human user both at the workstation and in the nearby work area. Workplace design is discussed first and many examples are provided. The design of the work space is then discussed. Following that, a design process is presented together with how to handle the inevitable trade-offs that occur. Anthropometry, the science of human body dimensions, completes the chapter. Although most situations are covered with design guidelines and recommendations, the designer will occasionally resort to the basic anthropometric data to solve design problems. Those data are good for use with the problem job model as well as for new design work.

7-1 INTRODUCTION

"Shorty is about to retire," said my client. "How can we get someone else that size to clean the equipment?" Shorty was a slightly built fellow, about 4 ft. 11 in. tall. He could fit almost anywhere it seemed, but his real value was cleaning behind some equipment on the shop floor. He not only could fit there, he could also do the work when he was back there. Larger people had trouble bending over and twisting to clean properly. "If only we had placed that equipment a few inches farther out. . . ."

The task for those of us in industrial ergonomics is how to design equipment and facilities to fit people properly. This is especially troublesome when there are multiple users, many tasks, and lots of equipment to consider.

A critical element in workplace and work space design is the use of data about human dimensions. This field is called anthropometry and is a classical area for ergonomists. Many excellent examples abound of successful use of these data, such as the designs for seating, chemical control rooms, aircraft cockpit and cabin seats, telephones, computer workstations, automobile driver's seats, kitchen work areas, hand tools, and eyeglasses.

From the design standpoint, it may seem economical to avoid the problems associated with designing for a variety of human operators. It will certainly be quicker to design for a limited range of people rather than to ensure more universal operability. Yet the money saved on faster design is often a false economy, particularly as people experience problems with operation. The more flexible the equipment and the more people who can use the equipment, the less expensive it will be to operate. Figure 7-1 may help to illustrate why operations management can become frustrated as their operating costs rise on an otherwise "perfect design."

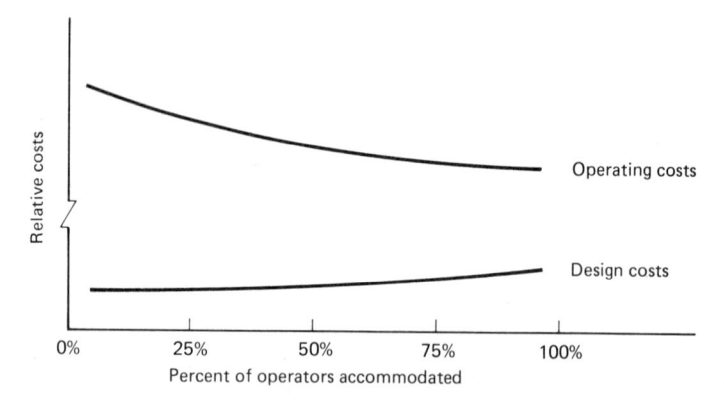

Figure 7-1 Design and operating cost trade-offs.

The degree of deliberate use of data about human beings as input for design is difficult to assess. Use of this input represents a range from a full integration of the human being into the design all the way to designs so poor that a redesign or retrofit is necessary before the design is usable. This range of human design input is illustrated in Table 7-1.

As one begins to think about design and how to design for human beings, it is very easy to become frustrated. "Where should I begin?" is the common response. There are three critical areas: (1) the person must use the equipment; (2) the person and the equipment must fit together into a usable workplace; and (3) the person must be able to move about the work space.

These elements represent a continuum for the designer (see Figure 7-2). At one end is the design of the equipment and how the user will interact with that single piece of equipment. Next comes the design of the workplace, which

Chap. 7   The Workplace and the Work Space   103

TABLE 7-1 Use of Anthropometric Data in Design Work

Case	Description	Example
Best design	Human fully integrated right from the start	The space program
Trial-and-error designs	Repeated designs that consider the human user more in each succeeding design	Personal computers
Marginal designs	Designs that have operating problems, yet are not severe enough to warrant redesign and/or retrofitting	Automobile maintainability
"Redesigns"	Designs that require redesign and/or retrofitting to be usable from the human standpoint (these may be very good designs after the retrofitting)	Nuclear power electrical generating plant control rooms

is the integration of the person with multiple pieces of equipment. The final element in this continuum is the design of the work space, which is the design for multiple users and/or movement around and about the workplace.

Equipment.   Normally, the designer in an industrial setting is faced with simply putting together "building blocks" of items that are already designed and fabricated. While the design of equipment is important, it is generally beyond the scope of this chapter. In the industrial setting, the designer/engineer more often works with purchased equipment and therefore has little control over the equipment design other than to influence purchase decisions.

Workplace.   The design of workplaces is usually the initial focus of the person practicing ergonomics in industry. Workplace design can be as simple as the workbench design for a person working from a chair, or it can be as complex as the design of a control room for a nuclear power generating station.

Work space.   The design of the work space represents the next major element in design. This is the design for multiple users and/or movement

Equipment ←→ Workplace ←→ Work space

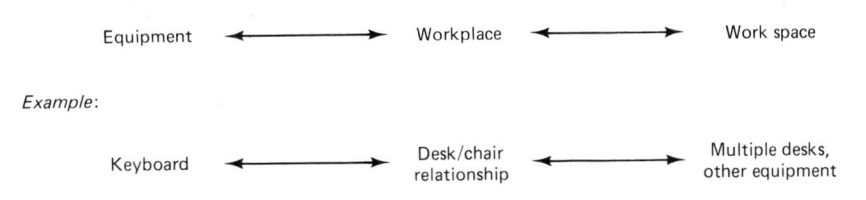

Example:

Keyboard ←→ Desk/chair relationship ←→ Multiple desks, other equipment

Figure 7-2 Ergonomic workplace design continuum.

around and about the workplace. The design of the work space is important from an ergonomics viewpoint, yet this is where the designer must have a fully developed appreciation of the many other variables that are important in engineering design. This stage of the design is often confounded with additional constraints, such as space limitations and cost considerations.

The focus of this chapter is on the middle of this continuum, simply because that is where more applications occur in the practice of ergonomics in industry.

7-2 WORKPLACE DESIGN

The design of workplaces is a very broad topic. Of course, it covers the situations where people work, but it also affects everyday activities as common as driving and cooking, and even recreational activities. The design of "workplaces" covers all situations where people are doing something. Even today, you have no doubt found yourself in many "workplaces," whether you have chosen to call them that or not.

It is important to have a workplace that is designed to accommodate the human user. The application of anthropometric data will allow the designer to "fit" the user into the design, without the extensive need for trial-and-error testing of the design. It will also allow the designer to "fit" a larger population than the one that is usually available for many trials. Equally important, the use of anthropometric data can allow the designer to predict those portions of the possible user population not being accommodated.

It is important to realize that workplace design is a very broad topic and that detailed applications require more information than is presented here. The focus of the rest of this section will be on the design of the industrial workplace and on the use of anthropometric data that are of value in the design of these industrial workplaces.

7-2-1 A Few Design Principles

Although this section is devoted to the illustration of workplace design, it is important to understand a few of the basic principles that support these design guidelines. The designer should strive to accommodate the largest user population possible (see, for example, Figure 7-3). A conventional standard was to design for at least 90% of the population, although it is becoming more and more common to design for accommodation rates of 95% or higher. In different cases, it will be important to emphasize either one end of the distribution or the middle portion of the distribution.

The smallest people in the design population establish the "reach" dimensions. If they can reach or touch a certain item, the larger people in the

Chap. 7 The Workplace and the Work Space   105

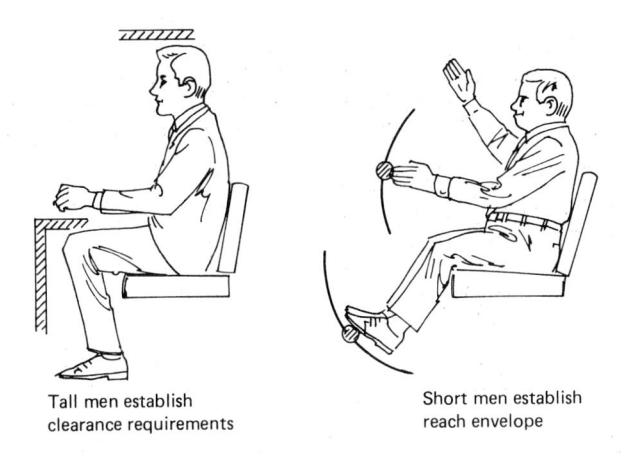

Tall men establish
clearance requirements

Short men establish
reach envelope

Figure 7-3 Design principles. *Source:* Adapted from H. P. VanCott and R. G. Kinkade, eds., *Human Engineering Guide to Equipment Design* (Washington, D.C.: U.S. Government Printing Office, 1972), p. 388.

population can also reach that item. This becomes apparent when thinking about the reach for items located on shelves. If the shorter people can reach the items, virtually everyone else can also. This logic is obvious in the location and/or design of many commonplace items such as the placement of fire extinguishers. In this case it is important for everyone to reach the fire extinguishers, so they have been placed low enough to accommodate virtually the entire population.

A contrasting situation can be found when there is a need to design clearance or access spaces. The larger people in the design population establish the "clearance" dimensions. If they can fit through an opening or pass in an aisle, others in that population can also. This logic is used in the design of doors, passageways, hatches, and emergency exits. Depending on the particular situation, it might seem proper to completely overdesign so that there is no question about use of the opening by the entire population. In many cases, however, there is a cost or penalty with overdesign. For example, with emergency hatches in aircraft, the size of emergency openings will affect the structural integrity of the craft. In most industrial situations, there is a significant cost associated with the addition of extra square footage to a layout. It is a constant struggle to provide adequate access space when that space simply looks empty on the blueprints.

The remaining situation occurs when it is important to accommodate the middle of the distribution rather than to favor one end of the distribution or the other. The midpart of the design population is used to establish the initial dimensions for the workplace. Although people at the ends of the design population may not be fully accommodated, this provides a usable workplace for the most people. These situations are also common, and demonstrate the requirements to trade off one portion of the population to accommodate another portion of the population. One common example is the height of a work desk which allows clearance for the legs of most people and at the same time allows most people to use the writing surface comfortably with their hands. But the work desk does not accommodate everyone—some people find it too small and others find it too high. A similar situation exists with the design for the automobile driver, where there is a need for clearance (especially during entry and exit), yet there is also a need for the driver to reach many controls during vehicle operation. Even with the adjustability that is built into most automobiles, it cannot accommodate the full range of the user population.

Usually, in a design situation, the designer has discretion in the actual design dimensions. The question then arises as to which design dimension to use. Many of the examples used in this text will list two dimensions, one that is preferable (it will accommodate more people) and one that is acceptable (it will accommodate approximately 90% of the population). The designer should use the preferable dimension if possible, and then default to the acceptable dimension if that is necessary due to other constraints. In establishing standards, this convention should be used, since it provides the designer with some additional information about the "criticality" of the design dimension.

7-2-2 Basic Elements

The application of anthropometry is usually a matter of finding the right dimension to fit each application. This section contains a number of situations that one may encounter with the appropriate workplace dimensions. These cover many of the typical needs of designers.

Clearances. Clearance dimensions are required for the human being to operate and maintain equipment. A frequent design flaw is the failure to allow adequate space for maintenance work. Figures 7-4 through 7-13 provide appropriate information about the clearance space needed for many different postures. Clearances are determined by the larger part of the expected work population.

Reaches. Reach dimensions help determine the maximum allowable spaces. Some typical reach positions are provided in Figures 7-14 through 7-17.

Workplaces. Some common dimensions for workplace designs are illustrated in Figures 7-18 through 7-20. These designs incorporate a number of design dimensions. They are based on the midband of the expected work population in order to accommodate the largest portion of people possible.

	Best	Minimum
A (Depth)	30 in.	20 in.
B (Height)	80 in.	74 in.

Figure 7-4 Stand-up operator workplace. *Source:* Wesley E. Woodson and Donald W. Conover, *Human Engineering Guide for Equipment Designers,* 2nd rev. ed. (Berkeley: University of California Press, 1964), pp. 2-123.

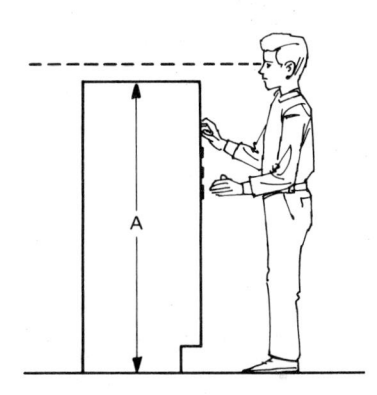

Standing height—55 in.
Sitting Height—Seat plus 25 in.

Figure 7-5 Visual clearance for a standing operator. *Source:* Wesley E. Woodson and Donald W. Conover, *Human Engineering Guide for Equipment Designers,* 2nd rev. ed. (Berkeley: University of California Press, 1964), pp. 2-123.

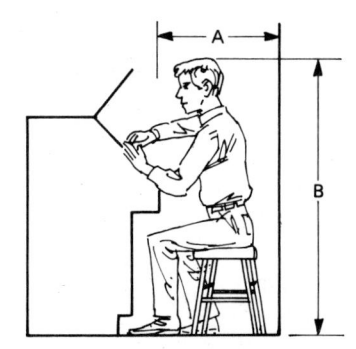

	Best	Minimum
A (Depth)	45 in.	36 in.
B (Height)	60 in.	52 in.

Figure 7-6 Maintenance performed in a seated operation. *Source:* Wesley E. Woodson and Donald W. Conover, *Human Engineering Guide for Equipment Designers,* 2nd rev. ed. (Berkeley: University of California Press, 1964), pp. 2-123.

	Best	Minimum
A (Depth)	40 in.	36 in.

Figure 7-7 Operator in a stooped position. *Source:* Wesley E. Woodson and Donald W. Conover, *Human Engineering Guide for Equipment Designers,* 2nd rev. ed. (Berkeley: University of California Press, 1964), pp. 2-123.

	Best	Minimum
A (Depth)	52 in.	46 in.
B (Height)	56 in.	48 in.

Figure 7-8 Work in a kneeling position. *Source: Standard-Human Engineering Design Criteria; MSFC-STD-267A* (Huntsville, Ala.: George C. Marshall Space Flight Center, National Aeronautics and Space Administration, 1966), p. 226.

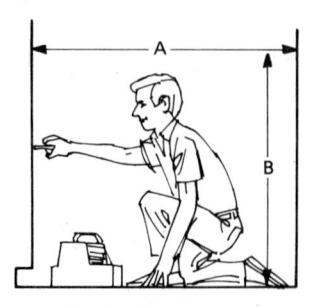

	Best	Minimum
A (Depth)	40 in.	36 in.
B (Height)	52 in.	48 in.

Figure 7-9 Work in a squatting position. *Source:* Wesley E. Woodson and Donald W. Conover, *Human Engineering Guide for Equipment Designers,* 2nd rev. ed. (Berkeley: University of California Press, 1964), pp. 2-123.

Chap. 7 The Workplace and the Work Space **109**

	Best	Minimum
A (Depth)	—	59 in.
B (Height)	36 in.	31 in.

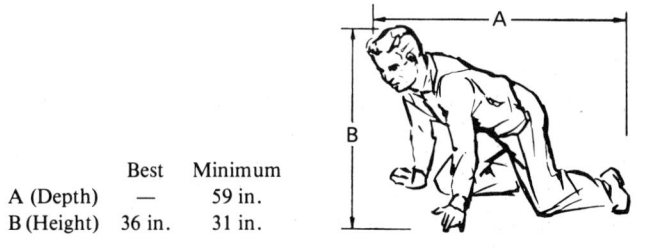

Figure 7-10 Maneuvering on all fours. *Source: Standard-Human Engineering Design Criteria; MSFC-STD-267A* (Huntsville, Ala.: George C. Marshall Space Flight Center, National Aeronautics and Space Administration, 1966), p. 226.

	Best	Minimum
A (Depth)	—	96 in.
B (Height)	20 in.	18 in.

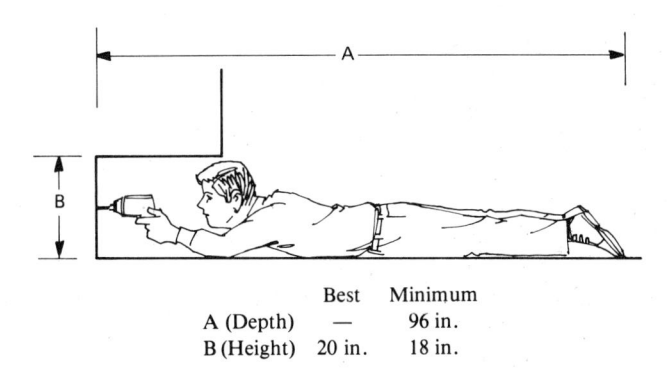

Figure 7-11 Working in a prone position. *Source:* Wesley E. Woodson and Donald W. Conover, *Human Engineering Guide for Equipment Designers,* 2nd rev. ed. (Berkeley: University of California Press, 1964), pp. 2-124.

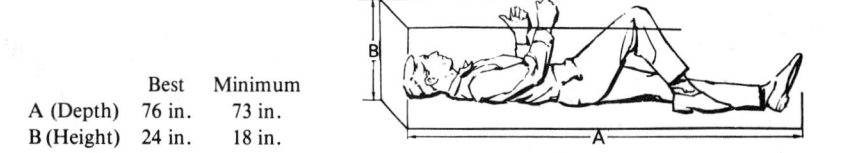

	Best	Minimum
A (Depth)	76 in.	73 in.
B (Height)	24 in.	18 in.

Figure 7-12 Working while laying on back. *Source: Standard-Human Engineering Design Criteria; MSFC-STD-267A* (Huntsville, Ala.: George C. Marshall Space Flight Center, National Aeronautics and Space Administration, 1966), p. 224.

Figure 7-13 Other clearance dimensions for maintenance tasks. *Source:* Wesley E. Woodson and Donald W. Conover, *Human Engineering Guide for Equipment Designers*, 2nd rev. ed. (Berkeley: University of California Press, 1964), pp. 2-124.

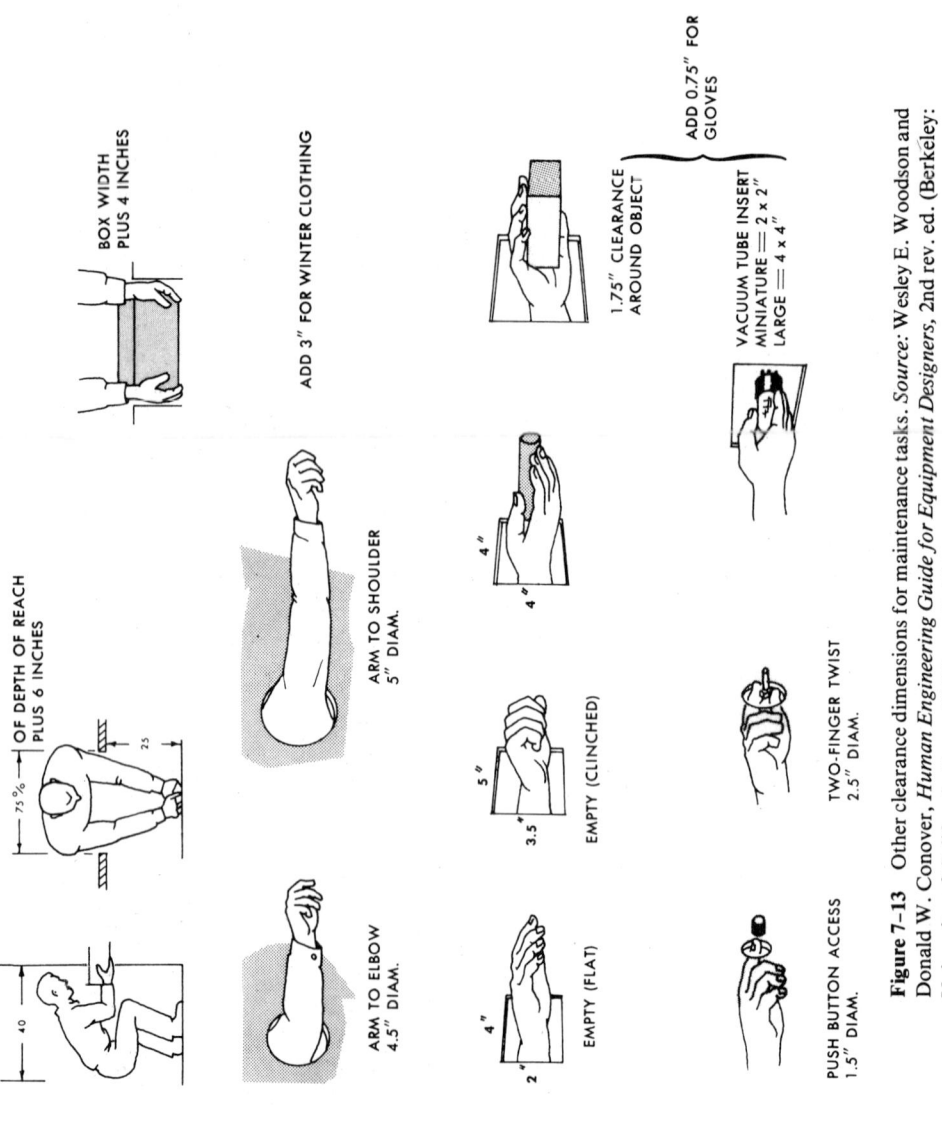

Chap. 7 The Workplace and the Work Space 111

95%	Male	88.5 in.
	Female	84.0 in.
5%	Male	76.8 in.
	Female	72.9 in.

Figure 7-14 Overhead reach. *Source:* Adapted from Julius Panero and Martin Zelnik, *Human Dimensions and Interior Space* (New York: The Whitney Library of Design, 1979), p. 100. Copyright © 1979 by Julius Panero and Martin Zelnik. Reprinted by permission of Watson-Guptill Publications.

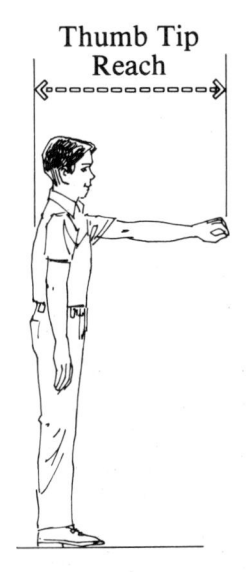

Thumb Tip Reach

95%	Male	35.0 in.
	Female	31.7 in.
5%	Male	29.7 in.
	Female	26.6 in.

Figure 7-15 Forward reach. *Source:* Adapted from Julius Panero and Martin Zelnik, *Human Dimensions and Interior Space* (New York: The Whitney Library of Design, 1979), p. 100. Copyright © 1979 by Julius Panero and Martin Zelnik. Reprinted by permission of Watson-Guptill Publications.

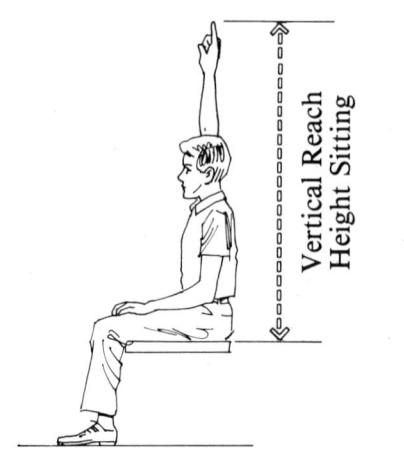

112   The Workplace and the Work Space   Chap. 7

5%	Male	51.6 in.
	Female	49.1 in.
95%	Male	59.0 in.
	Female	55.2 in.

Figure 7-16 Seated reach. *Source:* Adapted from Julius Panero and Martin Zelnik, *Human Dimensions and Interior Space* (New York: The Whitney Library of Design, 1979), p. 100. Copyright © 1979 by Julius Panero and Martin Zelnik. Reprinted by permission of Watson-Guptill Publications.

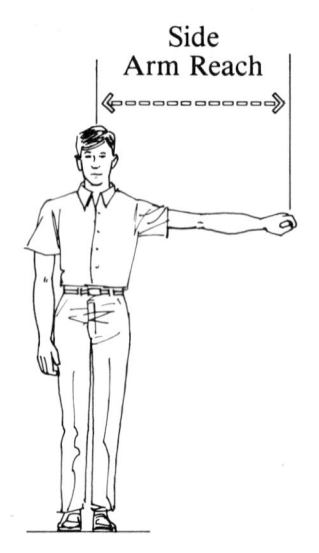

95%	Male	39.0 in.
	Female	38.0 in.
5%	Male	29.0 in.
	Female	27.0 in.

Figure 7-17 Sideways reach. *Source:* Adapted from Julius Panero and Martin Zelnik, *Human Dimensions and Interior Space* (New York: The Whitney Library of Design, 1979), p. 100. Copyright © 1979 by Julius Panero and Martin Zelnik. Reprinted by permission of Watson-Guptill Publications.

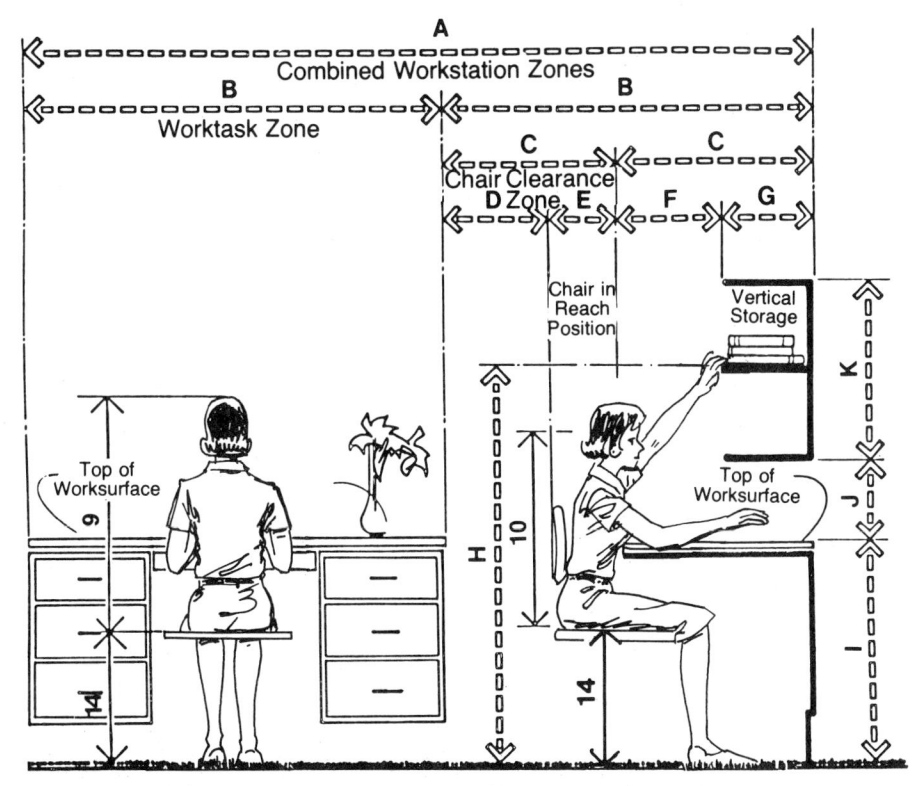

Chap. 7 The Workplace and the Work Space 113

BASIC WORKSTATION WITH VERTICAL STORAGE

A	120–144 in.
B	60–72 in.
C	30–36 in.
D	18–20 in.
E	12–16 in.
F	18–24 in.
G	12 in.
H	53–58 in.
I	29–30 in.
J	15 in. min.
K	25–31 in.

9 Sitting height (See item 9, Fig. 7-31)
10 Sitting eye height (See item 10, Fig. 7-31)
14 Popliteal height (See item 14, Fig. 7-31)

Figure 7-18 Office workstation. *Source:* Adapted from Julius Panero and Martin Zelnik, *Human Dimensions and Interior Space* (New York: The Whitney Library of Design, 1979), p. 181. Copyright © 1979 by Julius Panero and Martin Zelnik. Reprinted by permission of Watson-Guptill Publications.

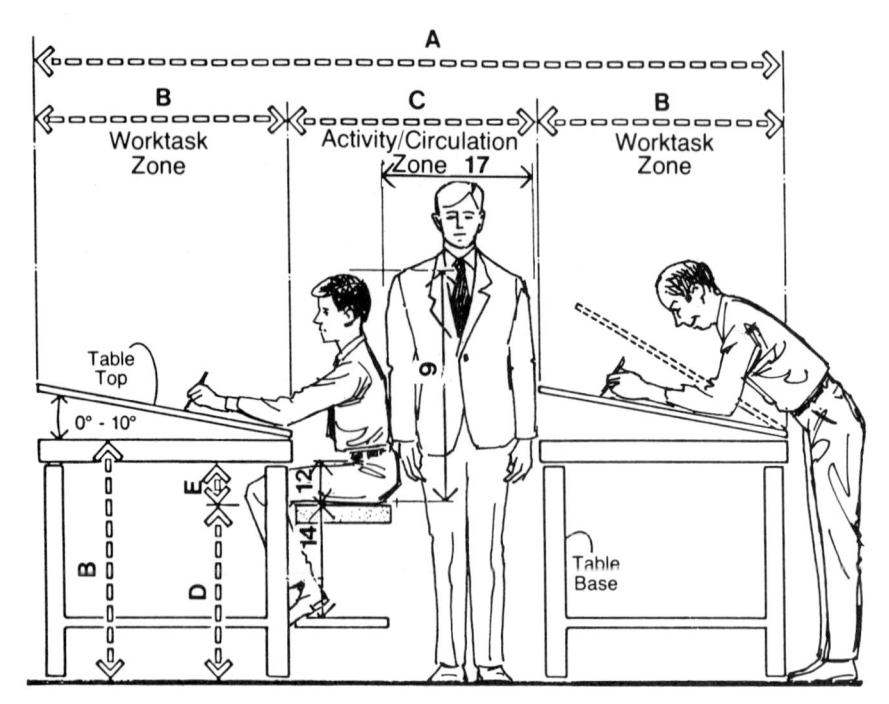

DRAFTING TABLES/CLEARANCE BETWEEN

A	108–120 in.		
B	36 in.	9	Sitting height (See item 9, Fig. 7-31)
C	36–48 in.	12	Thigh clearance (See item 12, Fig. 7-31)
D	21–27.5 in.	14	Popliteal height (See item 14, Fig. 7-31)
E	7.5 in.	17	Maximum body breadth (See item 17, Fig. 7-31)

Figure 7-19 Drafting table workstation. *Source:* Adapted from Julius Panero and Martin Zelnik, *Human Dimensions and Interior Space* (New York: The Whitney Library of Design, 1979), p. 261. Copyright © 1979 by Julius Panero and Martin Zelnik. Reprinted by permission of Watson-Guptill Publications.

7-2-3 Consolidated Designs

With experience, a designer soon finds that it is impossible to use all the recommended dimensions all of the time. For example, if a design calls for clearance for a taller person together with reach access for shorter people (see Figure 7-21), the designer begins to face a serious dilemma. Either one or both criteria will not be met.

These situations are probably more common (and more frustrating) to the designer than those shown previously. With the earlier situations, it is pos-

Chap. 7 The Workplace and the Work Space 115

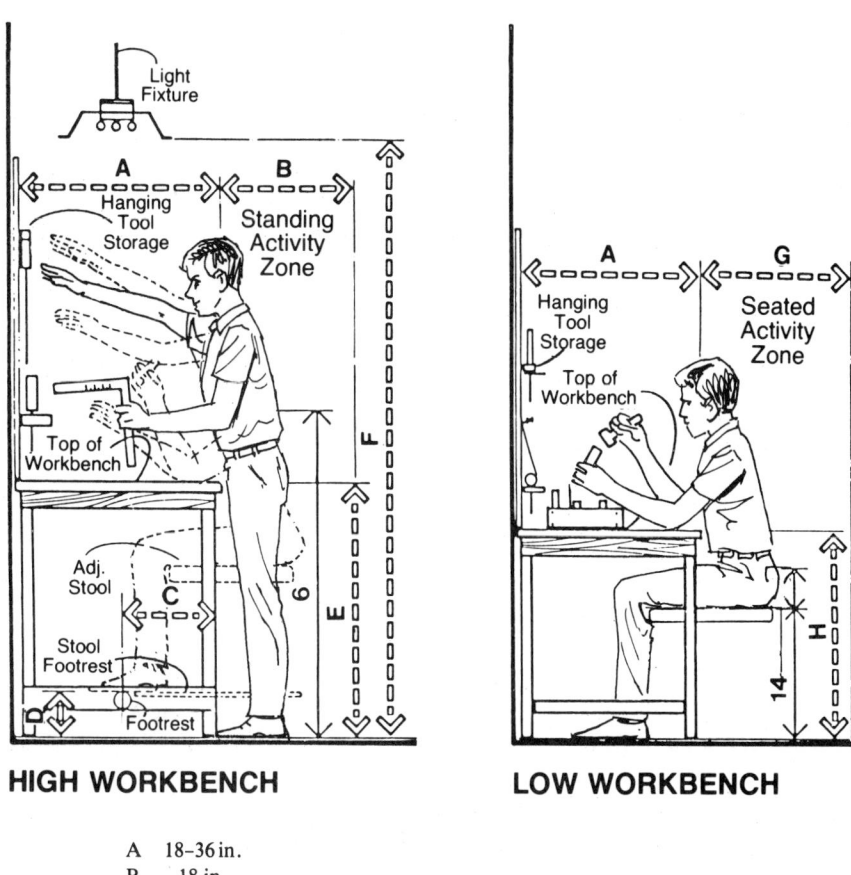

HIGH WORKBENCH   **LOW WORKBENCH**

A 18–36 in.
B 18 in.
C 6–9 in.
D 7–9 in.
E 34–36 in.
F 84 in.   6 Standing elbow height (See item 6, Fig. 7-31)
G 18–24 in.   12 Thigh clearance (See item 12, Fig. 7-31)
H 29–30 in.   14 Popliteal height (see item 14, Fig. 7-31)

Figure 7-20 Standing and seated workstations. *Source:* Adapted from Julius Panero and Martin Zelnik, *Human Dimensions and Interior Space* (New York: The Whitney Library of Design, 1979), p. 262. Copyright © 1979 by Julius Panero and Martin Zelnik. Reprinted by permission of Watson-Guptill Publications.

sible to provide design recommendations with confidence. Situations with complex criteria require more judgment. Since designs usually require the designer to balance several criteria, this section provides graphical illustrations of common situations. A more detailed discussion of design principles and trade-off situations is provided in the next section.

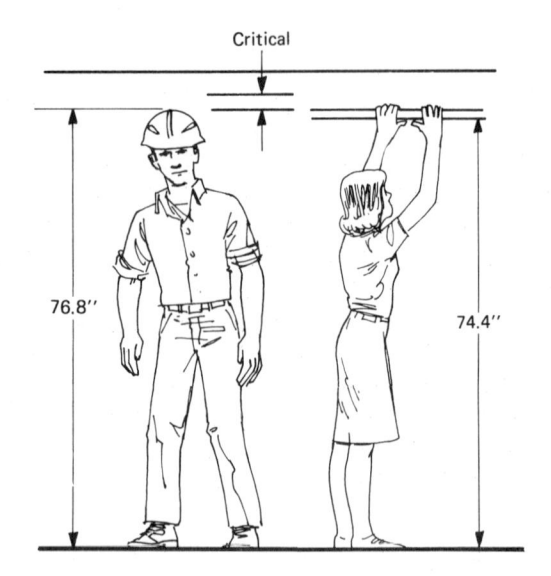

Tall male (95%) stature	72.8 in.		
Work shoes	1.5 in.	Short female (5%) vertical grasp reach	72.9 in.
Hard hat	2.5 in.	Work shoes	1.5 in.
Minimum clearance needed	76.8 in.	Maximum reach possible	74.4 in.

Figure 7-21 Reach and clearance tradeoffs. *Source:* Adapted from Wesley E. Woodson and Donald W. Conover, *Human Engineering Guide for Equipment Designers,* 2nd rev. ed. (Berkeley: University of California Press, 1964), pp. 5-16.

7-2-4 Design Processes and Concerns

Up to this point, there has been a lot of discussion about design and how to do it. It might be instructive to look at some more comprehensive examples of design, especially ones that illustrate the consolidation of many variables. A few instructive examples are a typical seated workplace, a standing workplace, and the workplace for someone using a video display workstation shown in Figures 7-22 through 7-24. These designs are inclusive and involve the use of many variables. Note the adjustability features to allow accommodation of more people.

7-3 WORK SPACE DESIGN

Work space design reflects the design for multiple users and/or movement around and about the workplace. The work space is more than the immediate work area because it includes:

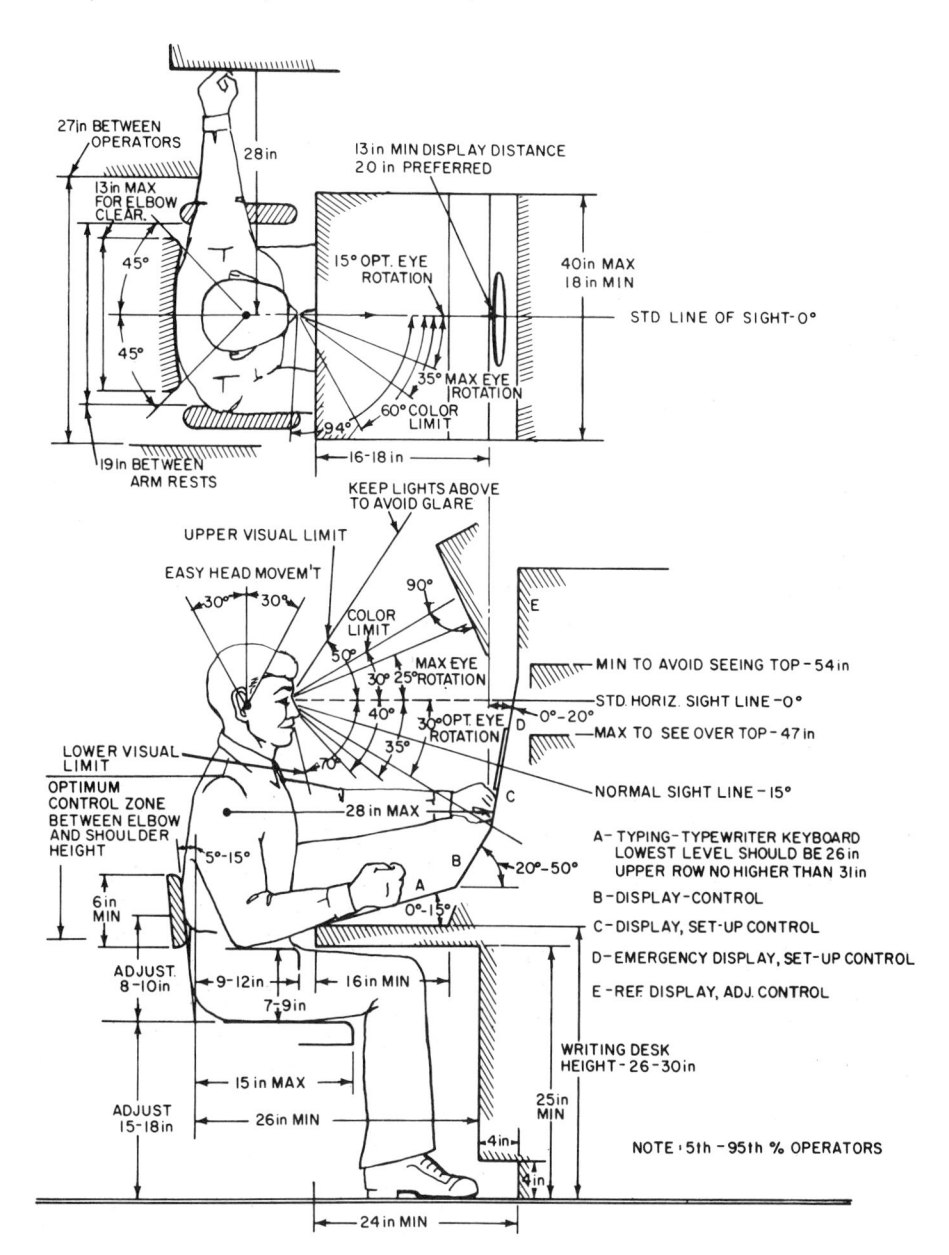

Figure 7-22 Suggested parameters for a seated operator. *Source:* H. P. VanCott and R. G. Kinkade, eds., *Human Engineering Guide to Equipment Design* (Washington, D.C.: U.S. Government Printing Office, 1972), p. 393.

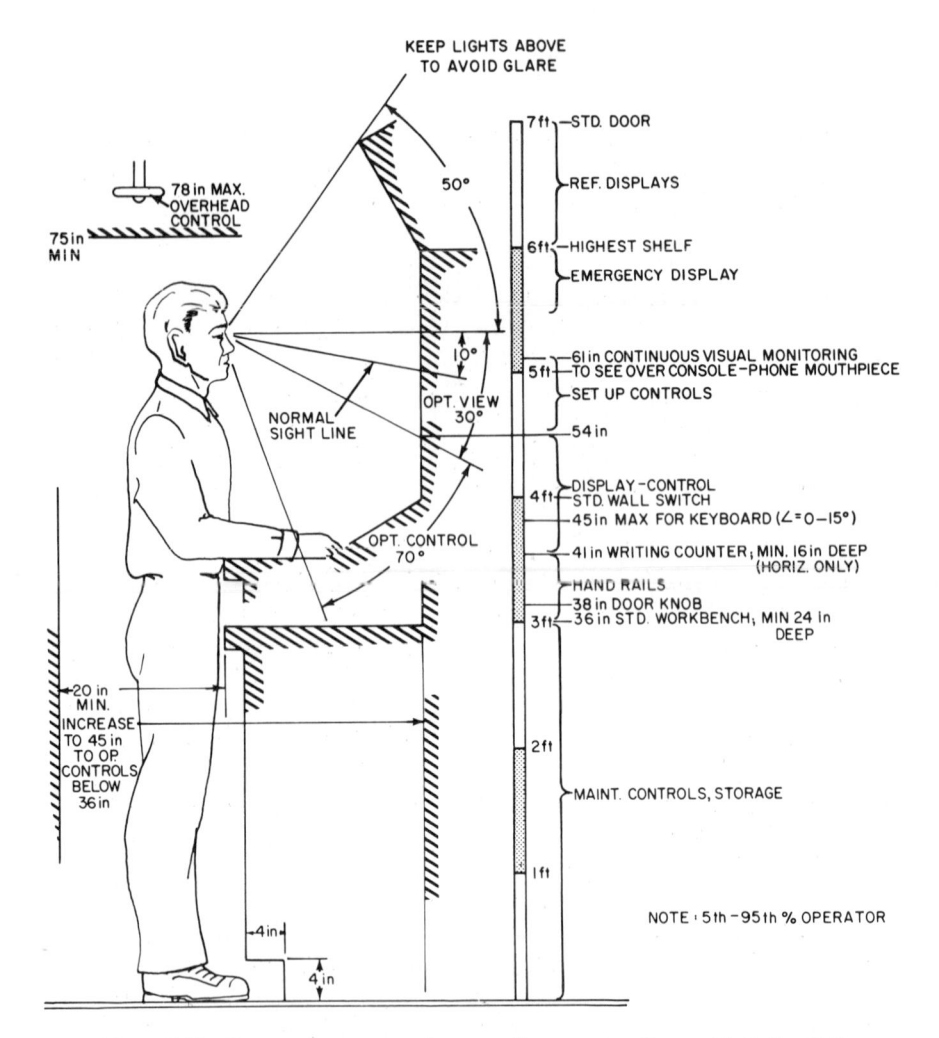

Figure 7-23 Suggested parameters for a standing operator. *Source:* H. P. VanCott and R. G. Kinkade, eds., *Human Engineering Guide to Equipment Design* (Washington, D.C.: U.S. Government Printing Office, 1972), p. 394.

- Access to and from the work area
- Movement in the work area
- Provisions for multiple persons at one work area
- The space requirements for visitors to the work area

There are more factors to deal with and these factors interact in more ways and in more complex ways.

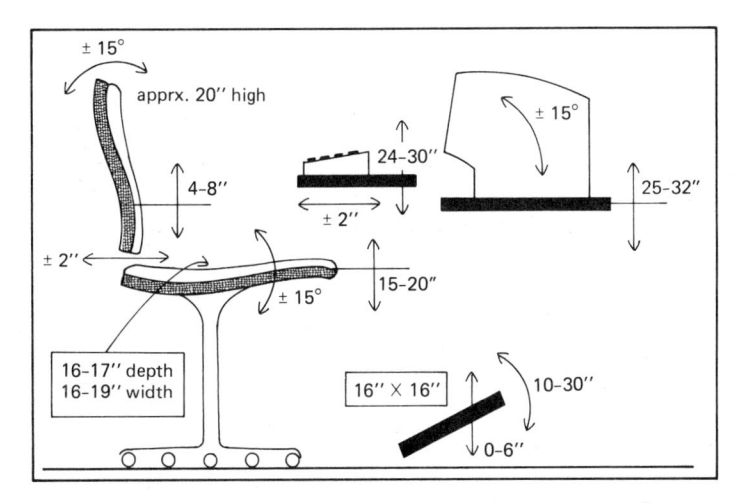

Figure 7-24 Dimensions (inches) for a VDT workstation. *Source:* K. H. E. Kroemer and D. L. Price, "Ergonomics in the Office: Comfortable Work Stations Allow Maximum Productivity, Part Six," *Industrial Engineering,* 14(7), July 1982, p. 26. Reprinted with permission. Copyright © Institute of Industrial Engineers, 25 Technology Park/Atlanta, Norcross, GA 30092.

7-3-1 Introduction

Work space design is less quantitative than workplace design. Since the events well be irregular, unpredictable, and infrequent, the need for them is not as easy to quantify. The design becomes more difficult:

- Should one allow room for one visitor or for two visitors?
- What about movement—how many people will walk abreast in the aisles?
- As one designs for handicapped personnel, should one also design so that two people in wheelchairs can meet in any area?

The questions are there, yet the answers are subject to speculation and judgment.

While work space design deals with clearance in the physical sense, it also covers a concept called "visual clearance"—the ability of operations personnel to see what is going on in the area. In work space design, visual clearance is often critical as a communications link allowing the personnel to work together effectively and/or as a way to monitor remote activities and instrumentation.

In addition, the design of the work space is sensitive to the dynamics of the work situation. In many designs, a static test of the appropriate people for the proper fit is a fair test of good design. In work space design, however, the movement of people in the work area is often variable, so a test for proper design often occurs only with a comprehensive mock-up or with the actual use of the design area. The "dynamics" of this type of design compared with a "static" design are very complicated.

Overall, the design of the work space may be the "softest" type of design work that the designer will do relative to anthropometry. In the work space design, the designer may have to make many assumptions about people and their activities in order to complete the design. These assumptions are required less with the design of the workplace and with simpler design issues.

7-3-2 Design Factors

When design trade-offs were discussed in the section on workplace design, the decisions were typically to favor one portion of the population over another. Since work space design situations usually involve only clearances (instead of the classical reach versus clearance trade-off), there are fewer decisions from the anthropometric standpoint. If enough space is provided, the design will be suitable for all people and for their movement in the design area. In work space design, bigger is better for the person (up to a point). The trade-offs in work space design are usually decisions between the person and traditional design criteria such as the cost of extra space or limitations on space imposed by fixed walls or other barriers.

Some important "human-oriented" design factors that need to be considered in the design of work spaces are the needs for:

- Production
- Service
- Maintenance

The production operator needs things to be convenient in order to have an efficient and effective operation. Work items and tools must be close and the operator should be able to work with a minimum of required movement. Of course, the operator should not be confined in one position for extended periods of time either, for fatigue from "static work" can be quite detrimental to good work.

Service work is often thought of as a need to bring in supplies and materials. It also covers routine service such as oiling and lubrication. Service work requires clearance needs for people and room to maneuver in the handling of material. These needs are not as regular as the needs of the production operator but still occur frequently and must be allowed for.

Maintenance needs are the least regular of the items mentioned above, but when needed, they are very important. Maintenance needs cover both access to the equipment and access within the equipment. Of course, there must be room to open the access ports and doors on the equipment and clearance for tools and parts to be brought in. Maintenance is a classical example of the "dynamics" of work space design, because it is necessary not only to leave room for the maintenance person, but to leave room for the person to move during the work.

There are many other criteria that also play an important part in design work. There are the costs involved, the need for proximity of equipment and people, and of course, the clearance and accessibility requirements of all the workplaces throughout the system. The integration of these design elements are discussed in Section 7-4.

7-3-3 Examples of Work Space Design

The use of anthropometry in the design of work spaces primarily uses whole body dimensions. The designs often need to reflect both a static design and a dynamic design. The static design is simply the placement of people in the area to see if they fit. The dynamic design will see if they have room to move together and to interact with each other. The dynamics of movement have not been as widely researched or published as the static human dimensions and more often must be assumed or tried with mock-ups of the design.

Usually, the design of the work space is based on clearance dimensions, and typically on the clearance of individuals. This means that when the design will accommodate the larger people, it will also accommodate the smaller people. Common situations for work space design involve the movement of several people, a single person moving among multiple pieces of equipment, and/or multiple person–multiple machine layouts.

Some common examples of this type of design are the passage of people in aisles and the design of a drafting table work space. A familiar example of a multiple person–multiple machine layout is a typical business office. Examples of these type of designs are shown in Figures 7-25 through 7-28.

7-4 DESIGN PROCESSES AND CONCERNS

In the design of a work system, the designers must be concerned not only with how the equipment works, but also with how a human operator works in the system. Designers must consider how the equipment will be used and how the operator will interface with the equipment. The economics, productivity, and quality of the whole system are important, not just optimizing one part of it.

The designers commonly use a standard method for each project: Determine design system objectives and criteria, determine constraints, gather appropriate design specifications and information, and then proceed with the design.

One critical aspect of the design information is information about people. The designer needs information about people: What can they do? What

**ONE PERSON PASSING
ONE STANDING AGAINST WALL**

36 in.
(30 in. MIN.)

TWO PERSONS PASSING

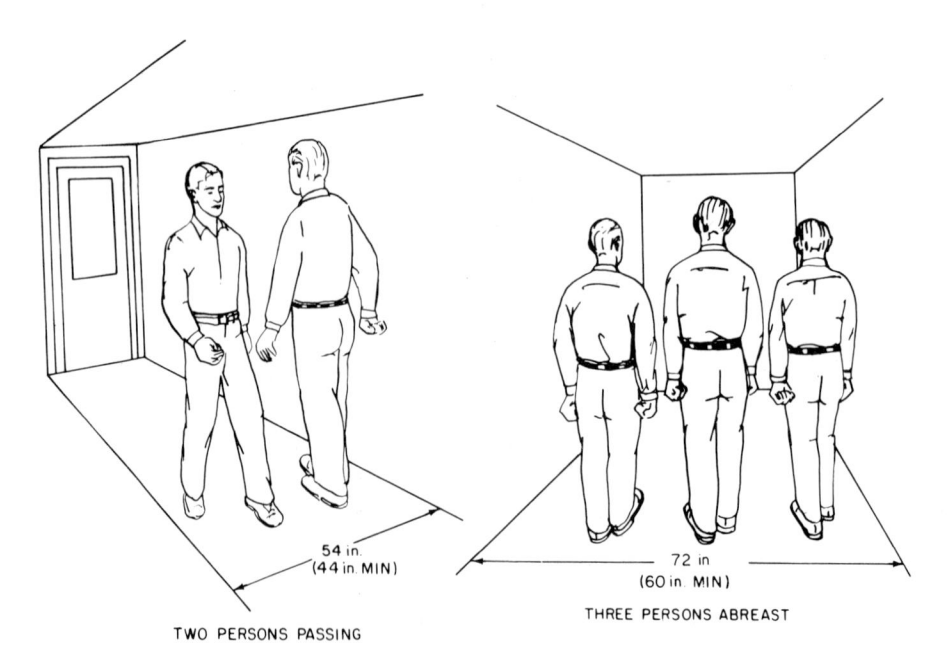

54 in.
(44 in. MIN)

THREE PERSONS ABREAST

72 in
(60 in. MIN)

Figure 7-25 Clearance for aisleways. *Source:* H. P. VanCott and R. G. Kinkade, eds., *Human Engineering Guide to Equipment Design* (Washington, D.C.: U.S. Government Printing Office, 1972), pp. 450-51.

Chap. 7　The Workplace and the Work Space　　　　　　　　　　　123

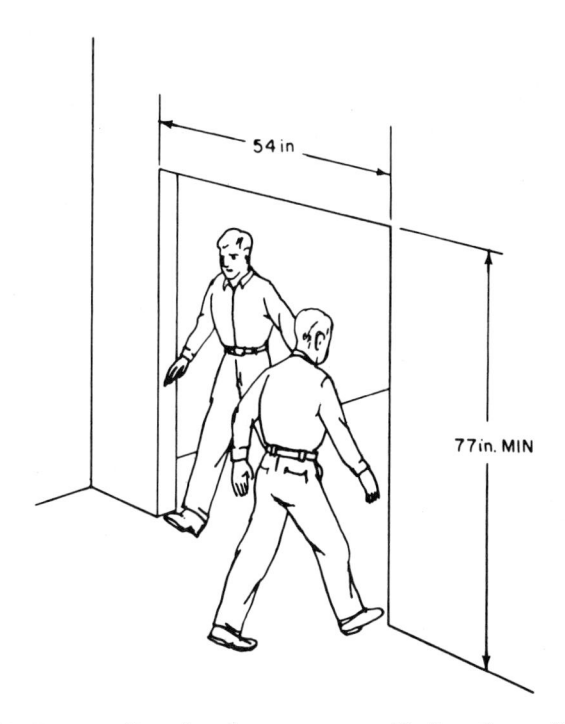

Figure 7-26　Doorway dimensions for two-person traffic flow. *Source:* H. P. Van-Cott and R. G. Kinkade, eds., *Human Engineering Guide to Equipment Design* (Washington, D.C.: U.S. Government Printing Office, 1972), p. 455.

are their capabilities? What are their limitations? And the designer needs to understand how to use that information.

"In principle, the logic of the design process is simple. In practice, many complex factors constrain the freedom of the designer and force him to make compromises. Nevertheless, poor work area arrangement dramatically affects system performance as well as human efficiency and safety."[1]

The designer needs to be able to solve the design decisions that come up relative to ergonomics. These decisions are almost always better when they are made during the design stages rather than later during a review or approval process by another person.

Design decisions are the providence of the designer. It would be presumptuous, for example, to maintain that the space allocations for the human being could not be adjusted. The fact of the matter is that in industrial design, the space allocations for the human being will be adjusted together with many other factors. By understanding the importance of allowing space

---

[1]Harold P. Van Cott and Robert G. Kinkade, eds., *Human Engineering Guide to Equipment Design,* (Washington, D.C.: U.S. Government Printing Office, 1972), p. 420.

DRAFTING CUBICLE

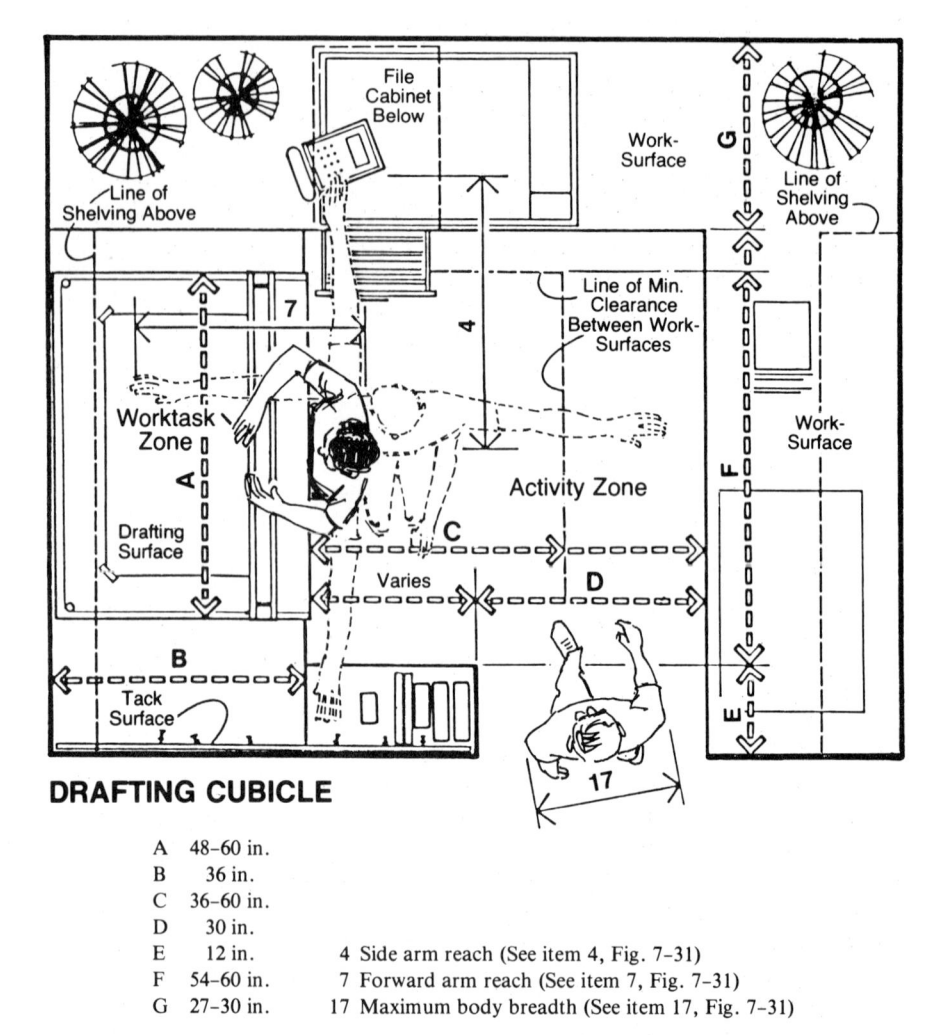

A	48–60 in.	
B	36 in.	
C	36–60 in.	
D	30 in.	
E	12 in.	4 Side arm reach (See item 4, Fig. 7-31)
F	54–60 in.	7 Forward arm reach (See item 7, Fig. 7-31)
G	27–30 in.	17 Maximum body breadth (See item 17, Fig. 7-31)

Figure 7-27 Drafting station workspace. *Source:* Adapted from Julius Panero and Martin Zelnik, *Human Dimensions and Interior Space* (New York: The Whitney Library of Design, 1979), p. 261. Copyright © 1979 by Julius Panero and Martin Zelnik. Reprinted by permission of Watson-Guptill Publications.

for the person and having information on the space needs of the person, a good designer will be able to incorporate the person with other criteria to reach a suitable layout that meets the needs of the human user.

7-4-1 Design Processes

There are certain guidelines that can be used to help the designer balance some of the multiple objectives of each major design. The compromises that

Chap. 7   The Workplace and the Work Space   125

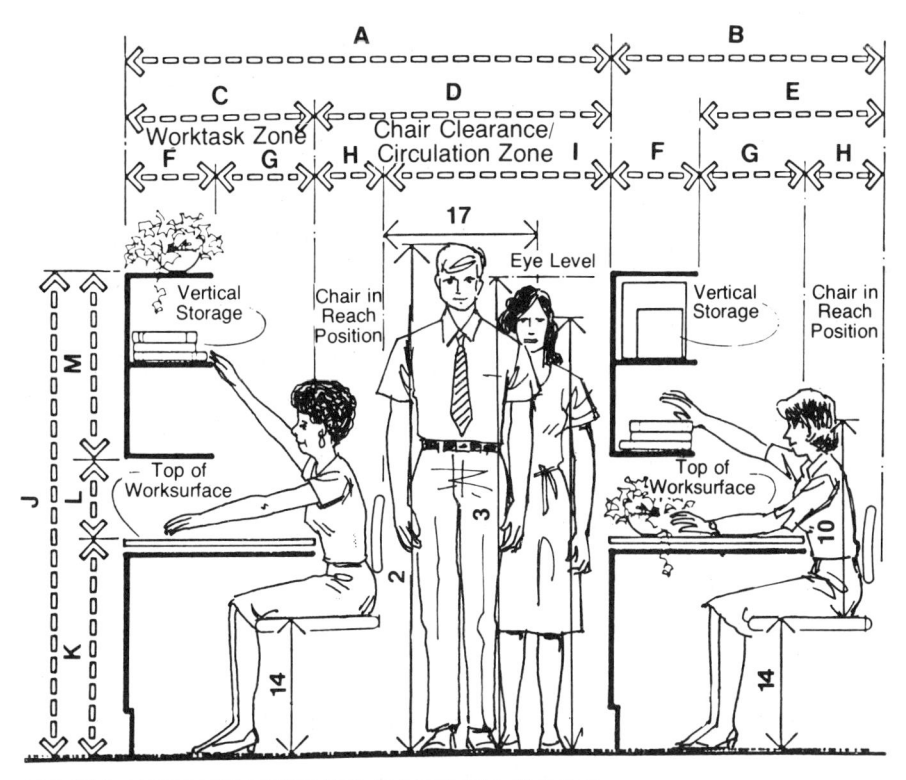

BASIC WORKSTATIONS BACK TO BACK
WITH VERTICAL STORAGE

A	78–94 in.
B	42–52 in.
C	30–36 in.
D	48–58 in.
E	30–40 in.
F	12 in.
G	18–24 in.
H	12–16 in.
I	36–42 in.
J	69–76 in.
K	29–30 in.
L	15 in. min.
M	25–31

2 Stature (See item 2, Fig. 7-31)
3 Standing eye height (See item 3, Fig. 7-31)
10 Sitting eye height (See item 10, Fig. 7-31)
14 Popliteal height (See item 14, Fig. 7-31)
17 Maximum body breadth (See item 17, Fig. 7-31)

Figure 7-28 Multiperson office workstation. *Source:* Adapted from Julius Panero and Martin Zelnik, *Human Dimensions and Interior Space* (New York: The Whitney Library of Design, 1979), p. 182. Copyright © 1979 by Julius Panero and Martin Zelnik. Reprinted by permission of Watson-Guptill Publications.

you make in designing are not easy choices. These general guidelines, used in sequence, ensure that the most critical elements are included.[2]

1. Visual tasks take priority. Design so that things can be seen from a normal eye position without stooping, craning or twisting the head, and so on. The eyes are in a fixed location, and moving them requires fatiguing body movements or head movements.
2. Major controls, hand tasks, or maintenance points should be placed after the visual tasks are located. They must be within easy reach of the hands. The hands have more movement flexibility than the eyes. Emergency controls count as major controls. The hands and arms can move easily, especially if the body does not have to move. Normal arm movements are not fatiguing.
3. Maintain a control/display relationship. Have controls near displays, and provide for compatible movements. The controls can be used more easily and reliably if they are near the appropriate displays.
4. Establish a sequence of operations. As the operator moves through a sequence of events, there should be a natural, orderly flow.
5. Emphasize convenience. Place minor controls, tools, and supplies near the hands.
6. Aim for consistency. Design properly rather than copying a bad design unless considerable changes in very critical items would be required and ingrained habits could cause misoperations.

Everyday examples can illustrate these points. "Tasks that require more visual acuity, such as fine assembly or writing, are normally done at a higher work surface than are tasks which require primarily hand work, such as routine assembly or typing. Tasks requiring strength more than vision are done at even lower working surfaces. A lawnmower is designed for easy arm and hand control, whereas a wall thermostat is positioned for visual priority."[3]

7-4-2 Trade-Offs

There are several types of trade-offs, all with their own particular ramifications. It is useful to look at the different kinds because of the impact that each has on accommodating the necessary population.

---

[2]Adapted from Van Cott and Kinkade, *Human Engineering Guide to Equipment Design,* pp. 387-389.

[3]David C. Alexander and L. A. Smith, "Evaluating Industrial Jobs, Determining Problem Areas and Making Modifications," *Industrial Engineering,* 14(6), June 1982, p. 47. Reprinted with permission from *Industrial Engineering* magazine, vol. 14, no. 6, June 1982. Copyright © Institute of Industrial Engineers, 25 Technology Park/Atlanta, Norcross, GA 30092.

Two different populations. The first type of trade-off is where one population is accommodated at the expense of another population. During the period when females were first beginning to join traditionally male dominated industries in large numbers, this type of trade-off was frequently used as the excuse to avoid the hiring of the females. It was often said, "Women can't do the job"; this was partially true, because the job had been designed for men and for the capabilities of men. In these cases, it was not malicious intent that caused the jobs to be designed for the men, but rather the lack of expectation that women would be joining the labor force in these particular jobs. Women were omitted from the design population, and in that sense, the trade-off had been made to favor the male population.

Another good example of this type of trade-off is the placement of an emergency shower handle. There are several important design criteria. First, people who may need to use the emergency shower must be able to reach the shower handle. Second, since the showers are usually placed in easily accessible locations, there is a need for clearance for a person to walk under the shower head and handle. When the design population was traditionally male, designing within these constraints was possible. With only the male population, the overhead reach of the shorter workers was enough to provide the clearance needed for the taller operators. However, as one began to include the significantly shorter female population, there were people who could not reach the shower handle. Attempts to lower the shower handle reduced the clearance needed for the taller males. This was clearly a dilemma that was accentuated with the addition of the second population. With this situation, there was a trade-off between two user populations, and each group could only benefit at the other's expense.

Fortunately, a suitable solution was found. A short rope was attached to the shower handle (see Figure 7-29), thus providing the necessary reach for the shorter people while maintaining the required clearance for the taller people. Had this solution not been found, the problem would still exist.

Within a population. A second trade-off is the trade within a population. This can be understood by thinking about accommodating everyone as an automobile driver. It cannot be done for the whole user population. This trade-off situation usually occurs when the designer decides to design for less than the whole population (typically 90 to 95% of the potential population). Whether this is done intentionally or not, the outcome is the same—some portion of the potential user population will be accommodated and the needs of another portion will not be accommodated.

Single person. The third trade-off is that of trading important functions that affect a single person. This situation can be understood by looking at the complex needs of a sewing machine operator. There are conflicting needs that simply cannot all be met. Since the operation is one that requires extensive

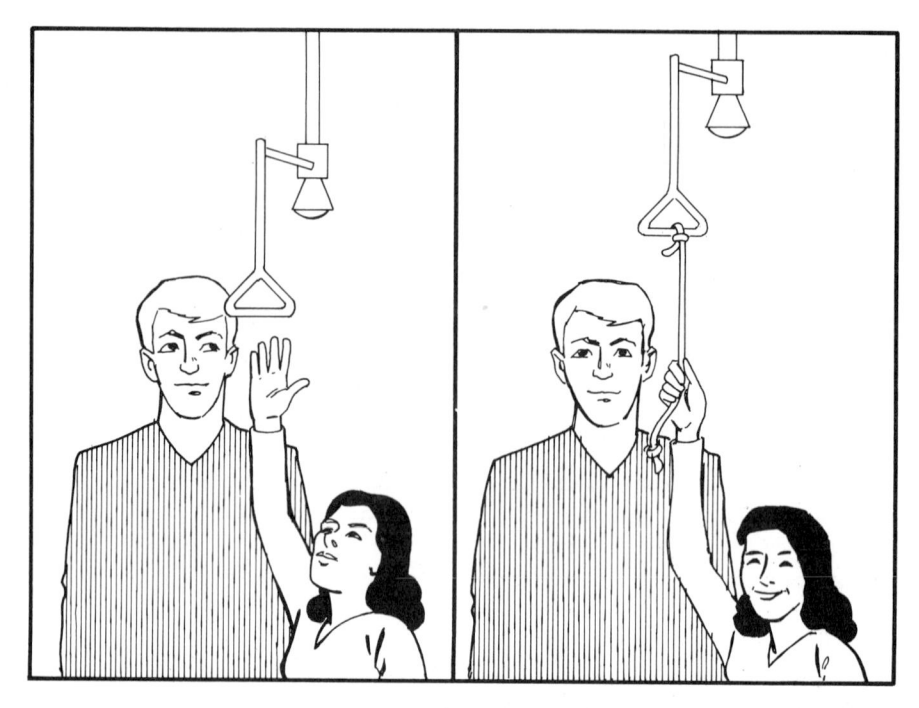

Figure 7-29 Trade-offs of reach and clearance. *Source:* David C. Alexander and L. A. Smith, "Evaluating Industrial Jobs, Determining Problem Areas and Making Modifications," *Industrial Engineer,* 14(6), June 1982, pp. 47–48. Reprinted by permission. Copyright © Institute of Industrial Engineers, 25 Technology Park/Atlanta, Norcross, GA 30092.

use of the hands, it is important to have the sewing head placed so that the operation can be done while the hands are at a convenient working height. This height usually approximates the seated elbow height for the operator. Another important function of the operator is to visually monitor the sewing task and, occasionally, to closely inspect the work being done. This, of course, suggests the placement of the sewing head closer to the eyes, to get proper visual acuity for the operators. Clearly, both of these needs cannot be met in the optimal manner with one design. In this case, it becomes a matter of trading off the comfort of a good hand height with the need for visual inspection. In these situations, there is no way that the optimal design can be achieved, even if the design is customized for only one person.

7-4-3 Solutions

These trade-offs present real problems for the designer. There are several methods of dealing with the trade-off problems discussed earlier. The most common methods are:

- Overdesigning
- Building in adjustability and/or flexibility
- Controlling the user population
- Creative solutions

Each of these methods is discussed below.

Overdesigning. Overdesigning may sound like poor design practice, but it is probably the most common method of avoiding the problems discussed earlier. In many cases, the design is one-sided; that is, it affects the group only if it is too big or too small. Examples of this design practice are so common that you may have to look for them before you become aware of them. The 7-foot doorways used in most commercial buildings accommodate almost everyone; hallways are wide enough for people to pass each other. This is the best approach if it can be employed in your design.

Adjustability. Probably, the classical method of dealing with the conflicting needs described earlier is to build adjustability and/or flexibility into the design. It is common to design for adjustability. This allows users with different anthropometric characteristics to use similar items and still have reasonably good accommodation for these users.

The use of adjustability is very common. Adjustability is designed into the automobile. In most offices, the chairs have built-in adjustability, providing comfortable seating for a wide range of people. In industry, where a fixed-height work table is critical, several thin movable platforms can be provided to create the correct individual work height. This type of design will accommodate the larger person, yet the built-in adjustability brings the shorter workers to the proper work height.

Controlling the user population. One rather dramatic solution to the problem of accommodating a large population is to limit the user population to one that is reasonable to accommodate. The U.S. military uses this strategy. To join the army, a male must be between 5 ft. and 6 ft., 8 in. tall; and a female must be between 4 ft., 10 in. and 6 ft. tall. This controls, to some extent, the wider range of statures found in the general population in the United States and limits both the wide design ranges on equipment and the extreme size ranges on clothing. Although this approach may not be possible or even desirable in many situations, it is reasonably effective in this case. It is an interesting way to limit the need for extreme adjustability and/or to avoid major misfits of the person-equipment interface. Selection is also done in industry, although it is becoming less common. It is used primarily in situations where extensive retrofitting of equipment is prohibitively expensive.

Win–win solutions. In some cases it may be possible to find a creative, win–win solution to the problem, like the addition of the rope to the shower handle mentioned earlier. How this is done is beyond the scope of this book, yet it is a common way of dealing with problems like these as they arise. Although many creative solutions come from the designer, that is not the only source of good ideas. The people doing the job often have good ideas and solutions to these same problems. If the situation is one that currently favors the taller portion of the population, yet now needs to be redesigned, go find the guy nicknamed "Shorty" and ask him how he is able to do the job. His homegrown accommodation may well be the answer to the problem that you are facing.

When designing to accommodate disabled persons, it is often good practice to consult with the individuals concerned. They probably have dealt with a similar situation on their own and may have reached a practical solution. One example comes to mind where a disabled person needed assistance with mobility on the job. The solution was a modification of an idea that had originated at home, the use of a rolling tea cart to provide stability during movement.

Design process. There are several ways to handle the design and tradeoff problems. Regardless of the situation, the following design process can be used to achieve a good solution.

1. Define the dimensional ranges needed and look for conflicts.
2. As necessary, use the concepts outlined above to develop a solution:
   a. Overdesign.
   b. Expand the design by including adjustability.
   c. Design for one end of the range and add adjusters to meet the needs of the remaining population.
3. If necessary, develop creative solutions, possibly in collaboration with the user population.
4. Allow people to adjust and accommodate themselves, to the extent possible.
5. Limit the user population to that which it is possible to accommodate.

7-4-4 Other Techniques

In the layout of the work space there are some other tools that can be of value to the designer. These are mentioned here to ensure that the design for the human being is done in a manner that allows not only for the physical space but for efficient and error-free operations as well. It is as important to accommodate the person physically as it is to design to help the person do the good job that they want to do.

In the layout, some good tools are link analysis, use sequence, and use frequency. Link analysis is a tool that illustrates the connection, or "link," between any two elements in the design. The link can be physical, where the person must travel from one element to another or it can be a "visual link" where one gets data from another location visually. Items with a high link count should be placed close together. The frequency of their associated use is enough to warrant this design decision.

Use sequence is a tool to illustrate the manner in which the use of one piece of equipment follows the use of another piece of equipment. This will provide a design order and layout for the pieces of equipment so that during operation, the person can move smoothly from one piece of equipment to the next with a minimum of backtracking and wasted time and effort. The idea is to place the items used in a large circuit so that in one complete loop, the person can do the job and then be ready to start again at the initial location.

Use frequency is a concept that places the most frequently used item in the most prominent location. This will allow a cluster of important items to be placed close together for easy access by the operator. These tools are valuable to the designer and can help to make the workplace and the work space as functional as possible.

7-5 ANTHROPOMETRY

"Anthropometry is the technology of measuring various human physical traits, primarily such factors as size, mobility, and strength. Engineering anthropology is the effort to apply such data to equipment, workplace, and clothing design to enhance the efficiency, safety, and comfort of the operator."[4]

Anthropometric measures vary considerably, with such factors as gender, race, and age playing a dominant role in this variability. The application of anthropometric data, then, is controlled largely by the anticipated user population. Engineering anthropology is shown most clearly in the workplace and work space design sections as the application of anthropometric data.

7-5-1 Importance of Human Body Dimensions

It is critical for the designer to consider the human being intentionally and thoroughly from the conception of the design rather than as an incidental or add-on part of the design. "The quality of the interface which connects man with his machines frequently determines the ability and ultimate performance of the man/machine unit.

"The beginning of any man/machine interface is objective knowledge of the full range of man's size, shape, composition and mechanical capacities.

[4]Van Cott and Kinkade, *Human Engineering Guide to Equipment Design*, p. 467.

Hence, anthropometry is fundamental to successful designs. . . . The only alternative is the costly process of trial and error."[5]

7-5-2 Data Basics

A few key concepts will help to explain the use of anthropometric data.

Data ranges. One fundamental concept that serves as a foundation for the application of anthropometry is that the data form a range or distribution. The range of the data is as important as the average point of the data, for the range must be accommodated, not just the mean. Often the data will approximate a normal distribution, although there are frequent cases, such as body weight, where a skewed distribution can be found.

The percentage of the range that can be accommodated is important since that determines the usability of the design or work areas by a population, not just a person. When the tails of a distribution are quite long, it becomes very expensive to accommodate the extreme ends of the distribution. At the same time, by designing for adjustability and providing flexibility to accommodate a larger percentage of the potential users, the design is more universal.

Figure 7-30 illustrates the concept of variability with seated eye height. Accommodations for a small percentage of the population are easy, whereas accommodations for larger percentages require more initial design work to build in adjustability.

Average person. There is no "average" person. There are distributions of all the human body attributes, but unfortunately, these distributions do not correlate well. For example, height is not an indicator of body weight since there are tall people who are thin and there are tall people who are heavy. This "scattered data pattern" holds for the relationships between most variables and rules out the use of an "average" person in the design process. The designer must determine the task requirements and then find the anthropometric data that fit those requirements. Only then can an effective design be completed.

It is important to realize that the designer is not average either. It may be useful for the designer to ask "Can I do that?," simply to verify that the design is not completely unusable. However, an affirmative answer to that question does not mean that the design is suitable for the entire user population.

Whom to design for. Designing for the average can lead to unexpected operational problems. For example, if a designer sized a vessel access port for

---

[5]*NASA Anthropometric Source Book, Vol. 2: A Handbook of Anthropometric Data* (Houston, Tex.: National Aeronautics and Space Administration, 1978), p. iii.

Chap. 7   The Workplace and the Work Space   133

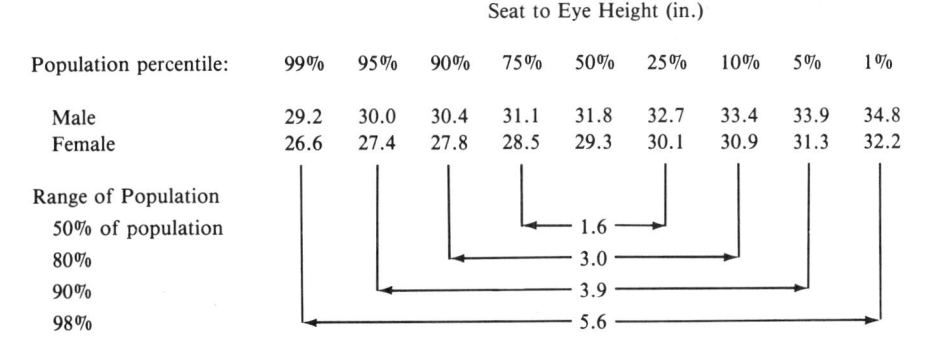

Seat to Eye Height (in.)

Population percentile:	99%	95%	90%	75%	50%	25%	10%	5%	1%
Male	29.2	30.0	30.4	31.1	31.8	32.7	33.4	33.9	34.8
Female	26.6	27.4	27.8	28.5	29.3	30.1	30.9	31.3	32.2

Range of Population
50% of population — 1.6
80% — 3.0
90% — 3.9
98% — 5.6

Figure 7-30 Anthropometric variations. *Source: NASA Anthropometric Source Book, Vol. 2: A Handbook of Anthropometric Data* (Houston, Tex.: National Aeronautics and Space Administration, 1978), p. 209, NASA RP-1024.

the average person, 50% of the population would not be able to enter the vessel or could do so only with great trouble.

The larger people in the design population establish the "clearance" dimensions. If they can fit through an opening, others in that population can also.

The smallest people in the design population establish the "reach" dimensions. If they can reach or touch a certain item, the larger people in the population can also reach that item.

The middle band of the design population establishes the initial dimensions for workplace design. Although people at the ends of the design population may not be fully accommodated, this provides a usable workplace for the most people.

There is an interesting story that circulates about the fact that most human body dimensions are normally distributed. So the story goes, the Quartermaster Corps during World War II asked what sizes of clothing to buy so that the soldiers could easily be fitted. They received a list of the dimensions and sizes that would be required and proceeded to order equal numbers of each size. Since the soldiers, of course, came in a normally distributed range of sizes and were forced to wear what was available, the often heard complaint of having to wear clothing that was either too large or too small could have been well founded.

7-5-3 Use of Anthropometric Data

The use of anthropometric data is not as straightforward as it looks. There are several ways to misuse the data without realizing it. The correct populations must be measured or referenced. For example, when designing toys, use data on the proper age group to determine correct design sizes. Determine whether it is appropriate to design for reaches, clearances, or for the middle of the population. Which limits the fewest people? How are trade-offs handled?

These steps help to use anthropometric data properly to satisfy the design specifications. A brief description of these steps follows this list.

1. Determine the user population and plan where to get the necessary data.
2. Determine the type of accommodation to be made—is it a reach situation, a clearance situation, or does the middle band of the population need to be accommodated?
3. Determine the percentage of the population to be accommodated.
4. Gather the design specifications.
5. Repeat for other design parameters as necessary.
6. As needed, plan trade-offs to meet the needs of the users and the needs of the design.
7. Design.

Population. The first step in the use of anthropometry is to select the desired population for the data source. Who will be using the equipment, the workstation, or the tool that is designed? Who could be the user of the design? Who must be accommodated, and who should be accommodated? For most design work, one can usually use a 50/50 mixed population of males and females for the design input. Once this decision has been made, the source of the design information can be specified or located.

Accommodation. The second step is to determine the type of accommodation to be made. Is it a reach situation or a clearance situation? Is it appropriate to design for the middle band of the population, and thereby limit the use of the design by those on the ends of the user distribution?

Whom to design for. The third step is to decide the percentage of the population to design for. This is a difficult step because, in most cases, it actually involves the specification of the percentage of people who will *not* readily be able to use that design. The determination of the percentage of the population to design for is important. It is common to try to accommodate as large a percentage of the population as possible. Frequently, a designer will try to accommodate 90, 95, or even 98 or 99% of the population. Generally, there is a small increase in initial design cost with increases in the percentage of the population accommodated. This cost should be offset later by reducing the need to modify the design and by the ease of use of the item by the user population.

Design specifications. The fourth step is to gather the design specifications. This involves looking up the design information and documenting it. At this time, it may be appropriate to modify the percentage of the population that is accommodated. Once the data on the user population are in hand, the designer may find that it is not difficult to accommodate a larger percentage of the population.

Repeat. The fifth step is to repeat this process as necessary, in order to obtain all the necessary design information. As this process is accomplished, it is increasingly useful to have full documentation of each parameter. The information on the extra space necessary to accommodate larger percentiles will be useful during the sixth step.

Conflicting situations. The sixth step involves planning how to accommodate possibly conflicting situations. Concurrent reach and clearance situations are the most troublesome. The earlier information on trade-off situations will be useful.

Proceed. The last step is to proceed with the design. Once the work to accommodate the human user has been accomplished, the designer can proceed with confidence regarding the usability of the design.

7-5-4 Anthropometric Data

This section provides general anthropometric data for use in design work to supplement the applied information already provided in the sections on workplace and work space designs.

As mentioned earlier, there are both gender and ethnic differences in the anthropometric data bases. These have generally been consolidated in the information presented, since the most useful information is that which represents a consolidation of the usual design populations. Rather than forcing the designer to integrate the male and female populations, for example, those populations have been consolidated in a single design distribution.

There are two methods of providing information on the human beings. The first is called a static or structural dimension. It provides the results of the measurements of a person in a standard position. The other method is called a dynamic or functional dimension. It provides the results of the measurements of a person in a working position. The static dimensions are the easiest to obtain, while for the ergonomist, the dynamic dimensions are probably the most useful. Unfortunately, most data are available only in static form.

Figure 7-31 and Table 7-2 are provided as general references on the human being. For more detailed information, some useful references are provided. For further anthropometric data and their applications, the following sources will be useful.

TABLE 7-2 Clothing Allowances

Clothing Type	Allowance (in.)	Most Important Body Dimensions Affected
Men's suit	0.50	Body depth
	0.75-1.0	Body breadth
Women's suit	0.25-0.50	Body depth
or dress	0.50-0.75	Body breadth
Winter outerwear,	2.0	Body depth
including basic suit	3.0-4.0	Body breadth
or dress	1.75-2.0	Thigh clearance
Men's heels	1.0-1.5	Stature, eye height, knee-height sitting, popliteal height
Women's heels	1.0-3.0	Stature, eye height, knee-height sitting, popliteal height
Men's shoes	1.25-1.5	Foot length
Women's shoes	0.5-0.75	Foot length
Gloves	0.25-0.50	Hand length, hand breadth
Hard hat	2.5-3.0	Stature

Source: Julius Panero and Martin Zelnik, *Human Dimensions and Interior Space* (New York: The Whitney Library of Design, 1979), p. 72. Copyright © 1979 by Julius Panero and Martin Zelnik. Reprinted by permission of Watson-Guptill Publications.

		Males 5%	Males 95%	Females 5%	Females 95%
1. Vertical grip reach	Reach	76.8	88.5	72.9	84.0
2. Stature	Clearance	63.6	72.8	59.0	67.1
3. Standing eye height	Reach	60.8	68.6	56.3	64.1
4. Side arm reach	Reach	29.0	39.0	27.0	38.0
5. Maximum body depth	Clearance	10.1	13.0	—	—
6. Standing elbow height	Adjustable	41.3	47.3	38.6	43.6
7. Forward arm reach	Reach	29.7	35.0	26.6	31.7
8. Sitting vertical reach height	Reach	51.6	59.0	49.1	55.2
9. Sitting height	Clearance	31.6	36.6	29.6	34.7
10. Sitting eye height	Reach	30.0	33.9	28.1	31.7
11. Seated elbow height	Adjustable	7.4	11.6	7.1	11.0
12. Thigh clearance	Clearance	4.3	6.9	4.1	6.9
13. Knee height	Clearance	19.3	23.4	17.9	21.5
14. Popliteal height	Reach	15.5	19.3	14.0	17.5
15. Buttock-toe length	Clearance	32.0	37.0	27.0	37.0
16. Hip breadth	Clearance	12.2	15.9	12.3	17.1
17. Maximum body breadth	Clearance	18.8	22.8	—	—

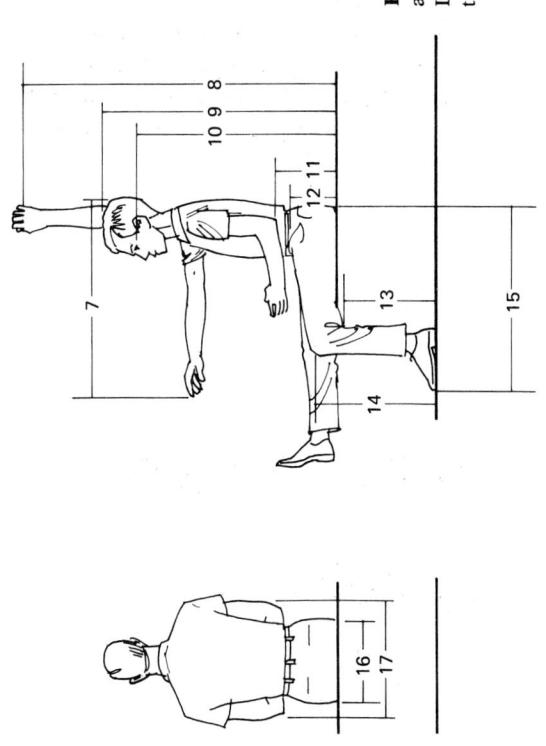

Figure 7-31 Anthropometric data (inches). *Source:* Adapted from Julius Panero and Martin Zelnik, *Human Dimensions and Interior Space* (New York: The Whitney Library of Design, 1979), pp. 86–104. Copyright © 1979 by Julius Panero and Martin Zelnik. Reprinted by permission of Watson-Guptill Publications.

Damon, Albert, Howard W. Stoudt, and Ross McFarland, *The Human Body in Equipment Design.* Cambridge, Mass.: Harvard University Press, 1971.

NASA Anthropometric Source Book, Vol. II: *A Handbook of Anthropometric Data.* Houston, Tex.: National Aeronautics and Space Administration, 1978.

Panero, Julius, and Martin Zelnik, *Human Dimension and Interior Space.* New York: The Whitney Library of Design, 1979.

Van Cott, Harold P., and Robert G. Kinkade, eds., *Human Engineering Guide to Equipment Design.* Washington, D.C.: U.S. Government Printing Office, 1972.

7-5-5 Design Aids

The concept of a "design person" is something that many people have been interested in for a long time. The design person is a collection of anthropometric measures that reflect the population that you are working with. This design person then becomes a useful design tool because it can be used to answer many questions about the dimensions that you will need for your designs. Some useful tools are "Humanscale 1/2/3" by Niels Diffrient and "The Measure of Man" by Henry Dreyfuss.

Although it is helpful to have these tools and to use them in design work, it is important to remember that you are dealing with consolidated anthropometric measures. They are certainly helpful as a guide and as a design tool, but for critical dimensions, it is important to check the actual data to ensure the proper usability by the population.

7-6 SUMMARY

One of the most common problems faced by the ergonomist is the design of work areas that "fit" the human operator. This chapter provides data on human dimensions and a background on how to use those data.

It is important to ensure that other designers have suitable data to enable them to design for the human user and that they understand the need for this design practice. Designing for people is most effective when all the designers see it as one of their responsibilities and then have the design data to carry it out.

QUESTIONS

1. Regarding the "fit" of the person on the job, what are three critical areas?
2. The designer must strive to accommodate the largest user population possible. How much of the user population should be accommodated?
3. What group of workers determines the reach dimensions in a workplace? Which group determines the clearance dimensions in a workplace?

Chap. 7 The Workplace and the Work Space   139

4. List five clearance needs and the appropriate design dimension.
5. List three reach needs and the appropriate design dimension.
6. In comprehensive designs, what is the key feature that allows the design to be used by most people?
7. What are the design needs of work space design?
8. Discuss the difference between a "static" and a "dynamic" test of a work space.
9. List the six design guidelines, in order, that let the designer balance some of the multiple objectives found in major designs.
10. Discuss three trade-off situations and give an example of each.
11. Is it more important to accommodate the average point in a distribution or the range of points in a data distribution? Why?
12. Examine the distribution of the male seated eye height. What amount of adjustability is required to accommodate 50% of the total male population? To accommodate 80% of this population? How much adjustability must be added to increase the accommodation from 90% to 98% of the population?
13. List the steps in the proper use of anthropometric data. Illustrate the steps with the design of a doorway in a home.
14. What are the critical raw anthropometric dimensions for the following situations?
    a. Grip an overhead valve?
    b. Doorway opening?
    c. Seat back to control panel?
    d. Seat to table distance?
    e. High density standing in an elevator?

Chapter 8

THE HEAVY JOBS

The material on "heavy jobs" will provide an overview of manual materials handling tasks and the detrimental effect they have on the human workers. Emphasis is placed first on analysis to determine whether or not there is a problem, then on pinpointing the specific problems, and finally on developing practical corrections to the problem.

There is information on the prevention of problems through proper design. Tables and graphs of acceptable manual materials handling limits are provided.

8-1 INTRODUCTION

"Tell me," asked the design engineer, "how many injuries will be prevented by the changes that you are recommending?" I hadn't expected this question, so I hesitated, and then suggested instead that the percentage of the population able to do the job would double. This was a significant improvement from my point of view, and I was proud of the changes I was recommending. "But," he protested, "how can I determine a rate of return from that information? I have to justify the expense of these proposed changes. . . ."

Often the ergonomist is confronted with those who want quantitative data in a human world. The designer is used to working with numbers and specifications, and data from the human worker may be too vague to be of appropriate value. When the designer is uncompromising, it is a difficult position to be in.

Yet injuries from manual materials handling are expensive and frequent. "Low back injuries are the largest single subset of musculoskeletal injuries. The Bureau of Labor Statistics recently reported that approximately 1 million workers sustained back injuries in 1980 and that back injuries account for one of every five injuries and illnesses in the workplace."[1] Of course, there are other injuries from manual materials handling which are revealed in these totals.

The costs of these injuries are difficult to determine, but "worker's compensation reports for Arkansas in 1976 . . . revealed an average of almost $3,000.00 per incident was spent in lost wage compensation and medical payments. . . . "[2] Unpublished data from private industry found costs to be as high as $6000 to $10,000 per incident in the early 1980s. The costs to industry begin at $3 billion per year and range much higher.

Included in these costs is the impact of handling injuries other than lifting injuries. One increasing cost results from cumulative trauma disorders. These injuries are primarily manipulation injuries that stem from twisting the hands, applying forces with the hands, using tools, and making fast or frequent bends of the wrist. While lifting injuries represent heavy objects, the cumulative trauma disorders reflect primarily lighter loads.

Three billion dollars is not an insignificant cost to industry, yet prevention of these injuries fails to play the strong role it should. One primary reason is the wide range of situations involved. Almost all jobs require some degree of lifting, manual materials handling, and manipulation of materials. The weights involved vary considerably, the frequency and pace of lifting are also highly variable, and personal factors such as fitness, size, and strength are also inconsistent. Table 8-1 is a list that clearly shows the wide range of factors that affect lifting. It is impossible to control all of them.

At the same time, there are "management problems" that also make it difficult to design properly for the human being in manual materials handling activities. A few of these are listed below.

- There are many existing jobs that require overexertion but cannot be improved. These jobs cause problems themselves and unfortunately serve as "models of acceptable design" for new jobs.
- There is not a clear relationship between redesign efforts and reductions in injuries. Thus it is difficult to convince designers and managers who "just want the facts" or require changes to be based on specified rates of financial return.
- The relatively low incidence of injuries on most jobs often makes it difficult to see improvements when they have occurred.

[1] *Journal of the American Medical Association,* 249(17), May 6, 1983, p. 2301.
[2] *Work Practices Guide for Manual Lifting* (Washington, D.C.: U.S. Government Printing Office, 1981), p. 2.

TABLE 8-1 Factors Comprising Each of the Four Major Manual Materials Handling System Components

Worker Characteristics

Physical: general worker measures, such as: age; sex; anthropometry; postures

Sensory: measures of worker sensory processing capabilities, such as: visual; auditory; tactual; kinesthetic; vestibular; propioceptive

Motor: measures of worker motor capabilities, such as: strength; endurance; range of movement; kinematic characteristics; muscle train state

Psychomotor: measures of worker capabilities interfacing mental and motor processes, such as: information processing; reaction/response time; coordination

Personality: measures of worker values and job satisfaction by attitude profiles; attribution; risk acceptance; perceived economic need

Training/experience: measures of the worker education level in terms of formal training or instruction in manual materials handling skills; informal training; work experience

Health status: measures from worker general health appraisal, such as: previous medical complaints; diagnosed medical status; emotional status; regular drug usage; pregnancy; diurnal variations; deconditioning

Leisure time activities: measures of the persons choosing to be involved in physical activities during leisure hours, such as: holding a second job or regular participation in sports

Material/Container Characteristics

Load: measure of mass; pushing/pulling force requirements; mass moment of inertia

Dimensions: measures of size of unit work load, such as: height; width; breadth when indicating the form of rectangular, cylindrical, spherical, etc.

Distribution of load: measure of the location of the unit load center of gravity (CG) with respect to the worker for one-handed and two-handed carrying

Couplings: measures of simple devices used to aid in grasping and manually manipulating the unit load, such as: texture; handle size, shape, and location

Stability of load: measure of load CG location consistency, for handling liquids, bulk materials

Task Characteristics

Workplace geometry: measures of the spatial properties of the task, such as: movement distance; direction and extent of path; obstacles; nature of destination

Frequency/duration/pace: measures of the time dimensions on the handling task, including frequency, duration, and required dynamics of activity over the short term and long term

Complexity: measures of combined or compounding demands of the load, such as: manipulative requirements of movement; objective of activity; precision of tolerance; number of kinestic components

Environment: measures of added deteriorative environmental factors, such as: temperature; humidity; lighting; noise; vibration; foot traction; seasonal toxic agents

Work Practices Characteristics

Individual: measures of operating practices under the control of the individual worker, such as: speed and accuracy in moving objects; postures (i.e., lifting techniques) used in moving objects

Organizational: measures of work organization, such as: physical plant size; staffing of medical, hygiene, engineering, and safety functions; and utilization of teamwork

Administrative: measures of administration of operating practices, such as: work and safety incentive system; compensation scheme; safety training and control; hygiene and safety surveys; and medical aid and rescue; long work shifts; rotation; personal protective devices

Source: Don B. Chaffin and M. M. Ayoub, "The Problem of Manual Materials Handling," *Industrial Engineering,* July 1975, p. 26. Copyright 1975 Institute of Industrial Engineers, 25 Technology Park/Atlanta, Norcross, GA 30092. Reprinted by permission.

- Since people are highly variable, some people can easily do the job when others are overstressed. The materials handling task is usually designed for the young male, but in actual practice will also be performed by females and older workers. Many designers are male, and they find it difficult to understand how much harder the lift is for females (the female has significantly lower shoulder and arm strength) or how endurance requirements could hamper older workers.
- Risk taking (wrong techniques, lifting too much, or using shortcuts) is frequently rewarded. These expectations are then built into new jobs.
- Accidents usually have a low probability, and have many other causes, so that poor design rarely sticks out as the only cause of a serious injury.
- There are no clear regulations with legal penalties.

There is a need to manage this area from the ergonomics standpoint. The costs are high and the results are variable; thus this area will take a lot of the ergonomist's time and effort. Accurate prediction of results will enhance the credibility of the ergonomist and allow changes to be made with only his or her recommendations. It is important to develop this credibility as soon as possible. The use of the NIOSH Guides and the tables in this chapter will help predict suitable results.

The process outlined in Chapter 6 is useful for heavy-job problems. In fact, the problem job model was originally developed as a means of instructing others in the thought processes required to analyze and correct lifting problems successfully. Only with repeated experience was the problem job model applied to other ergonomic problems and then developed into a more universal problem-solving tool.

The importance of correcting existing situations and of designing new jobs properly cannot be overemphasized. The remainder of this chapter presents the NIOSH lifting guidelines and discusses their application, provides guidelines for other manual materials handling tasks, discusses both engineering and administrative solutions, and discusses ways of managing manual materials handling efforts in your organization.

8-2 NIOSH LIFTING GUIDELINES

The technical report *Work Practices Guide for Manual Lifting*, published by the National Institute for Occupational Safety and Health (NIOSH), is a useful starting point for the prevention of lifting injuries. These guidelines were published in 1981 to aid in the identification of hazardous situations and to help evaluate both these jobs and possible solutions when problems are encountered.

Lifting is a very complex task, and the NIOSH Guides have brought four different methods of analysis into the development of these recommendations. Each of these four methods of analysis examines different lifting situations and how material can be safely handled. These four methods are:

1. Epidemiology, which surveys the experiential factors of risk that result in injuries, particularly in low back pain
2. Biomechanical, which predicts injuries from the complex stresses to the body while lifting
3. Physiological, which determines cardiovascular and fatigue limitations for lifting tasks
4. Psychophysical, which quantifies the subjective tolerance of people to the stresses of lifting

The work practices presented in the NIOSH Technical Report will be referred to as the NIOSH Guides.

8-2-1 Using the NIOSH Guides

The NIOSH Guides, although significant, do have limitations in their application. The range of lifting situations is quite broad and the guide was developed only for the two-handed lifts in front of the body. The guide assumes:

- A two-handed lift of a container with handles
- A symmetric lift performed directly in front of the body
- No twisting during the lift
- A lift done in a smooth manner
- A compact load no wider than 30 in.
- A load that does not shift during the lift
- No restrictions on the lifting posture
- A level floor with good friction and no obstructions
- Good environmental conditions

The guidelines are useful for predicting the acceptability of a lifting task. If a task, once analyzed, falls into the lowest level, there is a nominal risk of injury for most of the work force. This level, considered acceptable for over 75% of the females and over 99% of the males, is marked by the action limit (AL). It is permissible to have a wide range of people perform lifts below the AL without any changes to the task when the assumptions described above are met.

When lifts exceed the AL, they may be performed if administrative con-

Chap. 8   The Heavy Jobs   145

trols are used. The administrative controls include training of the worker in principles of lifting and selection or placement of workers based on their physical capabilities. Of course, the task could be redesigned to make the task fall below the AL.

The maximum permissible level (MPL) defines the highest permissible risk. The MPL is, by definition, three times the AL. Only about 25% of the males and less than 1% of the females have muscle strength to lift above the MPL, and thus their risk is substantial. When lifts fall above the MPL, they should not be performed. Engineering controls, like redesign, are the appropriate solution. Mechanical assistance is often used to make these lifts acceptable.

The action limit and the maximum permissible limit are calculated from formulas. Under ideal conditions, 90 pounds would define the AL. Figure 8-1 graphically depicts the AL and MPL.

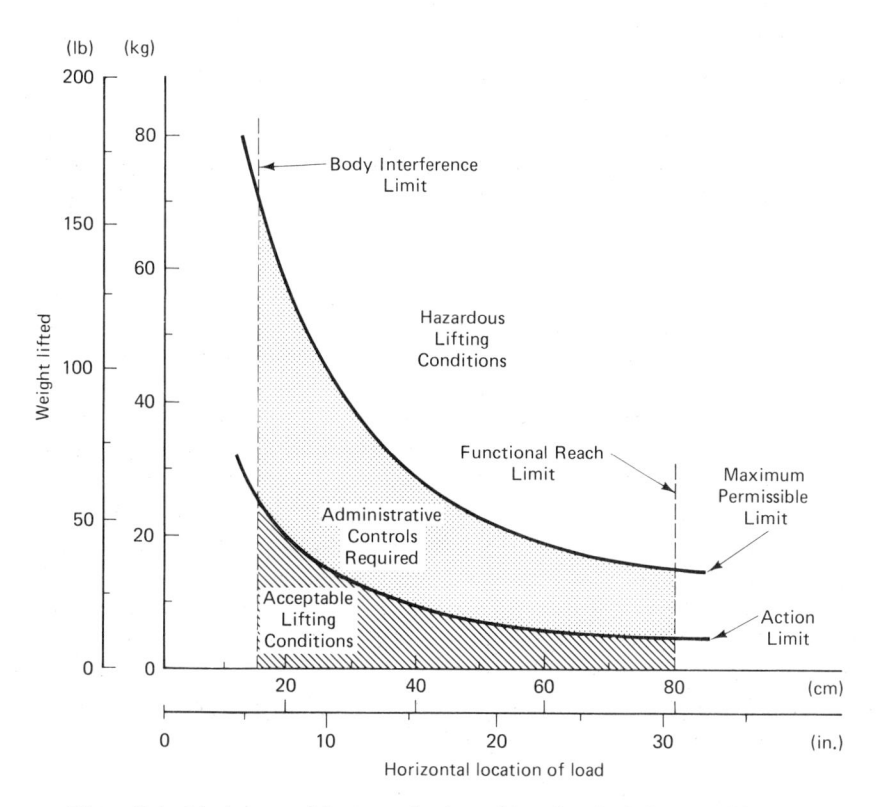

Figure 8-1 Maximum weight versus horizontal location for infrequent lifts from floor to knuckle heights. *Source: Work Practices Guide for Manual Lifting* (Washington, D.C.: U.S. Government Printing Office, 1981), p. 125.

$$AL(\text{pounds}) = 90 \times \frac{6}{H} \times (1 - 0.01|V - 30|)$$

$$\times \left(0.7 + \frac{3}{D}\right) \times \left(1 - \frac{F}{F_{\max}}\right)$$

$$MPL(\text{pounds}) = 3(AL)$$

Where *AL* is the action limit in pounds, *MPL* is the maximum permissible limit in pounds, and *H* (inches) is the horizontal location of the hands at the origin of the lift, measured from a point midway between the ankles. *H* is assumed to be at least 6 in. (otherwise, body interference will occur) and to be less than 32 in. so that most people can reach the load. Increasing the *H* factor will cause the *AL* to decline rapidly, as shown in Figure 8-2.

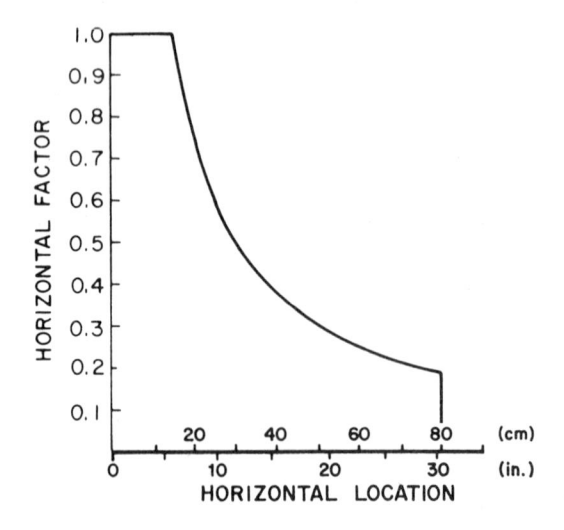

Figure 8-2 Horizontal factor nomogram. *Source: Work Practices Guide for Manual Lifting* (Washington, D.C.: U.S. Government Printing Office, 1981), p. 133.

V (inches) is the vertical location of the hands at the origin of the lift, measured from the floor. *V* will be between 0 (floor level) and 70, which is the vertical reach for many people. *V* is best at 30 inches (standing hand height) and declines in a linear fashion with higher or lower pickup points, as shown in Figure 8-3.

D (inches) is the vertical travel distance from the origin to the destination of the list. *D* will be from 10 in. up to (80 - *V*) in. For lifts less than 10 in.,

Chap. 8   The Heavy Jobs   147

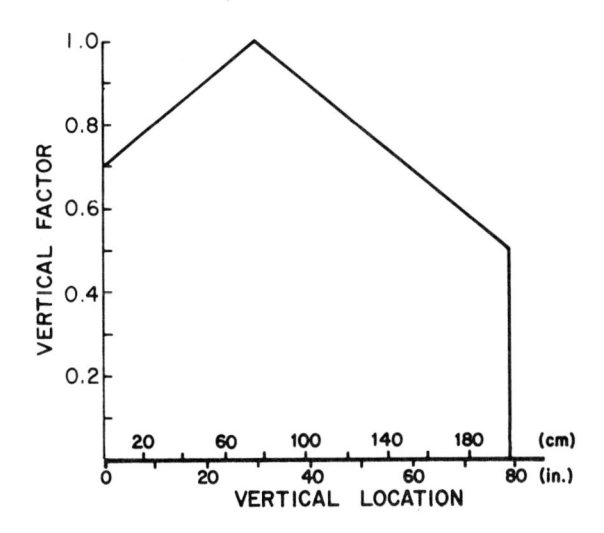

Figure 8-3 Vertical factor nomogram. *Source: Work Practices Guide for Manual Lifting* (Washington, D.C.: U.S. Government Printing Office, 1981), p. 133.

use $D = 10$. D is best with short lifts and declines gradually as the lift increases, as shown in Figure 8-4.

F (lifts/minute) is the average frequency of lifts. F is assumed to be between 0.2 (one lift every 5 minutes) and F_{max}. For lifting less than once every 5 minutes, set $F = 0$. The F factor will decline linearly with increasing fre-

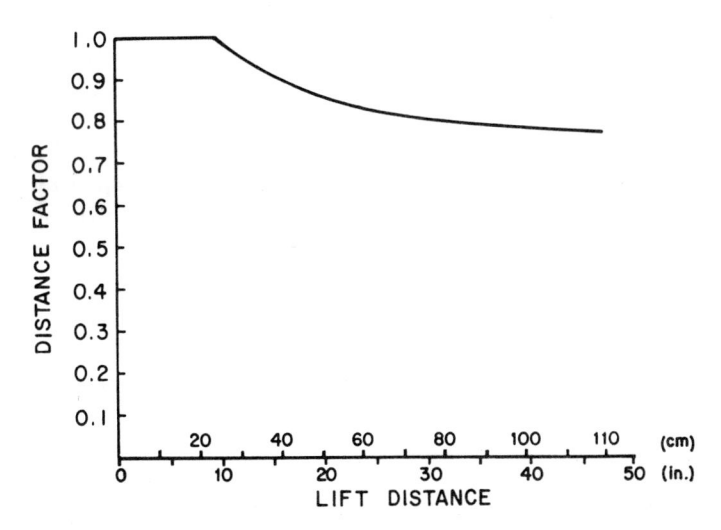

Figure 8-4 Vertical distance factor nomogram. *Source: Work Practices Guide for Manual Lifting* (Washington, D.C.: U.S. Government Printing Office, 1981), p. 134.

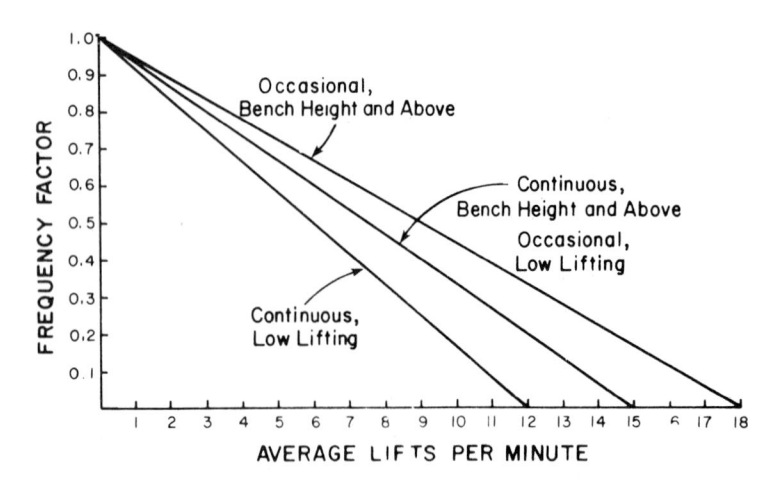

Figure 8-5 Frequency of lift factor nomogram. *Source: Work Practices Guide for Manual Lifting* (Washington, D.C.: U.S. Government Printing Office, 1981), p. 134.

quency, as shown in Figure 8-5. F_{max} (lifts per minute) is the maximum frequency that can be sustained and is shown in Table 8-2.

The maximum weight inside the action limit is 90 lb. This is for a lift next to the body ($H = 6$ in.), with a pickup point at the hands ($V = 30$ in.), a very short lift ($D = 10$ in. or less), and a frequency of less than once per 5 minutes ($F = 0$). As these conditions become less optimal, the weight at the action limit will decline.

One strength of this formula and the nomograms is that the sensitivity of the various factors can be assessed. Often, in industry, many changes are possible and they all have differing costs. It is desirable to predict the maximum improvement for the minimum cost.

An example can help to clarify the calculation of the action limit. A worker must lift a small electric motor from the floor to a repair bench. The motor weighs 26 lb.; H is 10 in.: V is 0 in.; D is 30 in.; and F is 0. Is this acceptable?

TABLE 8-2 Maximum Lifts per Minute (F_{max})

		Average Vertical Location (in.)	
		$V > 75 (30)$ *Standing*	$V \leq 75 (30)$ *Stooped*
Period	1 hr	18	15
	8 hr	15	12

Source: Work Practices Guide for Manual Lifting (Washington, D.C.: U.S. Government Printing Office, 1981), p. 127.

$$AL(\text{pounds}) = 90 \times \frac{6}{H} \times (1 - 0.01|V - 30|)$$

$$\times (0.7 + \frac{3}{D}) \times (1 - \frac{F}{F_{max}})$$

$$= 90 \times \frac{6}{10} \times (1 - 0.01|0 - 30|)$$

$$\times (0.7 + \frac{3}{30}) \times (1 - \frac{0}{15})$$

$$= 90 \times 0.6 \times 0.7 \times 0.8 \times 1$$

$$= 30.2 \text{ lb.}$$

Therefore, the task is acceptable.

If an obstruction kept the worker's feet 14 in. from the motor, is the lift still acceptable?

$$AL = 90 \times \frac{6}{14} \times (1 - 0.01|0 - 30|)$$

$$\times (0.7 + \frac{3}{30}) \times (1 - \frac{0}{15})$$

$$= 90 \times 0.43 \times 0.7 \times 0.8 \times 1$$

$$= 21.7 \text{ lb.}$$

Therefore, the task is acceptable only with administrative controls.

With the same obstruction ($H = 14$ in.), would mounting the motor on a 10-in. platform be of any help? Now $V = 10$ in. and $D = 20$ in.

$$AL = 90 \times \frac{6}{14} \times (1 - 0.01|10 - 30|)$$

$$\times (0.7 + \frac{3}{20}) \times (1 - \frac{0}{15})$$

$$= 90 \times 0.43 \times 0.8 \times 0.85 \times 1$$

$$= 26.3 \text{ lb.}$$

Therefore, the task is acceptable even with the obstruction if the 10-in. platform is added. If the 10-in. platform had not been sufficient, the required platform height could have been calculated through trial and error. A useful form for the collection of data is provided by NIOSH and is reproduced in Figure 8-6.

PHYSICAL STRESS JOB ANALYSIS SHEET

Department _____ Date _____

Job title _____ Analyst's name _____

Task description	Object weight Ave. Max.	Hand location Origin H (cm) V (cm)	Destination H (cm) V (cm)	Task freq.	AL	MPL	Remarks

Figure 8-6 Example coding form. *Source: Work Practices Guide for Manual Lifting* (Washington, D.C.: U.S. Government Printing Office, 1981), p. 131.

8-2-2 Solutions

The solutions to lifting problems are rarely easy. However, with the NIOSH Guides, the search for a proper solution becomes easier and has more predictable results. The solutions generally fall into two categories, engineering solutions and administrative solutions. Engineering solutions are required for lifts that exceed the MPL and may be used for lifts above the AL. Administrative solutions are useful only for lifts that fall between the AL and the MPL.

Engineering solutions require a modification to the task such as a change in the weight handled, changes in the lift configuration to bring the material in closer, or the use of mechanical assistance. Section 8-4 provides more details on the engineering solutions.

Administrative solutions do not involve changes to the task but require that the workers be selected or trained to enhance their ability to perform the task. Administrative solutions are detailed in Section 8-6.

It is important to analyze the lifting situations carefully, to determine the most effective solutions. By determining the sensitivity of the four major factors, it is possible to economically evaluate the benefit of a number of different solutions.

8-3 JOB SEVERITY INDEX

Another method of examining lifting tasks is the Job Severity Index (JSI). This index was developed to provide a measure relating the frequency and severity of injuries to the difficulty of manual material handling tasks. JSI is designed for analysis of jobs that require substantial lifting.

Chap. 8   The Heavy Jobs   151

The premise for the Job Severity Index is that lifting injuries are a function of two variables—the demands of the job and the capacities of the person or population performing the job. Information about these two variables can aid in the prediction of injury and severity rates, thus allowing appropriate job changes to be made early. The JSI, for given job conditions, is represented by the formula:

$$JSI = \frac{Job\ demand}{Operator\ capacity}$$

To quantify a particular JSI, both the specific job demands and the person's or work population's capacity must be determined. The details for determining job demand and worker capacity are covered later. Larger JSI values indicate more stressful jobs. Even so, the JSI ratio recognizes that high-capacity people can lift more with a suitable safety factor. To design a safer job, the JSI provides two separate options for the ergonomist: to change the job or to change the people assigned to the job.

The JSI is more specific than the NIOSH Guides. In addition to identifying hazardous tasks and jobs, the JSI identifies the portions of the work population that will experience problems. The JSI may explain why a job is suitable for the current work population but will not be suitable as that population begins to change. It can also be used to predict individual lifting capacities although specific data about the person is required. The NIOSH Guides deal with general work populations and provide guidelines for the design of tasks only from that standpoint.

The advantage of JSI is the additional decision-making capability. JSI allows the designer to change the job, the people, or both, in order to achieve a specific ratio of job demands to operator capabilities. It can answer questions in more detail and can pinpoint a problem or an answer more specifically. The JSI appears to be more useful for employee selection, especially with the validations studies performed. The disadvantages of JSI is the requirement for more information from the ergonomist.

The JSI was developed in 1978, and validations studies were conducted in 1978 and 1982. A total of 101 jobs involving 385 male and 68 female industrial workers from 28 private companies and government entities were used in the validation. This section is too brief to provide details of the validations studies, but interested readers should seek the original source: M. M. Ayoub, J. L. Selan, and B. C. Jiang, *A Mini-Guide For Lifting* (Lubbock: Texas Tech University, 1983).

8-3-1 Using the Job Severity Index

The JSI is used for lifting and lowering tasks associated with manual materials handling. It is not suitable for pushing, pulling, or other types of handling tasks. The JSI is designed specifically for those jobs requiring lifting as a substantial portion of the job.

The types of lifting injuries included in the JSI analysis are musculoskeletal injuries to the back, musculoskeletal injuries to other body parts, surface tissue injuries due to impact, other surface tissue injuries, and miscellaneous injuries. The information collected on these injuries included injury type, injury cause (lifting or nonlifting), number of days lost, medical expenses, wages paid during lost work days, worker's compensation paid, and extraordinary expense.

A strength of JSI is the predictive knowledge that higher JSI values result in higher expense and injury rates. As an example, the frequency of total injuries as a function of JSI for industry is shown in Figure 8-7. The total injury rate is shown for the equivalent of 100 full-time employees.

A JSI level of 1.5 is generally used as the cut-off for an acceptable job design. Above 1.5, there is a substantially increased risk of both injuries to workers and of increased expenses due to injuries.

The task demands. Each lifting task must be described in terms of the actual weight of lifts, the frequency of lifts, the container size, and the

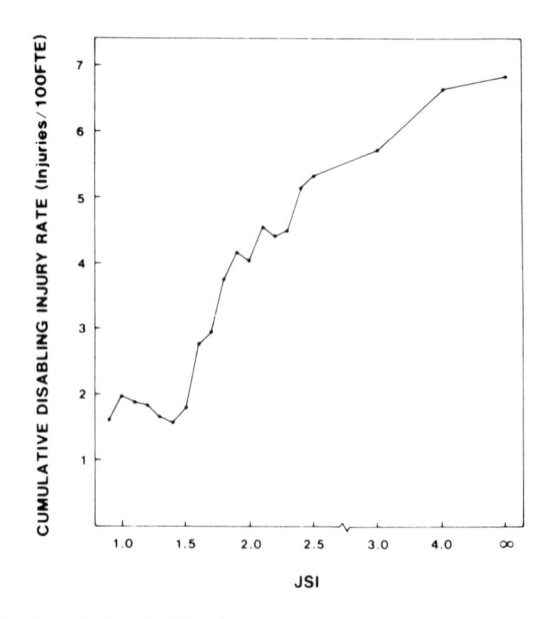

Figure 8-7 Cumulative disabling injury rate as a function of calculated JSI values. *Source:* D. H. Liles, S. Deivanayagam, M. M. Ayoub, and P. Mahajan, "A Job Severity Index for the Evaluation and Control of Lifting Injury," *Human Factors,* 26(6), 1984, p. 691. Copyright 1984 by The Human Factors Society, Inc., and reproduced by permission.

Chap. 8   The Heavy Jobs   153

range of the lifts. Each task should be defined to represent a set of relatively constant parameters. The tasks are eventually summarized into a job for the JSI calculations. The summation process is discussed briefly in the next section.

To define the tasks:

The weight of the lift is measured in pounds.

The frequency of the lift is the average number of lifts per minute required for the particular task.

The container size is the full depth of the container measured in inches perpendicularly from the front of the body.

The range of the lift is measured vertically in inches from initiation point of the lift to the termination point of the lift.

The total job demands. To evaluate the stress on the individual or work population, the task demands are summarized into the total job demands. Information on exposure hours for the work week is calculated first, and then the tasks are combined for total severity. The following formula is used for this purpose:

$$\text{JSI} = \sum_{i=1}^{n} \left(\frac{\text{Hours } i}{\text{Hours } t} \times \frac{\text{Days } i}{\text{Days } t} \right)_i \sum_{j=1}^{mi} \left(\frac{F_j}{F_i} \times \frac{\text{WT } j}{\text{CAP } j} \right)$$

Where:

n	= number of subtask groups
mi	= number of tasks in group i
Days i	= total days per week for group i
Days t	= total days per week for job
Hours i	= exposure hours per day for group i
Hours t	= number of hours per day that job is performed
$F j$	= lifting frequency for task j
$F i$	= total lifting frequency for group i
WT j	= maximum weight of lift required by task j
CAP j	= the smallest applicable maximum acceptable weight of lift adjusted for frequency of lift and box size.

When a person works at the same task all week, the time-base summation equals 1.0 as shown:

$$\sum_{i=1}^{n} \left(\frac{\text{Hours } i}{\text{Hours } t} \times \frac{\text{Days } i}{\text{Days } t} \right) = \frac{8}{8} \times \frac{5}{5} = 1.0$$

Only when the nonlifting portion of the job occupies substantial portions of the work day or work week will this factor change appreciably.

If only one task is performed, then the frequency term also equals 1.0, because the task frequency will equal the total lifting frequency.

These terms are most useful to isolate lifting from nonlifting tasks, or as a means to examine several different lifting tasks that are required of the same person. An example in Section 8-3-2 will illustrate the use of this formula. Remember that the JSI is useful only when lifting is a substantial portion of the job. For a single lift, the frequency essentially goes to zero. This is because the JSI uses psycho-physical lifting limits, not biomechanical limitations which occur with infrequent lifts.

Work population capacities. The lifting capacity models used in the JSI result from the psychophysically-based capacity data bases developed by Ayoub and Snook. The parameters used to predict lifting capacities are the range of the lift, the frequency of the lift, the gender of lifter, the container or box size, and the population percentage performing the task, as listed below.

The range of the lift is measured vertically in inches from the initiation point of the lift to the termination point of the lift. There are six ranges of lift normally used:

1. Floor to knuckle—FK
2. Floor to shoulder—FS
3. Floor to reach—FR
4. Knuckle to shoulder—KS
5. Knuckle to reach—KR
6. Shoulder to reach—SR

These points are used because the biomechanical strength changes at these points due to changes in the muscles used and the mechanical advantages of the muscles. The frequency of the lift is the average number of lifts per minute required for the particular task. Gender is determined for current or anticipated worker population. The container size is measured in inches perpendicularly from the front of the body. The population percentage is that portion of the population expected to perform the task.

The computation of the predicted lifting capacity is done in three consecutive steps:

1. An initial capacity is determined from the lifting range, the frequency of lift, and the gender of the worker population. These are found in Table 8-3.
2. The initial capacity is adjusted based on the size of the box or container.

Chap. 8   The Heavy Jobs   155

TABLE 8-3 Initial Prediction Equations for Lifting Capacity Based on the Lifting Range, the Lifting Frequency and Gender

Range of Lift	Frequency of Lift (Lifts/Minute)	
Male	$0.1 < FY < 1.0$	$1.0 \leq FY \leq 12.0$
1. FK	57.2 × (FY) ** (−0.184697)	57.2 − 2.0 × (FY − 1)
2. FS	51.2 × (FY) ** (−0.184697)	51.2 − 2.0 × (FY − 1)
3. FR	49.1 × (FY) ** (−0.184697)	49.1 − 2.0 × (FY − 1)
4. KS	52.8 × (FY) ** (−0.138650)	52.8 − 2.0 × (FY − 1)
5. KR	50.0 × (FY) ** (−0.138650)	50.0 − 2.0 × (FY − 1)
6. SR	48.4 × (FY) ** (−0.138650)	48.4 − 2.0 × (FY − 1)
Female	$0.1 < FY < 1.0$	$1.0 \leq FY \leq 12.0$
1. FK	37.4 × (FY) ** (−0.187818)	37.4 − 1.1 × (FY − 1)
2. FS	31.1 × (FY) ** (−0.187818)	31.1 − 1.1 × (FY − 1)
3. FR	28.1 × (FY) ** (−0.187818)	28.1 − 1.1 × (FY − 1)
4. KS	30.8 × (FY) ** (−0.156150)	30.8 − 1.1 × (FY − 1)
5. KR	27.3 × (FY) ** (−0.156150)	27.3 − 1.1 × (FY − 1)
6. SR	26.4 × (FY) ** (−0.156150)	26.4 − 1.1 × (FY − 1)

Where FY = Frequency of lift (lifts/minute)
   ** = exponentation (e.g., FY to the power of −0.184697)

The initial lifting capacity is 57.2 lbs. for males and 37.4 lbs. for females based on the mean capacity for lift based on published data from M. M. Ayoub and S. N. Snook for the various ranges of lift for the 50th percentage and 1.0 lift/minute.

Source: M. M. Ayoub, J. L. Selan, and B. C. Jiang, *A Mini-Guide For Lifting* (Lubbock: Texas Tech University, 1983), Table 8, pp. 45-46. Research supported by NIOSH Grants 5R010H00545-02 (1978) and 5R010H00798-04 (1983).

The full box depth is measured in inches perpendicular from the body. The adjustment factors are found in Table 8-4.

3. The final adjustment is made for the percentage of the population expected to perform the task. If many people will perform the task, use high population percentage (to include most people); if the job is expected to be performed only by the stronger individual selected for the job, a lower population percentage may be used. The population percentage adjustment factors are found in Tables 8-5 and 8-6.

In this example to determine lifting capacity, a full-time worker will be required to handle and inspect boxes from several conveyors, and then pack them into a larger carton. The incoming boxes arrive at an average of 2 per minute, weigh 14 lbs. and are 15 in. across. The incoming conveyors are at a height of 25 in., and the packing station is at a height of 40 in. If the task is to be performed by 75% of the female work population, what is the predicted capacity of the lifters?

Table 8-4 The Adjustment to the Initial Lifting Capacity Based on the Box Size

Range of Lift	Box Size (In. in the Sagittal Plane)	
Male	*$12" \leq BX \leq 18"$*	*$BX > 18"$*
1. FK	CAP + 1.65 × (18 − BX)	CAP + 0.8 × (18 − BX)
2. FS	CAP + 1.65 × (18 − BX)	CAP + 0.8 × (18 − BX)
3. FR	CAP + 1.65 × (18 − BX)	CAP + 0.8 × (18 − BX)
4. KS	CAP + 1.10 × (18 − BX)	CAP + 0.8 × (18 − BX)
5. KR	CAP + 1.10 × (18 − BX)	CAP + 0.8 × (18 − BX)
6. SR	CAP + 1.10 × (18 − BX)	CAP + 0.8 × (18 − BX)
Female	*$12" \leq BX \leq 18"$*	*$BX > 18"$*
1. FK	CAP + 1.10 × (18 − BX)	CAP + 0.4 × (18 − BX)
2. FS	CAP + 1.10 × (18 − BX)	CAP + 0.4 × (18 − BX)
3. FR	CAP + 1.10 × (18 − BX)	CAP + 0.4 × (18 − BX)
4. KS	CAP + 0.55 × (18 − BX)	CAP + 0.2 × (18 − BX)
5. KR	CAP + 0.55 × (18 − BX)	CAP + 0.2 × (18 − BX)
6. SR	CAP + 0.55 × (18 − BX)	CAP + 0.2 × (18 − BX)

Where CAP = Capacity of the lift as determined in Table 8-3
BX = Box size (in in.)

Source: M. M. Ayoub, J. L. Selan, and B. C. Jiang, *A Mini-Guide For Lifting* (Lubbock: Texas Tech University, 1983), Table 9, pp. 47–48. Research supported by NIOSH Grants 5R01OH00545-02 (1978) and 5R01OH00798-04 (1983).

The lifting range is in the knuckle-to-shoulder range (25 to 40 in.), the lifting frequency is 2 lifts/minute, the gender is female, the box size is 15 in., and the population percentage is 75%.

From Table 8-3, the knuckle-to-shoulder lifting capacity for females at lift frequencies between 1 and 12 lifts/minute is shown below. Capacity[a] is the adjusted capacity for the frequency of lift.

Capacity[a] (lbs.) = 30.8 − 1.1 × (FY − 1)
Capacity (lbs.) = 30.8 − 1.1 × (2 − 1)
Capacity (lbs.) = 30.8 − 1.1 × 1
Capacity (lbs.) = 30.8 − 1.1
Capacity (lbs.) = 29.7

Capacity[b] includes the adjustments for the box size as shown in Table 8-4:

Capacity[b] (lbs.) = Capacity[a] + 0.55 × (18 − BX)
Capacity (lbs.) = 29.7 + 0.55 × (18 − 15)
Capacity (lbs.) = 29.7 + 0.55 × 3

TABLE 8-5 The Final Adjustment to the Lifting Capacity Based on Work Population Percentiles and Lifting Frequency

Range of Lift	Frequency	
Male	$0.1 \leq FY < 1.0$	$1.0 \leq FY \leq 12.0$
1. FK	CAP + Z × 16.86 × (FY)**(−0.174197)	CAP + Z × (16.86 − 0.5964 × (FY − 1))
2. FS	CAP + Z × 15.09 × (FY)**(−0.174197)	CAP + Z × (15.09 − 0.5338 × (FY − 1))
3. FR	CAP + Z × 14.47 × (FY)**(−0.174197)	CAP + Z × (14.47 − 0.5119 × (FY − 1))
4. KS	CAP + Z × 14.67 × (FY)**(−0.156762)	CAP + Z × (14.67 − 0.5534 × (FY − 1))
5. KR	CAP + Z × 13.89 × (FY)**(−0.156762)	CAP + Z × (13.89 − 0.5240 × (FY − 1))
6. SR	CAP + Z × 13.45 × (FY)**(−0.156762)	CAP + Z × (13.45 − 0.5074 × (FY − 1))
Female	$0.1 \leq FY < 1.0$	$1.0 \leq FY \leq 12.0$
1. FK	CAP + Z × 6.87 × (FY)**(−0.251605)	CAP + Z × (6.87 − 0.1564 × (FY − 1))
2. FS	CAP + Z × 5.71 × (FY)**(−0.251605)	CAP + Z × (5.71 − 0.1300 × (FY − 1))
3. FR	CAP + Z × 5.16 × (FY)**(−0.251605)	CAP + Z × (5.16 − 0.1175 × (FY − 1))
4. KS	CAP + Z × 5.66 × (FY)**(−0.258700)	CAP + Z × (5.66 − 0.1289 × (FY − 1))
5. KR	CAP + Z × 5.01 × (FY)**(−0.258700)	CAP + Z × (5.01 − 0.1141 × (FY − 1))
6. SR	CAP + Z × 4.85 × (FY)**(−0.258700)	CAP + Z × (4.85 − 0.1104 × (FY − 1))

Where CAP = Capacity of lift as determined in Table 8-4
Z = Z score of population percentage (From normal distribution tables, and as shown in Figure 8-4)
FY = Frequency of lift (Lifts/minute)
** = Exponent

Source: M. M. Ayoub, J. L. Selan, and B. C. Jiang, *A Mini-Guide For Lifting* (Lubbock: Texas Tech University, 1983), Table 10, pp. 49-50. Research supported by NIOSH Grants 5R010H00545-02 (1978) and 5R010H00798-04 (1983).

Capacity (lbs.) = 29.7 + 1.65
Capacity (lbs.) = 31.35

Capacity^c is the adjustment for the population percentage as shown in Tables 8-5 and 8-6:

TABLE 8-6 Z Scores for Various Population Percentages

Population Percentage	Z Score
95	−1.6449
90	−1.2816
85	−1.0364
75	−0.6745
50	0.0
25	0.6745
15	1.0364
10	1.2816
5	1.6449

Source: M. M. Ayoub, J. L. Selan, and B. C. Jiang, *A Mini-Guide For Lifting* (Lubbock: Texas Tech University, 1983), Table 11, p. 51. Research supported by NIOSH Grants 5R010H00545-02 (1978) and 5R010H00798-04 (1983).

Capacity^c (lbs.) = Capacity^b + Z × (5.66 − 0.1289 × (FY − 1))
Capacity (lbs.) = 31.35 + (−0.6745) × (5.66 − 0.1289 × 2 − 1))
Capacity (lbs.) = 31.35 + (−0.6745) × (5.66 − 0.1289 × 1)
Capacity (lbs.) = 31.35 + (−0.6745) × (5.53)
Capacity (lbs.) = 31.35 − 3.73
Capacity (lbs.) = 27.62

The capacity of 75% of the female work population, lifting a 15-in. box in the knuckle-to-shoulder range twice each minute is 27.62 lbs.

Adjustments to the JSI calculations. It is common for the lifts being analyzed to fall outside the criteria presented earlier. Several adjustments are available to the Job Severity Index calculations to accommodate some of these real-world situations.

Ayoub cites research that recommends adjustments for twisting and the use of handles while lifting. If twisting occurs while lifting, the maximum acceptable weight of lift is decreased by 5%.[3] Handles make lifting easier, and the absence of handles reduces the lifting capacity by 7.2%.[4] The range of lift seldom falls within the precise ranges specified in the initial model. Ayoub has provided adjustments for the range of lift as shown in Table 8-7. These ad-

[3] S. S. Asfour "Energy Cost Prediction Models for Manual Lifting and Lowering Tasks." Ph.D. Dissertation, Texas Tech University, Lubbock, Texas, 1980.

[4] A. Garg and U. Saxena, "Container Characteristics and Maximum Weight of Lift," *Human Factors,* 22(4), 1980, pp. 487–495.

TABLE 8-7 Lifting Range Assignment

Point of Lift Initiation	Point of Lift Termination	Range Assignment
0" to KL/2	0" to KL + 10"	1. FK
	KL + 10" to KL + 30"	2. FS
	KL + 30" and above	3. FR
KL/2 to KL	KL/2 to KL	1. FK
	KL to KL + 30"	4. KS
	KL + 30" and above	5. KR
KL to KL + 10"	KL to KL + 30"	4. KS
	KL + 30" and above	5. KR
KL + 10" to KL + 20"	KL + 10" to KL + 20"	4. KS
	KL + 20" and above	6. SR
KL + 20" and above	KL + 20" and above	6. SR

Where KL Knuckle level

Range 1. FK is the floor-to-knuckle range
      2. FS is the floor-to-shoulder range
      3. FR is the floor-to-reach range
      4. KS is the knuckle-to-shoulder range
      5. KR is the knuckle-to-reach range
      6. SR is the shoulder-to-reach range

Source: M. M. Ayoub, J. L. Selan, and B. C. Jiang, *A Mini-Guide for Lifting* (Lubbock: Texas Tech University, 1983), Table 7, p. 44. Research supported by NIOSH Grants 5R010H00545-02 (1978) and 5R010H00798-04 (1983).

justments allow computations for lifts that cross the boundaries specified earlier.

Individual capacities. The Job Severity Index is useful for predicting the lifting capacity for an individual, once strength and anthropometric measures have been determined. It is not the intent of this book to provide the means to assess individual lifting capacity, but merely to provide an overview of the model and its possible use in industry. The interested reader is again encouraged to seek the original source, *A Mini-Guide for Lifting,* that describes the measurement techniques for the required variables.

The formula used to compute the lifting capacity is shown below. It is based on regression analysis and uses many factors. The factors change with the range of lift required, with the gender of the lifter, and with the body weight of the lifter. Additional adjustments are made based on the lift frequency, and with the box size.

    Capacity = Constant + (Arm Strength × factor)
    + (Shoulder Height × factor) + (Back Strength × factor)
    + (Abdominal Depth × factor) + (Dynamic Endurance × factor)
    − Body Weight

The units for the variables are:

Constant — lbs. median, and 1 if the body weight is above the median. The median weight is 135 lbs. for females and 170 lbs. for males.
Arm Strength — lbs.
Shoulder Height — centimeters
Back Strength — lbs.
Abdominal Depth — centimeters
Dynamic Endurance — minutes

The prediction of individual lifting capacities has two useful results. First, the predicted capacities may be used to select people for specific jobs and tasks requiring lifting. Second, the use of the predicted capacities can be used to identify those jobs where there is substantial risk of injury for the people on the job. This naturally leads to the redesign of the job, either by changing equipment and facilities or by changing work rates and work times.

The risk of injury is a function of two factors: the capacity of the individual and the requirements of the job. By changing either factor, the risk of injury is changed. The capacity of the individual on the job is most easily changed by selection, although physical training is another alternative. The job requirements are changed through job and equipment redesign.

8-3-2 Examples of the Use of the Job Severity Index

A person works 6 hours per day (30 hours per week) inspecting and packing boxes from several conveyors into larger shipping containers. The last 2 hours each day is spent on shipping, record-keeping, and other miscellaneous duties.

The boxes arrive on several different conveyors that are 25 in. high. The boxes weigh 14 or 28 lbs. each; the lighter ones arrive at an average of 3 per minute and the heavier ones arrive once per minute. The lighter boxes are 15 in. deep, while the heavier boxes are 18 in. deep. The boxes are packed in larger shipping containers which are located on a table 40 in. high.

As the ergonomist, you want the job to be acceptable to a wide range of the work force, and decide that 75% of the female work population should be able to perform the job with a JSI of 1.5 or less. The needed information is:

The lifting range is in the knuckle-to-shoulder range (25 to 40 in.), the total lifting frequency is 4 lifts/minute, the gender is female, and the population percentage is 75%.

Two boxes are lifted. The first weighs 14 lbs., is 15 in. deep, and is handled at a frequency of 3 per minute. The second box weighs 28 lbs., is 18 in. deep, and is handled at a frequency of 1 per minute.

Chap. 8   The Heavy Jobs   161

The first step in the analysis is to determine the capacity of the proposed population for each of the two lifts. From Table 8-3, the knuckle-to-shoulder lifting capacity for females at lift frequencies between 1 and 12 lifts/minute is:

$$\text{Capacity}^a \text{ (lbs.)} = 30.8 - 1.1 \times (FY - 1)$$

To make the adjustments for the box size, use Table 8-4:

$$\text{Capacity}^b \text{ (lbs.)} = \text{Capacity} + 0.55 \times (18 - BX)$$

To make the adjustments for the population percentage, use Tables 8-5 and 8-6:

$$\text{Capacity}^c \text{ (lbs.)} = \text{Capacity} + Z \times (5.66 - 0.1289 \times (FY - 1))$$

The calculations for the smaller box (15 in. deep, 3 lifts/min) are:

$$\text{Capacity}^a \text{ (lbs.)} = 30.8 - 1.1 \times (3 - 1) = 28.6$$

$$\text{Capacity}^b \text{ (lbs.)} = 28.6 + 0.55 \times (18 - 15) = 30.25$$

$$\text{Capacity}^c \text{ (lbs.)} = 30.25 + (-0.6745) \times (5.66 - 0.1289) \times (3 - 1)$$

$$= 26.61$$

The capacity of 75% of the female work population, lifting a 15-in. box in the knuckle-to-shoulder range 3 times per minute is 26.61 lbs. Further adjustments should be made for twisting during the lift, or the absence of handles on the container, as appropriate.

The calculations for the larger box (18 in. deep, 1 lift/min) are:

$$\text{Capacity}^a \text{ (lbs.)} = 30.8 - 1.1 \times (1 - 1) = 30.8$$

$$\text{Capacity}^b \text{ (lbs.)} = 30.8 + 0.55 \times (18 - 18) = 30.8$$

$$\text{Capacity}^c \text{ (lbs.)} = 30.8 + (-0.6745) \times (5.66 - 0.1289 \times (1 - 1))$$

$$= 26.98$$

The capacity of 75% of the female work population, lifting an 18-in. box in the knuckle-to-shoulder range once per minute is 26.98 lbs. Again, adjustments for twisting and container handles are appropriate.

The next step to complete the summation process is to calculate the JSI. The initial formula is used:

$$\text{JSI} = \sum_{i=1}^{n} \left(\frac{\text{Hours } i}{\text{Hours } t} \times \frac{\text{Days } i}{\text{Days } t} \right) \sum_{j=1}^{mi} \left(\frac{Fj}{Fi} \times \frac{WT j}{CAP j} \right)$$

n = 1 subtask
mi = 2 tasks as described
Days i = 5 days per week for the task

Days t = 5 days per week for the job
Hours i = 6 hours per day
Hours t = 8 hours per day for the job
F 1 = 3 lifts per minute for the lighter box
F 2 = 1 lift per minute for the heavier box
F i = 4 lifts per minute
WT 1 = 14 lbs. for the individual boxes
WT 2 = 28 lbs. for the individual boxes
CAP 1 = 26.61 lbs.
CAP 2 = 26.98 lbs.

$$JSI = \sum_{i=1}^{n} \left(\frac{\text{Hours } i}{\text{Hours } t} \times \frac{\text{Days } i}{\text{Days } t}\right) \sum_{j=1}^{mi} \left(\frac{Fj}{Fi} \times \frac{WT j}{CAP j}\right)$$

$$JSI = \sum_{i=1}^{1} \left(\frac{6}{8} \times \frac{5}{5}\right) \sum_{j=1}^{2} \left(\frac{3}{4} \times \frac{14}{26.61}\right) + \left(\frac{1}{4} \times \frac{28}{26.98}\right)$$

$$= (0.75) \times (0.3946 + 0.2595)$$

$$= 0.4905$$

This level of JSI indicates a safe level of manual material handling.

Now suppose that the level of the work has increased, so that the lighter boxes are arriving at 4 per minute, and an additional box weighing 45 lbs. also arrives once per minute. The 28-lb. boxes are still packed at the initial rate. The packing has increased to 8 hours each work day.

The calculations for the smaller box (15 in. deep, 4 lifts/min) are:

Capacity[a] (lbs.) = 30.8 − 1.1 × (4 − 1) = 27.5

Capacity[b] (lbs.) = 27.5 + 0.55 × (18 − 15) = 29.15

Capacity[c] (lbs.) = 29.15 + (−0.6745) × (5.66 − 0.1289 × 4 − 1))

= 25.59

The capacity of 75% of the female work population, lifting a 15-in. box in the knuckle-to-shoulder range 4 times per minute is 25.59 lbs.

The calculations for the larger box (18 in. deep, 1 lift/min) remain unchanged. The capacity is 26.98 lbs. The calculations for the largest box (18 in. deep, 1 lift/min) are:

Capacity[a] (lbs.) = 30.8 − 1.1 × (1 − 1) = 30.8

Capacity[b] (lbs.) = 30.8 + 0.55 × (18 − 18) = 30.8

Capacity[c] (lbs.) = 30.8 + (−0.6745) × (5.66 − 0.1289) × (1 − 1))

= 26.98

Chap. 8   The Heavy Jobs   163

The capacity of 75% of the female work population, lifting an 18-in. box in the knuckle-to-shoulder range once per minute is 26.98 lbs.

The next step is to complete the summation process to calculate the JSI. The initial formula is used:

$$\text{JSI} = \sum_{i=1}^{n} \left(\frac{\text{Hours } i}{\text{Hours } t} \times \frac{\text{Days } i}{\text{Days } t} \right) \sum_{j=1}^{mi} \left(\frac{Fj}{Fi} \times \frac{\text{WT } j}{\text{CAP } j} \right)$$

n = 1 subtask
mi = 2 tasks as described
Days i = 5 days per week for the task
Days t = 5 days per week for the job
Hours i = 8 hours per day
Hours t = 8 hours per day for the job
F 1 = 4 lifts per minute for the lighter box
F 2 = 1 lift per minute for the heavier box
F 3 = 1 lift per minute for the heaviest box
F i = 6 lifts per minute
WT 1 = 14 lbs. for the individual boxes
WT 2 = 28 lbs. for the individual boxes
WT 3 = 45 lbs. for the individual boxes
CAP 1 = 25.59 lbs.
CAP 2 = 26.98 lbs.
CAP 3 = 26.98 lbs.

$$\text{JSI} = \sum_{i=1}^{n} \left(\frac{\text{Hours } i}{\text{Hours } t} \times \frac{\text{Days } i}{\text{Days } t} \right) \sum_{j=1}^{mi} \left(\frac{Fj}{Fi} \times \frac{\text{WT } j}{\text{CAP } j} \right)$$

$$\text{JSI} = \sum_{i=1}^{1} \left(\frac{8}{8} \times \frac{5}{5} \right) \sum_{j=1}^{3} \left(\frac{4}{6} \times \frac{14}{25.59} \right) + \left(\frac{1}{6} \times \frac{28}{26.98} \right) + \left(\frac{1}{6} \times \frac{45}{26.98} \right)$$

$= (1.0) \times (0.3647 + 0.1730 + 0.2780)$

$= 0.8157$

Again, this level of JSI indicates a safe level of manual material handling.

In one final job analysis, a job requires all people in the work population to palletize 50 bags of material. The range of lifting is in the floor-to-shoulder range, the container size is about 12 in. (the flat side of the bag faces the lifter), and the lift frequency is 3 lifts/minute. What is the JSI?

The calculations for the lifting capacity for the bag (12 in. deep, 3 lifts/min) are:

Capacity^a (lbs.) = 31.1 − 1.1 × (3 − 1) = 28.9

Capacity^b (lbs.) = 28.9 + 1.1 × (18 − 12) = 35.5

Capacity^c (lbs.) = 35.5 + (−1.6449) × (5.71 − 0.1300 × (3 − 1))

= 26.54

The capacity of 95% of the female work population, lifting a 12-in. bag in the floor-to-shoulder range 3 times per minute is 26.54 lbs.

The next step is to complete the summation process to calculate the JSI. The initial formula is used:

$$\text{JSI} = \sum_{i=1}^{n}\left(\frac{\text{Hours }i}{\text{Hours }t} \times \frac{\text{Days }i}{\text{Days }t}\right) \sum_{j=1}^{mi}\left(\frac{Fj}{Fi} \times \frac{\text{WT }j}{\text{CAP }j}\right)$$

n = 1 subtask
mi = 1 task as described
Days i = 5 days per week for the task
Days t = 5 days per week for the job
Hours i = 8 hours per day
Hours t = 8 hours per day for the job
F 1 = 3 lifts per minute for the lighter box
F i = 3 lifts per minute
WT 1 = 50 lbs. for the individual boxes
CAP 1 = 26.54 lbs.

$$\text{JSI} = \sum_{i=1}^{n}\left(\frac{\text{Hours }i}{\text{Hours }t} \times \frac{\text{Days }i}{\text{Days }t}\right) \sum_{j=1}^{mi}\left(\frac{Fj}{Fi} \times \frac{\text{WT }j}{\text{CAP }j}\right)$$

$$\text{JSI} = \sum_{i=1}^{1}\left(\frac{8}{8} \times \frac{5}{5}\right) \sum_{j=1}^{1}\left(\frac{3}{3} \times \frac{50}{26.54}\right)$$

= 1.88

This level of JSI indicates an unsafe level of manual material handling. There are several possible ways to reduce the JSI, including altering the lift frequency, the lift height, the population (percentile or gender), the weight, or making fundamental changes to eliminate the whole job.

8-4 OTHER MANUAL MATERIALS HANDLING GUIDELINES

In industry, there are many tasks that involve the manual handling of material other than lifting. Material is carried, pushed, pulled, shoveled, rolled, and heaved. Some containers, like drums, are moved manually but not lifted.

Chap. 8   The Heavy Jobs

This section provides available guidelines for those areas. Typically, these other handling methods have not been studied as extensively as lifting tasks.

One clear consistency with the lifting guidelines used earlier is that there is no absolute limit that is safe for everyone, just as there is not one weight that is safe for everyone. By designing within the guidelines, the worst situations are avoided and the probability of injuries is lessened.

Also, the emphasis of solutions will remain on engineering solutions, such as task redesign, and on administrative solutions that promote selection, training, and the use of proper techniques when they are known.

8-4-1 Carrying

The loads found to be acceptable for carrying are listed in Figure 8-8. The weight involved is a function of the distance of the carry, as well as factors such as size, handles, and stability. Bent-elbow carrys are less efficient than

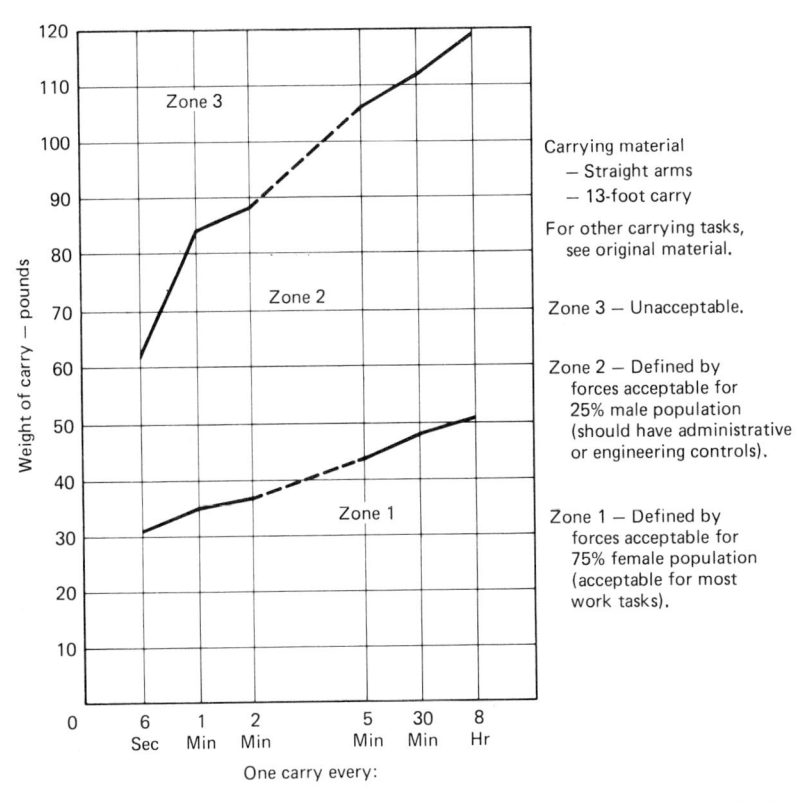

Figure 8-8 Graph of acceptable loads for carrying versus the frequencies involved. *Source:* Adapted from Stover H. Snook, "The Design of Manual Handling Tasks," *Ergonomics,* 21(12), 1978, p. 980.

straight-elbow carrys, so containers should be shallow enough not to bump on the thighs during a carry. Carrying on stairs is not recommended because the line of sight to the stairs may be blocked and because the load will tend to interfere with the movement of the legs.

Some objects, such as pails and suitcases, are designed to be carried with only one hand. One-handed lifts are awkward because they unbalance the carrier. It is preferable to carry two lighter containers, rather than one heavy one, to improve the person's balance. Continual carrying of heavy loads, such as a letter carrier's heavy mailbag, on the same side of the body should be discouraged since it can result in scoliosis of the back from constantly resisting the load.

Two-person carrys are not recommended because the load is usually not equally shared, and the close coordination required for safe pickup is difficult.

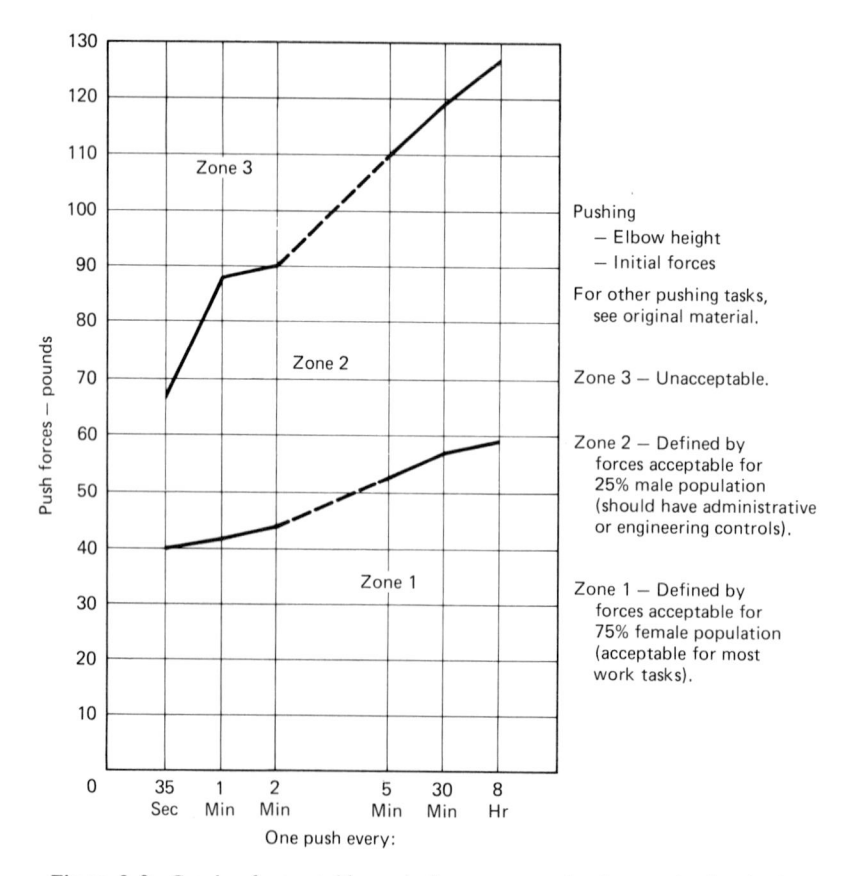

Figure 8-9 Graph of acceptable push forces versus the frequencies involved. *Source:* Adapted from Stover H. Snook, "The Design of Manual Handling Tasks," *Ergonomics,* 21(12), 1978, p. 979.

Chap. 8 The Heavy Jobs   167

Limited measurements of two-person loads show that one person can end up with as much as 70% of the load.

8-4-2 Pushing

Acceptable push forces are shown in Figure 8-9 as a function of the frequency required. These push forces are for short distances, and with longer push distances the acceptable forces will increase slightly. The initial forces reflect the typical peak forces normally incurred when starting a load from a standing stop.

The height of the hands is important during the push, with a handle height between elbow and shoulder height for the person producing the highest forces. A cart design using vertical bars will allow each person to achieve their maximum push force, as shown in Figure 8-10. If vertical bars are not possible, the handle should be set at 35 in., since that height better fits the proper push height of the female.

The designer should be able to determine the approximate push forces required based on the load to be pushed, the caster diameter, the caster material, and the floor conditions. Increasing the caster diameter or caster material hardness are usually inexpensive methods of reducing the push forces.

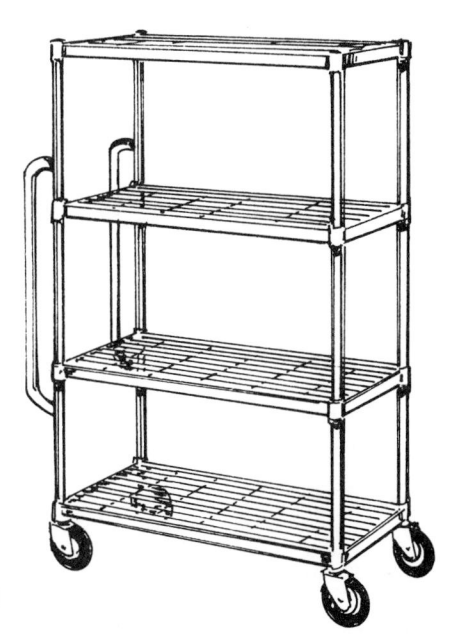

Figure 8-10 Sketch of industrial cart with vertical push bars.

8-4-3 Pulling

Pulling is not usually recommended as a method of moving material, since the load is more difficult to control. The load can easily get out of control and injure the worker. It is difficult to watch the load being moved and the path to be followed at the same time.

The acceptable forces for pulling loads are shown in Figure 8-11. Only short pulling distances are shown because loads moved longer distances are usually pushed.

8-4-4 Shoveling

Shoveling is a task that requires skill as well as strength. Shoveling guidelines are difficult to list, although certain recommendations are always in order.

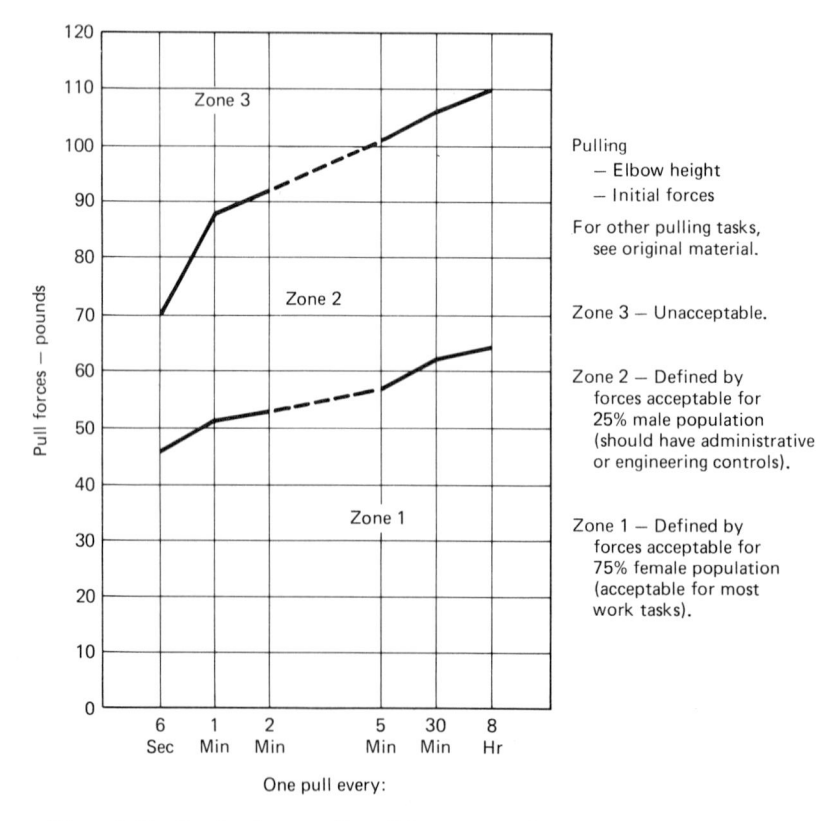

Figure 8-11 Graph of acceptable pull forces versus frequencies involved. *Source:* Adapted from Stover H. Snook, "The Design of Manual Handling Tasks," *Ergonomics,* 21(12), 1978, pp. 975-978.

With the wide variety of shovels available, the shovel should fit the job and materials to be handled. Scoop shovels are good for light, fluffy materials and smaller shovels are best for dense materials. A good rule of thumb for material weight to be handled is 10 lb per shovelful.

The job should be self-paced to allow an appropriate rhythm and use of the muscle dynamics. Learning should take place with lighter loads and at a slower pace. Stability is critical—the feet should be spaced at shoulder width and staggered front to back.

8-4-5 Drum Handling

The moving and palletizing of drums is common in industry. Like shoveling, it requires skill and strength. Drum heights from 30 to 36 in. are standard, which is a convenient height for handling. Rolling drums on the floor and up on pallets is best done using the weight of the drum to provide the momentum. The worker is there to control the drum by balancing it on its edge. Practicing with lightweight drums helps develop good drum-handling skills and should be required before heavy drums are moved. Good rules of thumb for manually moving drums are limiting drum weight to 300 lb and restricting transport distances to 30 ft. Heavier drums should be moved with mechanical equipment. Figure 8-12 shows proper positioning for drum handling.

8-5 ENGINEERING SOLUTIONS

The use of engineering solutions is required when the lifting task exceeds the maximum permissible limit (MPL) and is an alternative when the task exceeds the action limit (AL). Thus, once a problem is identified, engineering solutions are always acceptable. This is a good opportunity to work with designers in the correction of problems, and at the same time open some doors for the prevention of problems in the future.

Engineering solutions are better than administrative solutions since they tend to be more permanent. They usually involve only one application to correct the problem. Administrative solutions, on the other hand, require continual updating and monitoring to be effective.

The use of the NIOSH Guides and of the other guidelines will help to diagnose the problem situation and determine the major cause of the problem. By first determining the major cause, the solutions can be more rapidly identified, justified, and implemented with confidence.

Often, the major portion of the problem is the identification of the problem. A solution can usually be found if the problem is specific enough. The ergonomist must be conscious of his or her limited resources for solving problems of this nature, and be sure to include the designers in the solution process.

a. Person in position for breaking drum.

b. Person with hands on during a move.

Figure 8-12 Drum handling. *Source:* Eastman Chemicals Division of Eastman Kodak Company.

Although many situations seem to be problems waiting for solutions, there are so many design tricks available that the usual situation is one of having solutions waiting for suitable problems to be identified.

8-5-1 Types of Design

There are a number of ways in which design can be used to correct problems. This section discusses some of the general ways in which this can happen. Specifically, the design solutions discussed in the problem job model are all appropriate for solving manual materials handling problems. When you are stumped for a solution to a particularly difficult problem, the thought processes outlined in that model will be beneficial. Some specific solutions are presented in the sections that follow.

Workplace redesign. With the NIOSH Guides, the four factors that modify the load lifted are the distance to the load center, the pickup point, the height of the lift, and the lift frequency. The first three can all be modified by workplace configuration. The workplace can be redesigned to bring the load closer to the worker or to remove obstructions between the person and the load.

The height of the pickup point is often easy to change, for example by placing empty pallets under the load or by mounting motors on raised platforms instead of on the floor. Changing the pickup point will usually reduce the height of the lift, further improving the lifting task.

Other changes in the workplace include improving the footing, the elimination of twists required during the task, the reduction of transport distances, and reducing the need to manipulate the load.

Job redesign. The frequency of the lift can be reduced through job redesign. Rotation of people or changes in staffing can change the lift frequency. A realignment of the tasks comprising the job can change the lift frequency as well.

Changing the work methods can also improve the task. By picking up the load from a different side, the distance to the load center may be reduced. Methods of grasping the load can also make the task safer. The use of two people can change the weight lifted, as well as changing the stability required to lift and move the load.

Container redesign. Engineering changes to the container can make substantial improvements. The center of gravity of the load can be brought closer to the worker. The load can be stabilized to prevent shifting, such as adding baffles to containers containing liquids.

The shape and/or size of the container can be altered. Handles can be added or modified to provide clearance and comfort of the human hand.

The weight of the load can be reduced to the point where it is within the AL or MPL. Or the weight of the load can be increased so that it is so heavy that people do not try to lift it.

Mechanization and automation. There are a number of situations where automation can be used to correct manual materials handling problems. Conveyors, lift trucks, industrial robots, cranes, and hoists are all good substitutes for human muscle. Often the job can be done much cheaper with equipment than by human beings.

The equipment used need not be exotic to result in an improvement in the operation. The use of hoists can be just as effective a substitute for human beings as industrial robots. The newer technology does seem to be more flexible, thus allowing its use in more places or allowing it to be moved from one place to another.

The use of mechanization and automation starts when the problem tasks are identified. Then means are identified to replace the human being in the system.

8-5-2 Examples of Materials Handling Equipment and Aids

There are many ways to improve tasks that prove to be difficult for the human being. Recently, a task was identified that resulted in a high rate of injuries and redesign ideas were explored. The task and the possible solutions are discussed below.

The task of handling bags filled with material is a common task in many industries. Food products, agricultural products, plastics, animal feed, chemicals, and fertilizer are all handled in bags. Suppose that lifting the bags from the bag sealer to the pallet was determined to be an unacceptably hazardous situation. What alternatives are available?

Workplace redesign

Modify the height of the loading/unloading area to change the lift.

Slide the bag instead of lifting.

    Use a smooth metal.

    Use an air table.

Use a raised work surface so that the operator only has to lower material instead of lifting.

New layout to avoid twisting with a load.

Equipment redesign

Use a variable-height conveyer to mechanically raise bags.

Use a spring-loaded (cafeteria-style) leveler to position the material at the correct height.

Chap. 8   The Heavy Jobs   173

Use a stacker's fork to raise and lower the pallet to the correct height.
Automate the task with bag palletizers or use industrial robots to handle the bags.
Add mechanical assists such as a powered conveyer to move the material.

Redesigned methods
Use two people to handle the bags.
Use a deadstack instead of palletizing.
Use ballistic motions in handling the bags.
Keep the load in close.
Avoid twisting with the load.

Job composition
Rotate personnel to distribute the materials handling work among more people.
Consolidate all the materials handling into only a few jobs.

Work environment
Control heat, dust, noise, and foot traction.

Process
Purchase/sell in bulk quantities.

Product
Change the bag size and weight (lighter or heavier).
Give it handles.
Use a rougher bag surface for better grip.

There is no question that additional ideas could have been generated, but this served to point out the wide variety of methods available to improve a task once a problem situation has been initiated.

The ideas available to correct situations like this run all the way from high tech to low tech to no tech at all. Technical innovations can sometimes be used to solve a problem that has resisted solution before. But often a solution is available if one has read the right catalog or talked to the right designer.

There are many materials handling aids available featured in plant equipment and maintenance catalogs. Figure 8-13 features some of the equipment available. These items are not intended as product references, but merely serve to preview the vast array of these products.

The economics of these changes are often very favorable. There are reductions in injuries, which are quite expensive when the costs are properly

(a)

BIG WHEEL CART

Over-sized Wheels for Ease in Carrying Heavy Loads

Unique service cart has sturdy oversized 10" x 1¾" rubber tired wheels (two rigid, two swivel) for easier handling of loads. 500 lbs. capacity. Flat 30" x 20" top projects over cart base for easier loading of heavy or oversized stock, packages, parts, machinery, etc. Ample 17⅞" between shelves. 1 tubular steel handle. 33" high. All welded steel construction. Durable gray finish.

Stock No. FM2270

$172.00 LOW AS

ELEVATING WORK TABLES

2 Post Lift Table
500 Or 1000 Lb. Capacity
30" Lowered Ht. / 18" Lift

4 Post Lift Table
Up to 6000 Lb. Capacity
36" Lowered Ht. / 24" Lift

Lift and Transport Dies, Fixtures, and Machine Parts. Foot-Operated Pump on Heavy-Duty Steel Lift Tables. Large Phenolic Casters and Floor Lock are Standard

(b)

EASY WAY TO PALLETIZE YOUR HEAVY DRUMS

- Two-Wheel Brakes For Control On Inclines
- Shoe Grips Pallet And Eases Drum Transfer
- Steel Or Aluminum Construction, 53" High
- Solid Or Pneumatic Wheel Models

Easy To Load

Free Standing When Loaded

1 Sliding, Spring Loaded Chime Hook
2 Replaceable Steel Toes
3 Two Wheel Brake
SOLID RUBBER TIRED MODELS 152-D and 153-D
PNEUMATIC TIRED MODELS 204-D AND 205-D

(c)

STACKER HAND TRUCK (d)

Allows One Person to Handle Difficult Moving and Lifting Jobs

Capacities to 750 Lbs.
Two Lift Heights, 40" and 54"

Foot-Operated Hydraulic Lift Pump
Pedal Downstroke Lifts Platform 1¼"
Automatic Leveling Platform
Fold-Away Foot Pedal

8" x 2" Mold-on Rubber Wheels

Welded Steel Construction
or
Lightweight Magnesium

Beveled Platform Edge Tilts to Absolute Floor Level for Ease of Loading

(e) **DRUM SLINGS**

750 Lb. Capacity
Lightweight-Safe
Fits 55-Gallon Drums
Economy Steel
or
Non-Sparking Bronze Construction

REMOTE CONTROL DRUM CARRIER

For Safe, High Level Controlled Pouring

Manually powered remote control drum carrier pours materials at any height or plant location with all the accuracy and smooth control of hand-tipping. Saves labor, reduces costly spillage and improves plant safety. Easy one-man chain operation. Just hook carrier to your monorail, hoist, crane or chain block, raise and tilt to desired position. Will handle 800 lbs. when full, 500 lbs. when half full. 31"W x 31"H.

(f)

ELECTRICAL INPUT (OPTIONAL)

Jib cranes are designed to be floor/foundation mounted and self-supporting. All units rotate 360°.

(g)

Floor Level Loading Place Container in Load Position... Lock And...

Dump Granular Material Into Mixing Bins...

(h)

(i)

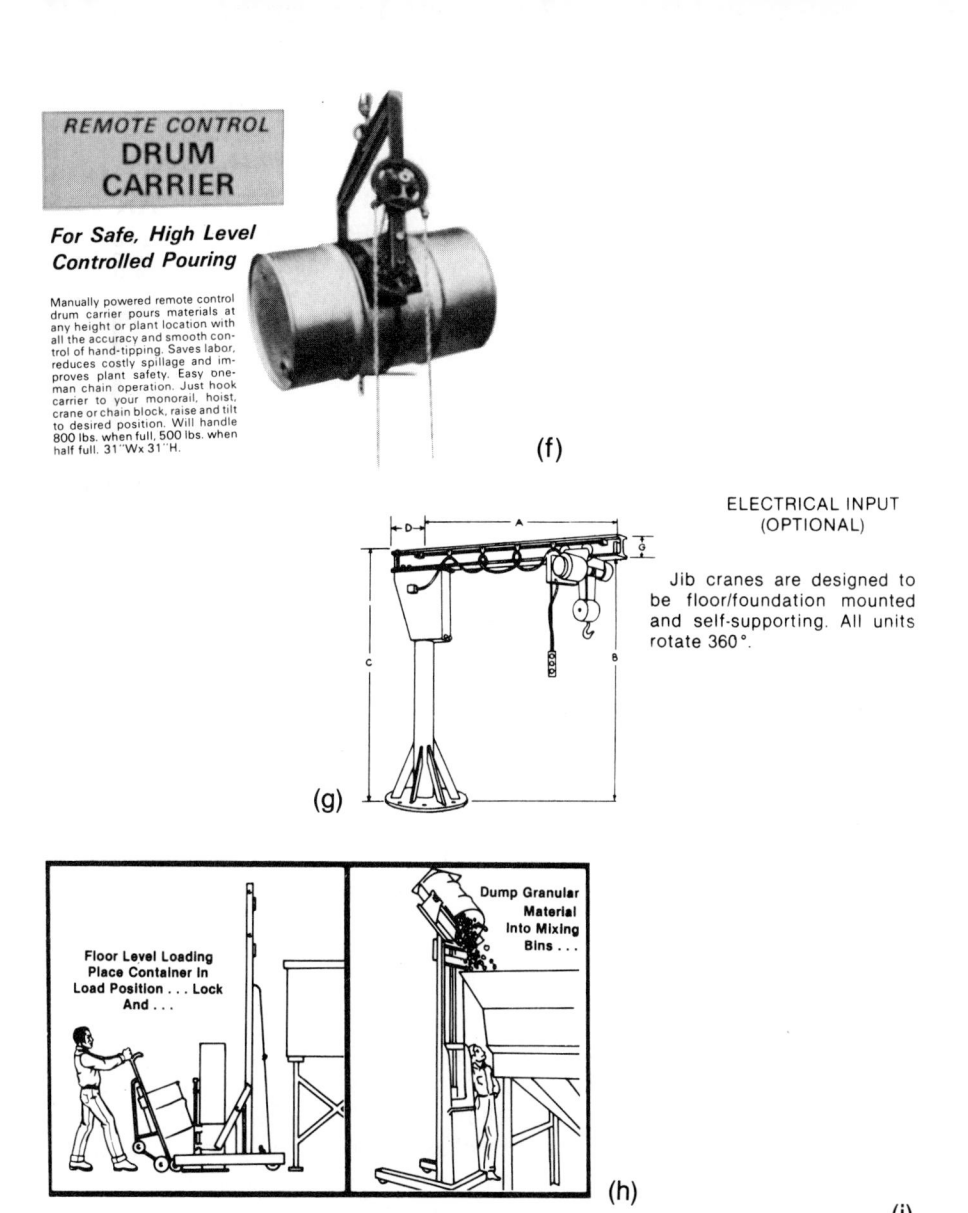

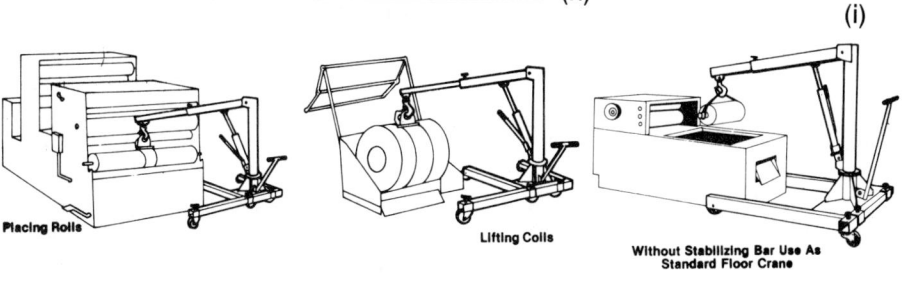

Placing Rolls

Lifting Coils

Without Stabilizing Bar Use As Standard Floor Crane

Figure 8-13 Examples of materials handling aids available from equipment catalogs. *Source:* Arrow Star, Inc., Lynbrook, NY; Dozier Equipment International, Nashville, TN; Morse Manufacturing Co., East Syracuse, NY; Valley Craft, Inc., Lake City, MN; and Wesco Manufacturing Co., Lansdale, PA.

documented. There may be labor savings. The speed of the operation is often likely to increase. There may be a more consistent package and/or pallet configuration, and the rate of product damage may decline. The author is aware of numerous situations where a project was initiated to help control the injury rate, but after careful cost analysis, the project could easily justify itself on tangible cost savings alone.

8-6 ADMINISTRATIVE SOLUTIONS

Administrative solutions are an acceptable alternative when lifting tasks fall between the AL and the MPL on the NIOSH Guides. They are not acceptable when the lifting task exceeds the MPL. These types of solutions are useful when part of the population can be expected to perform the task and when there are risks for the remainder of the population.

Administrative solutions should be explored when it is found that engineering solutions are not available or are prohibitively expensive. The engineering solutions tend to require only one time consideration of the problem, while administrative solutions must be audited for effectiveness and reinstituted periodically. The continued enforced use of administrative solutions is their major drawback. Their low cost and rapid implementation are major attractions. Three major types of administrative solutions will be discussed: (1) job redesign, (2) selection of personnel, and (3) training of personnel.

8-6-1 Job Redesign

Job redesign is sometimes thought of as an engineering solution, but it is described here to reflect the transitory nature of some of the redesigns. When the changes can be made permanently, they should be considered engineering changes, but when they can be easily reversed, they need to be thought of as administrative changes with all the follow-up requirements.

An example of this situation involved the cleanup of scrap material. Conventional "50-lb" bags were used to hold the scrap material, but since the material was unusually dense during this period of processing, the full bag weight was closer to 90 lb than to 50 lb. Lifting from the floor, this easily violated the AL and approached the MPL. Something needed to be done. The logical solution was to fill the bags only half full. This was implemented, but a follow-up later revealed that the new limitation was not being followed. This was a classic case of "the lazy man's load," carrying something twice as heavy to reduce the total number of trips. It is a common syndrome in industry and one that is very hard to fight. Repeated occurrences finally led to another solution, a rolling scrap cart that eliminated the lift altogether. The types of job redesigns that work well are creative staffing proposals to reduce lifting frequency and methods changes to alter lifting weights and configurations.

Staffing changes are designed primarily to reduce the frequency of the lift. The task may be shared by more people through rotation or by changing job assignments. Since the lifting tasks are performed by more people, the lift frequency is reduced. Neither of these solutions requires increases in staffing.

Some job redesigns involve the consolidation of the tougher lifting tasks into one job, which is then staffed by natural selection. This is a good example of using job redesign and selection in combination.

Methods changes focus on the lifting configurations used. By changing the pickup points with extra empty pallets stuck under the load, significant changes in the AL can occur. This change not only affects the height of the pickup and the distance the load is lifted, but it can also affect F_{max}, resulting in changes in the frequency factor as well.

The method of pickup can also be changed, for example by changing the side of the container faced during pickup, to allow the person to get closer to the center of gravity of the load. Changes to get the load closer to the body by allowing it to fit between the legs are also appropriate. Changing a one-handed lift to two hands or going from a one-person lift to two persons are also suitable methods changes.

The characteristic of the administrative changes is that they are easily reversible and may be reversed without management's or the ergonomist's knowledge. Caution must be taken to ensure that the workers themselves know the reasons for the changes and accept the changes as being in their best interest. Each succeeding generation of workers must develop the same understanding.

8-6-2 Selection

There continues to be a strong inclination to use selection techniques by line supervision. Requests to use selection methods are among the most frequent from the line organization and among the most difficult to answer. Conceptually, selection of workers to fit the job makes a lot of sense. Practically, however, selection may not even meet the needs originally identified, and the side effects can be very harmful. The appeal of selection is so powerful that once suggested, it may be difficult to get the client to consider other solutions, such as engineering design.

The major benefit of selection is supplying a fully capable worker for the job while excluding only those who cannot perform the job safely. There is a very fine line between performing the job and not performing acceptably. In fact, with strength variations, the same person may be able to perform the job one day and not the next. It is very hard to find the line that distinguishes acceptable performance from unacceptable, and consequently selection sounds better than it actually works.

One major problem with selection is that the attributes normally mentioned for personnel selection involve strength and size. Both of these factors have gender biases and strength has an age bias as well. If the selection test is done improperly, it will discriminate against females and older people.

A number of traditional selection tests used to prevent the occurrence of low back pain have been found to be unreliable. Snook found that "no significant reduction in low back injuries was found in employers who used medical histories, medical examinations, or low back x-rays in selecting the worker for the job."[5]

Due to the complexity of developing and proving validity of selection tests, the effort that goes into the development of selection tests could be used to provide redesigns, often with better results.

Before selection is used, it is recommended that other steps be investigated first. Since this is not what the client has requested, it is appropriate to detail a plan that involves selection as a solution, after other possible solutions have been explored. As the ergonomist, the expert on the resolution of these work site problems, you may have to "stick to your guns" to get this process accepted. A process for utilizing selection methods is listed in Table 8-8.

This process is not easy, nor should it be. Management selection of people for a job is serious business and it should be done carefully.

TABLE 8-8 Processes of Developing Employee Selection Procedures

1. Request for selection from client.
2. Client and ergonomist agree on the process, after discussions of this process and its rationale.
3. The problem is clearly defined—exactly what aspects of the job result in injuries and problems.
4. Engineering redesign is attempted. If successful, the process is terminated.
5. Job redesign is attempted. If successful, the process is terminated.
6. If earlier processes are unsuccessful, appropriate selection criteria are agreed upon. The criteria should simulate the actual task as much as possible. Pass and fail criteria for people taking the test are agreed upon.
7. The test methodology is developed and means to implement it are planned.
8. Blind testing is done of all job candidates, but placement is made in the old manner. The test results are stored for later analysis.
9. After a suitable number of candidates (approximately 20) and a suitable length of time on the job (at least 6 months), the test results are correlated with job and injury performance. A valid test will produce statistically significant results. If the test is not valid, additional selection criteria must be developed and tested.
10. With suitable validity, the test is adopted and used for all placements to the job. Monitoring for continued effectiveness should continue.

---

[5]Stover H. Snook, "The Design of Manual Handling Tasks," *Ergonomics*, 21(12), 1978, p. 983.

The process of "natural selection" or of self-selection for a job is preferred from the legal standpoint. Even with scientific testing, the person is normally the best judge of whether or not the job is suitable for them.

Natural selection has a limitation where there is a strong incentive to accept a new job. For example, to obtain a job that pays appreciably more, people may say that the job is suitable when, in fact, they realize that it exceeds their capacity.

Strength testing has an implicit legal agreement where the person being tested may recover medical and related costs for injuries during the testing. These are similar to workers' compensation claims and probably would be handled the same way. However, if strength testing is done prior to employment, the limitations of workers' compensation may not apply. In other words, in legal action for a preemployment testing injury, the employer may not be protected with the limited liabilities normally provided by workers' compensation.

8-6-3 Training

There are two types of training that are used in industry. The first type, and by far the most common, is training on the correct method of lifting. The second type, and most comprehensive, is the awareness training that allows the worker to understand the human body, the lifting process, and injury prevention techniques.

Training on the correct lifting method has been going on for the better part of this century. Normally, this emphasizes the classical lifting method, where the worker lifts with the knees bent and the back straight. Despite extensive use of this technique, it has not been proven to reduce the injury rate. Snook stated that "no significant reduction in low back injuries was found in employers who trained their workers to lift properly."[6]

Training on manual materials handling. NIOSH recommends that training should be directed toward awareness of the dangers of careless and unskilled MMH, showing people how to avoid unnecessary stresses, and teaching them individually to be aware of what they can safely handle. An outline of the material that they suggest for a training course is shown in Table 8-9.

A practical method of leading training in lifting is outlined in the "Lift School," a training methodology developed by the author and utilized extensively. The Lift School emphasizes four principles and the need to observe them during the lifting process. The principles were developed from a review of primary accident causes and biomechanical needs. The four principles are:

1. Keep the load in close to the body (to avoid large moments on the spine).

[6]Ibid.

TABLE 8-9 Topics for MMH Training Recommended by NIOSH

1. The risks to health of unskilled MMH
2. The basic physics of MMH
3. The effects of MMH on the body
4. Individal awareness of the body's strengths and weaknesses
5. How to avoid the unexpected
6. Handling skill
7. Handling aids

Source: Work Practices Guide for Manual Lifting (Washington, D.C.: U.S. Government Printing Office, 1981), pp. 100–101.

2. Keep the hips and shoulders aligned (to avoid twisting the spinal column while it is under load).
3. Maintain stability (to avoid sudden movements that can increase biomechanical forces or will cause a fall).
4. Think and plan (first, to plan unusual activities such as stairs or corners, and second, to avoid getting into a situation that is difficult to get out of, such as setting down an awkward load or having to place an object on a high shelf).

The preferred method of presenting the Lift School is on the shop floor, using materials commonly lifted by the workers and with extensive interaction with the workers. Classroom training is generally ineffectual, because it is difficult for workers to take the abstract class material and apply it to their own lifting tasks. It is better to have an experienced person make the connection between the general lifting techniques and the specific lifting tasks required on the job. The techniques can be taught and demonstrated for the actual items that are lifted and at the real work site.

Table 8-10 presents an overview of the Lift School. Table 8-11 presents a method of teaching others to lead the Lift School. This has proven an effective method of providing lifting experts at many plant sites. Candidates for instructors for the Lift School do not have to have prior technical experience with manual materials handling. They should, however, be familiar with actual lifting tasks and they should be comfortable working with small groups.

Skill acquisition. Training is also appropriate for MMH situations where the task is significantly skill dependent. Drum handling is an excellent example of a situation where skill, rather than the person's size, weight, or strength, is the best predictor of job suitability.

In these situations, lighter weights should be used in training. This develops proper motion patterns and the fine sensitivity necessary for control. Often, without practice, the problems that arise are ones of overcontrol and

TABLE 8-10 The Lift School

The Lift School
Participation and discussion are requested and required.
(Total time of session is 60 minutes.)

A. Introduction
   1. Sell the idea—prevent the injury
   2. Let them talk about people they know with back problems
      a. Severity
      b. Limited activity
      c. High personal cost
B. Proper handling techniques
   1. Techniques for any lift
      a. Load in close
      b. Hips and shoulders lined up
      c. Maintain stability
      d. Think and plan
   2. Demonstration and discussions
      a. Straight lift
      b. Lift above shoulders
      c. Lift from floor
      d. Carrying material
      e. Stair climbing
      f. Turns
      g. Two-person lifts
      h. When to get assistance
      i. Special cases as needed
         (1) Pushing-pulling
         (2) Bag handling
         (3) Shoveling
         (4) Drum handling
         (5) Others
C. What to do if an injury occurs
   (*Note:* This will be specific to each organization and probably to each plant. It should be tailored to that need.)
   1. Contact supervision
      a. Get help
      b. Prevent further injury
      c. Establish it as an on-the-job injury
   2. Medical department
      a. Examination
      b. Medication
      c. Therapy
      d. Light-duty work
      e. Records kept by the medical department
      f. Entry on work record?
D. Flexibility exercise program
   1. Prevention (prevent the injury)
   2. Rehabilitation (use doctor's prescription for exercise; it may include some of these)
   3. Demonstration of flexibility exercises by instructor
   4. Participation in the flexibility exercises by class

TABLE 8-11 Training Others to Lead the Lift School

The Lift School
"Train the Trainers"
(Approximate time is 6 hours.)

I. Introduction (30 minutes)
   A. Purpose
      1. To teach participants how to lead the Lift School
      2. To give them an understanding of lifting, job analysis, the back, and medical therapy to answer questions
   B. Purpose of the Lift School
      1. To demonstrate and discuss proper handling techniques for various types of lifts
      2. To demonstrate principles of lifting which enable an employee to cope effectively and safely with nonstandard lifts
   C. Assignment—come up with a tough lifting situation from their own work area that they want analyzed later in the session
   D. General discussion on lifting

II. Demonstration—"The Lift School" (60 minutes)
    (A live demonstration of the Lift School—just like this class was an actual work group. Participation and discussion are requested and required.)
    A. Introduction
       1. Sell the idea—prevent the injury
       2. Let them talk about people they know with back problems
          a. Severity
          b. Limited activity
          c. High personal cost
    B. Proper handling techniques
       1. Techniques for any lift
          a. Load in close
          b. Hips and shoulders lined up
          c. Maintain stability
          d. Think and plan
       2. Demonstration and discussions
          a. Straight lift
          b. Lift above shoulders
          c. Lift from floor
          d. Carrying material
          e. Stair climbing
          f. Turns
          g. Two-person lifts
          h. When to get assistance
          i. Special cases as needed
             (1) Pushing-pulling
             (2) Bag handling
             (3) Shoveling
             (4) Drum handling
             (5) Others
    C. What to do if an injury occurs
       (*Note:* This will be specific to each organization and probably to each plant. It should be tailored to that need.)
       1. Contact supervision

Chap. 8   The Heavy Jobs   183

TABLE 8-11   Training Others to Lead the Lift School (cont'd)

   a. Get help
   b. Prevent further injury
   c. Establish it as an on-the-job injury
  2. Medical department
     a. Examination
     b. Medication
     c. Therapy
     d. Light-duty work
     e. Records kept by the medical department
     f. Entry on work record?
  D. Flexercise program
     1. Prevention (prevent the injury)
     2. Rehabilitation (use doctor's prescription for exercise, it may include some of these)
     3. Demonstration of flexercises by instructor
     4. Participation in the flexercises by class

III. The Back and How It Works (45 minutes)
  (This is an extensive discussion of the back, how it works, the muscle support systems, the spine and how it interlocks, and the need for the abdominal support system. This section will cover some reasons for muscle tone, flexibility, special postures, and support of the back.)
  A. Muscle systems (overhead cell)
  B. The spine (overhead cell)
     1. Vertebrae
     2. Discs
     3. Nerves
  C. Abdominal support (overhead cell)

IV. Medical Therapy (30 minutes)
  A. Visit the on-site medical facilities
     1. How an injury is processed
     2. The examination
     3. Types of medication
        a. Dosage
        b. What it does
     4. Therapy
        a. Ultrasound
        b. Moist heat
        c. Whirlpool
        d. Traction
        e. Other
           (*Note:* Discuss and demonstrate each.)
     5. Prescription for light-duty work
        a. Who makes it
        b. For how long
  B. Discussion
     1. Records kept by the medical department
        a. Purpose
        b. Duration of retention
        c. Who has access

TABLE 8-11 Training Others to Lead the Lift School (cont'd)

---

  2. Entry on person's work record
      a. Is entry made on work record?
      b. Does line supervision have access to medical file?
V. How Much Can a Person Lift? (90 minutes)
  [Based on the author's personal experiences, plus the article "What Criteria Exist for Determining How Much Load Can Be Lifted Safely?" Arun Garg and M. M. Ayoub, *Human Factors,* 1980, 22(4), pp. 475-486.]
  A. Introduction
      1. How much can a person lift?
      2. How much should a person lift?
      3. How should people lift?
      4. List of lifting factors
      5. Four most important variables
      6. Very complex problem; no one answer; we don't know yet
  B. What basis exists for determining how much people can safely lift?
      1. Subjective—psychophysical
      2. Metabolic/cardiovascular
      3. Biomechanical
      4. Epidemiological—50 lb bag, 300 lb drum
      5. Types of limitations
          a. What you think you can lift
          b. Stamina—what your heart can take—snow shoveling
          c. Physical stress—what your physical system can take—structural approach, destructive testing
  C. Psychophysical approach
      1. Definition and approach
      2. Types of data gathered
      3. Proposed guidelines
      4. Example of bag handling
  D. Biomechanical approach
      1. Definition
      2. Example
  E. How to analyze lifts?
      1. Solutions and techniques
      2. Twenty factors; four most important
      3. Changes that can be made—example: bag handling
  F. Other factors
      1. Training and exercise
      2. Selection
VI. Special Cases—Lifting Assignments Analyzed (60 minutes)
  (The cases brought up by the class members are analyzed interactively; that is, the instructor will analyze the situations, but will ask questions of the class to identify and list the most likely causes. Solutions are generated, and they need to be as practical as possible. It is difficult to describe this section much better, since it is very dependent on the class makeup and the questions asked.)
VII. Practice—The Lift School (60 minutes)
  (This section is optional, depending on the confidence and skill of the class. If done, I like to ask different members of the class to lead different sections. Critiques are necessary to improve future performance.)

TABLE 8-11 Training Others to Lead the Lift School (cont'd)

VIII. Review—The Lift School (15 minutes)
    A. Review the four sections
    B. Summary—this is important
End of session

---

of strong muscle groups working in opposition to one another. Some other tasks that will benefit from skill training are those where ballistics are used extensively, such as for tossing material, or those where dynamic use of the body weight will ease the handling task.

Physical training. Physical training is probably not a good method to enable someone to do a task quickly that they cannot perform currently. However, over a longer time period, physical training can certainly be effective. The capacity of a person can be noticeably changed.

Physical training is also useful for situations where the heavy tasks are intermittent, so that training does not continue to occur as one performs the job. Firefighters find that they must consciously work out in order to be ready for the physical stress and lifting tasks that occur during fire fights.

Physical training will work best when the individual is self-motivated to perform the training. The person must clearly see the need and the benefit of the training. Otherwise, the training itself must be monitored or periodic tests must be administered to ensure that strength and/or conditioning are maintained.

8-7 MANAGING MATERIALS HANDLING

In any industry, there are many tasks that require materials handling. The costs resulting from manual materials handling injuries are usually large, especially temporary lost time and long-term disability. The management of MMH to control costs and injuries is an important part of the ergonomist's job. Managing requires an emphasis in the areas of design to prevent injuries, of cost allocation to highlight the true costs of injuries, and of constantly seeking the ideas of others to reduce injuries.

8-7-1 Design Focus

In the design area, the focus will be on aiding the designer and on ensuring that the designer accommodates the workers' limitations. Design guidelines and reference standards should be readily available. The guidelines/standards should indicate:

1. When people can be expected to perform the tasks and when they cannot
2. How to diagnose the major problem areas
3. Acceptable methods of dealing with the problems uncovered
4. How the designer can get assistance for additional help

Manuals of "solution tricks" can be compiled. These "tricks" show useful handling aids and mechanical equipment for the types of tasks that cause trouble. Some specific tasks where these books work well are drum handling and bag handling. These manuals are less rigorous than design standards and are easier to update with helpful-looking items. The manuals can be started with a few brainstormed ideas and then be added to as design solutions and new catalogs become available. Hands-on designers are excellent contributors to something like this.

It is important to ask designers regularly for their ideas. These ideas are useful for enhancing the "solution tricks" manuals and for creating additional guidelines. An ergonomist should also be available to assist the designers with their questions. Lack of assistance is such a frequent excuse for improper design that it must be avoided.

8-7-2 Costs of Injuries

One main reason that injuries are allowed to continue is that only a few people know how high the costs actually are. Often, these people are not in the best position to solve the problems. Injury costs are often hidden from view or ignored. The injury costs should be grouped together to establish their magnitude to the whole organization. The individual costs for each injury should also be sent to the appropriate operating unit. The actual injury costs will usually demonstrate that the costs of prevention are really not that high.

To supplement cost information, actual cost allocation is a powerful tool. During the startup of new designs, the costs of injuries should actually be charged to that design to emphasize inappropriate trade-offs of tasks to people and machines. Certainly, costs can be allocated back to the injured person's work department or section to further enhance control of the injuries. This serves two purposes: first, an awareness of the injury seems to help prevent additional injuries from occurring, and second, continuing cost allocation develops a more rapid placement of people returning to work.

The total costs of all injuries should be determined, but if that is not possible, the actual costs of some representative injuries should be calculated. These costs may be difficult to identify and harder still to quantify. Yet this lack of information is precisely what allows injuries to continue without extensive risk control. In comparison with the injury costs (between $5000 and $10,000 lost for each injury), the costs of preventive measures are often very small. Costs are an effective method of mobilizing management interest and attention. This will make the prevention of manual material handling injuries a higher priority.

8-7-3 Personal Actions

The ergonomist has a strong responsibility to be fair when examining MMH situations. If situations require changes, the ergonomist is obligated to pursue those changes. At the same time, if the situation really does not require any changes (even though some are proposed or approved), the ergonomist should point out that those resources can be better used elsewhere. This fairness is essential to develop a high degree of credibility.

The ergonomist should strive to publicize good results. Both the design and the designer should receive credit for an effort well done. Appropriate means of recognition are to notify supervision, highlight the work in case studies and examples, discuss it in classes, and use organizational newspapers. Use all means possible to make champions of good designers.

The ergonomist should strive to develop the relationship with management so that critical problems can be discussed. Management should not accept a poor design or continue to use one already in existence. The ergonomist is most influential through personal communications and by ensuring that early awareness training allows management to identify poor designs. The managers should be educated about the costs of manual materials handling at the same time that the designers are educated about correcting those problems.

The ergonomist should try to "make a friend" whenever dealing with a new designer. This means not to win the battle only later to lose the war. The ergonomist may choose to mute criticism on minor problems, provided that the designer has learned how to handle the situation next time, to ensure that the designer will ask for assistance next time it is needed.

Contacts with design personnel should be developed and maintained. Often, the ergonomist is in a position to identify those tough problems that seem to defy solution by conventional means. Yet it is surprising how often there is some new technology or equipment on the market that will correct the situation. By maintaining these contacts, the situation of "a solution looking for a problem" can be avoided.

The ergonomist should develop a network of contacts throughout the industry to discuss common MMH problems and their resolution. Ergonomists and safety personnel are good places to begin building this network. Find out how others are able to sell their ideas and redesigns and how they are able to stretch limited budgets for improvements in this area. One should also inquire how to get designers to understand, appreciate, and use lifting and other MMH design guidelines.

8-8 CUMULATIVE TRAUMA DISORDERS

Cumulative trauma disorders are just that—they are the injury that results from repeated use of the hand or other body part. They are normally seen as carpal tunnel syndrome, tendinitis, bursitis, and tenosynovitis. Not all people are equally susceptible to these injuries, and people working side by side will experience different responses to the same job stress. Nonjob activities will contribute to cumulative disorders, and in fact some disorders are known better by their nonwork names—tennis elbow and trigger finger.

8-8-1 Causes of Cumulative Trauma Disorders

With a reasonably low incidence of injuries (compared to the total exposure), there are no absolute causes of cumulative trauma disorders (CTDs). Some of the contributors to CTD, however, are large forces, repeated and severe bending, and low temperatures.

Large forces can contribute to injuries to the delicate structures of the hands and wrist. Forces from scraping paint or pounding on things with the palm of the hand will injure the palm and wrist areas.

Extreme wrist movements are hazardous, as are highly repetitive movements. The forces on the wrist at the limits of flexion and extension are severe. Highly repetitive movements are also harmful on the hand and wrist joints. When extreme joint motions are combined with high repetitions and with large forces, problems are sure to follow.

Low working temperatures are also believed to be a contributor, so working in a cold and/or wet environment will contribute to CTD problems. This environment severely reduces blood flow to the area, causing fatigue and inhibiting healing of any injuries.

For the elbow and shoulder joints, the extreme motions and repetitive movements are also believed to be a strong contributing source of CTD. The extreme motions can come from fully extended elbows or working with the arms overhead. Highly repetitive movements can come from a variety of sources, most in assembly where actions are frequently repeated.

8-8-2 Correcting and Preventing Cumulative Trauma Disorders

Many times the CTD injuries will heal themselves with a period of rest. Surgery may be necessary for some injuries. The more severe the case, the less likely it is that the person will later be able to perform the troublesome job.

A correction to a diagnosed CTD problem will be more effective when there are corresponding changes at the work site or to the job. The same types of improvements will also help prevent the injuries to start with.

In the prevention of CTD, proper workplace design and proper job design become even more critical. Seemingly minor changes in design can cause substantial changes at the work site. For example, a workbench that is slightly too high will cause people to raise their arms as they work. During the workday, the forces on the shoulders can lead to CTD. Working in a natural relaxed position is the best design.

One of this author's principles of design has been "anything in moderation." With CTD this certainly holds true. The best designs are those that limit fast, hard, and extreme movements by the person.

Work site design techniques can provide a workplace that is the correct height and that "fits" the person well. This helps to avoid the stress of supporting body parts all shift long.

Proper work site design techniques avoid cold and/or wet work environments. Any tools used should be free from vibration, should not require large forces to operate, and should not blow air at the person's hands. The comments in Section 8-8-3 are helpful in preventing CTD.

One good principle of work site design that always works is use machinery whenever possible to avoid highly repetitive operations by people. It is especially true with CTD problems.

Job design technology is also useful in controlling and reducing cumulative trauma disorders. The work pace should be operator controlled, if possible. When identical movements are required during an entire shift, rotation should be considered to other jobs with different required movements.

If rotation is not possible, adequate recovery pauses should be implemented. The recovery periods are better when they are shorter and more frequent than when the they are longer at less frequent intervals. About 3 to 6 minutes per hour will provide suitable recovery. The person does not have to be idle during this period, but may instead perform other needed tasks.

People new to a job should be broken in slowly, to allow their bodies a chance to adjust to the task requirements. The same holds true for a person who has been off the job for a while on another assignment, or home with a cumulative trauma disorder. Coming back gradually is preferable to having to perform a new task at full pace without any buildup of strength and job skill.

8-8-3 Tool Design

One of the common contributors to cumulative trauma disorders is the inappropriate use of hand tools or the use of improperly designed hand tools. Hand tools are marvelous extensions of the human being. They allow the person to do many things quickly and easily that would otherwise be difficult or impossible.

Unfortunately, tools can also be a source of injury. Direct injuries from the tools as well as overuse syndromes result in lost time, pain, and excessive costs. Well-designed tools can prevent these problems. Since many tools are purchased, rather than designed and fabricated internal to the user organization, the chapter on purchasing ergonomically designed products will be valuable.

In her article, "Handtool Design," Nemeth provides 10 principles for the design and use of handtools. These principles help prevent injuries, and enhance performance and quality of workmanship. The principles discuss not only the design of the tool, but also the use of the tool, the design of the workstation, and the design of the job. This is consistent with the systems-type approach followed in this book. The principles and brief comments are listed below.[7]

1. Maintain straight wrists. Bent wrists encourage carpal tunnel syndrome. Any wrist deviation is further aggravated by repetitive motions or large forces. The tool should be held and used in a neutral position.
2. Avoid static muscle loading. The work should be done with the arm and shoulder in a normal position to avoid excessive fatigue. This is especially true when tool weights are large or the tool is used for extended periods of time. Counterbalancing tools is a common solution, as noted in Figure 8-14.
3. Avoid stress concentrations over the soft tissue of the hand. Pressures on these tissues can obstruct blood flow and nerve function. Figure 8-15 shows a tool that concentrates the forces on the hand; these tools should be avoided.
4. Reduce grip force requirements. The grip forces can put pressure on the hands or result in tool slippage. Of special note is the distribution of force on the hand, resulting in the same problems mentioned in item 3.
5. Maintain optimal grip span. The optimal power grip with the fingers, palm, and thumb should span 2.5 to 3.5 in., as noted in Fig. 8-16. For circular tool handles, like screwdrivers, the optimum power grip is 1.25 to 2.0 in., and the optimum precision grip is 0.3 to 0.6 in. for fingertip use.
6. Avoid sharp edges, pinch points, and awkward movements. Sharp edges cause blisters and pressure points. Pinch points can make a tool almost unusable. Awkward movements are easily found through observation or use. One common problem is the movement required to open a tool, especially when it is used on a repetitive basis.
7. Avoid repetitive finger trigger actions. Using a single finger to operate a trigger is bad, especially with frequent use. Use of the thumb is preferred since the thumb muscles are in the hand, not in the forearm (this avoids

---

[7]Susan E. Nemeth, "Handtool Design," in *Industrial Ergonomics: A Practitioner's Guide,* eds. David C. Alexander and B. Mustafa Pulat (Atlanta, Ga.: Industrial Engineering and Management Press, 1985). Reprinted with permission. Copyright Institute of Industrial Engineers, 25 Technology Park/Atlanta, Norcross, Ga. 30092.

Chap. 8　The Heavy Jobs

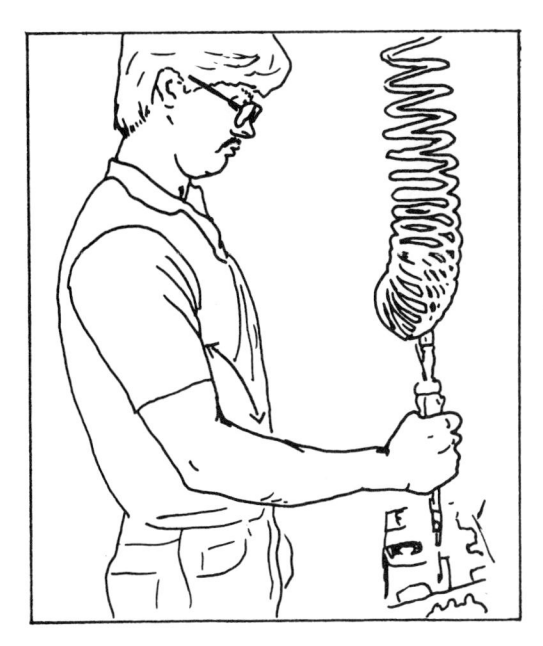

Figure 8-14 Counterbalance tools to avoid static loading. *Source:* Susan E. Nemeth, "Handtool Design," in *Industrial Ergonomics: A Practitioner's Guide,* eds. David C. Alexander and B. Mustafa Pulat (Atlanta, Ga.: Industrial Engineering and Management Press, 1985). Reprinted by permission, Copyright © Institute of Industrial Engineers, 25 Technology Park/Atlanta, Norcross, Ga. 30092.

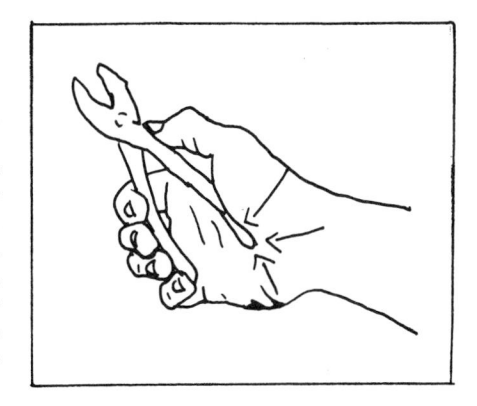

carpal tunnel syndrome). Other triggering mechanisms, like proximity or pressure switches, are possible.

8. Protect the hands from heat or cold. Tools with motors can produce and transfer heat, as can soldering irons or other heat-generating tools. Cold comes most often from air-powered tools with an exhaust near the hand grips.

Figure 8-15 Avoid stress concentrations from tool handles. *Source:* Susan E. Nemeth, "Handtool Design," in *Industrial Ergonomics: A Practitioner's Guide,* eds. David C. Alexander and B. Mustafa Pulat (Atlanta, Ga.: Industrial Engineering and Management Press, 1985). Reprinted by permission. Copyright © Institute of Industrial Engineers, 25 Technology Park/Atlanta, Norcross, Ga. 30092.

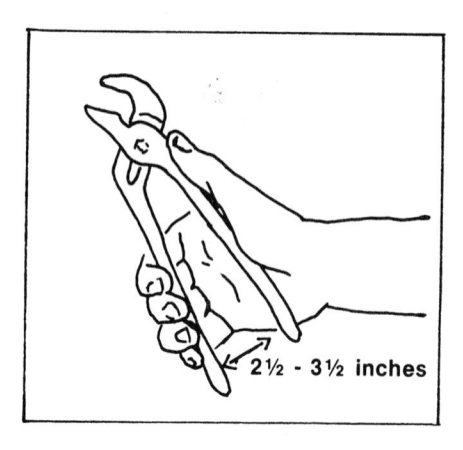

Figure 8-16 Utilize optimal power grip. *Source:* Susan E. Nemeth, "Handtool Design," in *Industrial Ergonomics: A Practitioner's Guide,* eds. David C. Alexander and B. Mustafa Pulat (Atlanta, Ga.: Industrial Engineering and Management Press, 1985). Reprinted by permission. Copyright © Institute of Industrial Engineers, 25 Technology Park/Atlanta, Norcross, Ga. 30092.

9. Avoid excessive vibration. From these vibrations can come Reynaud's syndrome, or dead fingers. Damping or isolating the vibration are the best solutions, although job rotation is a possible way to limit exposure.
10. Use gloves that fit. Gloves that are too big or too thick reduce control of tools because of a reduction of the person's strength and dexterity. Gloves of the correct "weight" for the job and gloves in several sizes are the most practical remedies.

8-9 SUMMARY

There are many jobs in traditional industry that require manual handling of materials. Even in office and service environments, materials are handled often enough to cause problems. The correction and prevention of handling injuries is important for the ergonomist.

In this chapter, means of identifying and diagnosing problems were provided. Tables of handling limits were provided. Many types of practical solutions were discussed, relative to specific diagnosed problems. When a "heavy job" is required, whether it involves lifting, manipulation, or other handling duties, the material in this chapter provides the information to diagnose and correct the problems that occur.

QUESTIONS

1. What is the frequency and cost of back injuries in the U.S.?
2. Why are industrial back injuries so difficult to control?
3. What are the "management problems" associated with lifting tasks?
4. Discuss the four methods of determining safe lifting limits.

Chap. 8   The Heavy Jobs

5. List the constraints of lifting tasks when using the NIOSH lifting guidelines.
6. Describe the three levels of the NIOSH lifting guidelines, and the solutions required at each level.
7. A worker must lift boxes of copier paper from a pallet to a nearby truck bed. Each box weighs 50 lbs. The boxes are picked up from a pallet. The truck bed is 30 in. from the floor. During loading, 10 boxes will be lifted in 2 minutes. Is the box within the action limit? Which factor has the greatest effect on reducing the action limit?
8. In the above situation, which change would have the greater impact, to add an extra pallet to raise the load, or to increase the loading time to 5 minutes?
9. In the above situation, what do both changes do to the AL?
10. How much can a person lift when working all day? The following conditions apply: The job is to feed and position castings on a drilling machine. The horizontal location is 12 in., the average pick-up point is 24 in., the lift is 16 in., the pace is 1 lift per minute.
11. What is the largest acceptable push force without administrative controls when pushing is required every 35 seconds? Every 5 minutes? Every 8 hours?
12. What is the advantage of vertical push bars on carts?
13. What is the largest acceptable pull force without administrative controls when pulling is required every 6 seconds? Every 5 minutes? Every 8 hours?
14. What is the "rule of thumb" for the amount of weight to be handled with each shovelful? The amount of weight to be handled in a drum?
15. List and discuss the methods of redesign for unacceptable lifting tasks.
16. In a grocery store, 25-lb. bags of dog food are laid on the bottom shelf. The shopper must lift one of these 25-lb. bags from the low shelf to the cart. List ideas for resolving this hazardous lifting situation using the methods for redesign.
17. List and discuss the three major types of administrative control solutions.
18. What is "the lazy man's load"?
19. When selection is requested, what types of solutions are always explored first?
20. Distinguish between "natural selection" and "management selection" of personnel.
21. What are three types of training associated with material handling?
22. List the major contributors to cumulative trauma disorders.

Chapter 9

THE HARD AND HOT JOBS

This chapter discusses those jobs that create cardiovascular stress on the worker. Both hard physical labor and heat stress are covered.

The chapter begins with a discussion of a "fair day's work" and how to determine it. This leads to an overview of the field of work physiology. Then the information on work physiology is used to evaluate work stress and heat stress problems and to develop practical solutions.

9-1 A FAIR DAY'S WORK

One day I received a telephone call from the midwest, describing an employee death from apparent heat stress. "How," the caller asked, "can I avoid this in the future? Is there a way to identify those jobs that are likely to overstress workers? And is there a way to predict the effects of changes that we might make?"

There are two major types of tough jobs, those that are tough on the musculoskeletal system and those that are tough on the cardiovascular system. This chapter focuses on jobs that tax the cardiovascular system.

Heavy physical work and/or work in hot environments requires a large blood supply to provide oxygen to the muscles and to carry heat from the muscles to the skin for dissipation. The blood supply is dependent on the cardiovascular system, which becomes the limiting factor in such jobs.

The effects of the overstressed cardiovascular system can be more serious than the results of an overworked musculoskeletal system. Cardiovascular problems show up as shortness of breath, fainting, strokes, or heart attacks. The ultimate outcome can be death or a serious injury from stroke, such as paralysis.

Specifically, some of the jobs that can be analyzed using a work physiology approach include:

- A job paced by a conveyor, such as palletizing boxes off the end of a line
- Emergency repair work
- A job in a hot work environment, such as a foundry
- Boiler repair work, when the boiler is still cooling down
- Firefighting

It is difficult to initiate a study of jobs like this without understanding the process involved. Why open up questions when one is unsure of how to develop the answers? Yet there are scientific methods to analyze and evaluate these situations. Typical questions are:

- Is the job too difficult?
- Is it too hot?
- Can everyone do the job, or do I need to select people?
- What is the cheapest way to modify the job?
- Is this a fair day's work?

An understanding of work physiology can help to answer these questions. Section 9-2 provides an overview of work physiology as it relates to industry. Following that, the two major areas of application are discussed. Section 9-3 will discuss the stamina applications, and Section 9-4 discusses jobs involving heat stress.

Section 9-2 on work physiology is not intended to be definitive, but rather should introduce the topic in an applications sense. Those readers with more interest are urged to consult *Physiology in Industry* by Lucien Brouha, *Textbook on Work Physiology* by Per-Olof Astrand and Kaare Rodahl, and the American Industrial Hygiene Association pamphlets "Ergonomics Guide to Assessment of Metabolic and Cardiac Costs of Physical Work" and "Ergonomics Guide to Assessment of Physical Work Capacity."

9-2 A BRIEF DESCRIPTION OF WORK PHYSIOLOGY

Work physiology is the scientific, repeatable method used to evaluate and analyze those jobs requiring stamina or involving heat stress. This approach assesses the difficulty of the work and the body's response to this work by measuring the heart rate (HR), the rate of oxygen consumption (O_2), and the kilocalories (Kcal) of energy required to perform the work.

The range of useful questions that this technique can answer are extensive and have been around for some time:

- What constitutes a fair day's work?
- If a job is too difficult, what should be done to change that job?
- When have enough changes been made?
- What are the most effective methods to combat heat stress?
- If rest is needed during the work shift, how many rest periods are necessary, how frequently, and for how long?

9-2-1 Assessing Difficult Jobs

The assessment of difficult jobs can be done by three methods. Preliminary assessment of job difficulty can be done with tables that describe tasks and their difficulty. More comprehensive assessments are provided through measurement of heart rates as the jobs are being performed. Even more comprehensive assessments come from measuring the oxygen required while performing jobs.

The unit of measurement used is kilocalories of work energy (1 kilocalorie of work energy is commonly referred to as 1 "calorie" when speaking of food intake). Thus work is simply a measure of the amount of equivalent food "calories" needed to perform the required task. It is possible, although quite difficult, to measure energy usage by monitoring food intake.

The energy required for a given task will vary from one person to the next, based on the individual's weight, fitness, age, gender, and other factors. Within reason, however, there are safe ranges to use for the industrial worker.

There is a relationship between kilocalories (or food calories), the amount of oxygen used by a person, and the heart rate during the work period. Although this relationship will vary from one person to the next, there are general relationships that suit our purposes here. Table 9-1 displays this relationship over a range of work loads. Figure 9-1 depicts the relationship between these items.

Comparison tables. For initial estimates of work load, comparison tables may be satisfactory. Table 9-2 provides a list of activities and the associated work loads. Since most jobs consist of several tasks, time-weighted assessments are required to learn the true cost of the whole job. Rest breaks are factored in the same manner as any other separate task.

The energy costs listed in Table 9-2 will not be exact, because people vary in their work pace and job skills, and the tasks will vary somewhat. Body weight is also a significant factor; "It may be necessary to make correction

Chap. 9   The Hard and Hot Jobs   197

TABLE 9-1   Level of Activity, Kilocalorie Load, Heart Rate, and Oxygen Uptake

Classification of Work Loads in Terms of Physiological Reactions

Work Load	Oxygen Consumption (liters/min)	Energy Expenditure (kcal/min)	Heart Rate during Work (beats/min)
Light	0.5–1.0	2.5–5.0	60–100
Moderate	1.0–1.5	5.0–7.5	100–125
Heavy	1.5–2.0	7.5–10.0	125–150
Very heavy	2.0–2.5	10.0–12.5	150–175

Source: Lucien Brouha, *Physiology in Industry* (Oxford: Pergamon Press Ltd., 1960), p. 96.

for the body weight of the worker, if it is much different from the 70 kg (154 lb) of the reference Standard Man. It is usually satisfactory to make proportional adjustment for actual body weight in such cases."[1]

Oxygen consumption. A more accurate method of determining the energy requirements of a task is to measure oxygen consumption during the performance of the task. There is a linear relationship between the work load and oxygen consumption, so it is possible to determine the caloric requirements by monitoring oxygen usage. Astrand states: "In many types of muscular exercise the oxygen uptake increases roughly linearly with an increase in work load."[2] The quantitative relationship is identified: "For each liter of oxygen consumed, about 5 kcal (4.7 to 5.05) will be delivered; hence, the higher

---

Energy* for task = *f*(oxygen available, food stored)
Oxygen available = *f*(% oxygen in blood, blood flow)
Percent oxygen in blood = *f*[oxygen intake,† fitness]
Blood flow = *f*[heart rate,† stroke volume]

---

*Desired information.
†Reasonably easy to measure.

Note: Heart rate also influenced by emotional stress and heat load.

Figure 9-1   Relationships between energy for tasks, heart rate, and oxygen consumption.

[1]*Ergonomics Guide to Assessment of Metabolic and Cardiac Costs of Physical Work*, American Industrial Hygiene Association, Ergonomics Guides Series, August 1971, p. 562.
[2]Per-Olof Astrand and Kaare Rodahl, *Textbook of Work Physiology* (New York: McGraw-Hill Book Company, 1970), p. 286.

TABLE 9-2 Average Energy Costs While Performing Selected Activities[1]

Body Position and Activity	Total Energy Cost, Typical	Kcal/min Range
Heavy activity at fast to maximum pace		10.0–20.0
Jogging, level, 4.5 mph	7.5	
Lifting, 44 lb, 10 cycles/min		
Floor to waist	8.2	
Floor to shoulder	10.8	
Reclining, at rest	1.3	
Running, level, 7.5 mph	12.7	
Shoveling, 18-lb load 1 yd with 1-yd lift, 10 times/min	8.0	
Sitting, at ease		
Light hand work (writing, typing)	1.7	1.6–1.8
Moderate hand and arm work (drafting, light drill press, light assembly, tailoring)		
Light arm and leg work (driving car on open road, machine sewing)	2.8	2.5–3.2
Heavy hand and arm work (nailing, shaping stones, filing)	3.5	3.0–4.0
Moderate arm and leg work (local driving of truck or bus)	3.6	3.0–4.0
Standing, at ease	1.9	
Moderate arm and trunk work (nailing, filing, ironing)	3.7	3.0–4.0
Heavy arm and trunk work (hand sewing, chiseling)	6.0	4.0–8.0
Walking, casual (foreman, lecturing)	3.0	2.5–3.5
Moderate arm work (sweeping, stockroom work)	4.5	4.0–5.0
Carrying heavy loads or with heavy arm movements (carrying suitcases, scything, hand-mowing lawn)	7.0	6.0–8.0
Transferring 35-lb sheet materials 2 yd at trunk level, three times per minute	3.7	
Pushing wheelbarrow on level with 220-lb load	5.5	5.0–6.0
Walking on level		
2 mph	3.2	
3 mph	4.0	
4 mph	5.9	
Walking up 5-degree grade at 3 mph	8.5	
Mailman climbing stairs	12.0	
Walking down 5-degree grade at 3 mph	3.4	

[1] Values apply for 70-kg (154-lb) man. For most activities adjustment is proportional to body weight.

Source: Ergonomics Guide to Assessment of Metabolic and Cardiac Costs of Physical Work, American Industrial Hygiene Association, Ergonomics Guides Series, August 1971, p. 562.

the oxygen uptake, the higher the energy output."[3] Therefore, by multiplying the oxygen uptake by 5, an approximation of the kcal expenditure can be obtained.

If this is used as the method of determining work loads, the following precautions are appropriate:

1. Use an average worker on the task.
2. Use a standard work pace.
3. Use the standard work methods.
4. Measure at steady-state working conditions; oxygen uptake will lag the actual requirements by a few minutes.
5. Consider whether there is any anaerobic work occurring and building up an oxygen debt that will not show up in the measurements.

To measure the oxygen used, the expired air must be collected and analyzed. The equipment used in the collection of the expired air is awkward and uncomfortable to wear. Its bulk may inhibit normal work. Usually, samples of only about 5 minutes' duration are taken to verify the work loads predicted by the heart rate data.

Measurement of oxygen consumption is not easy and mistakes in data collection, in analysis, and in calculation are easy. It is best to request additional support from a health professional experienced in this area if oxygen consumption measures are required.

Heart rate. The heart rate (HR) will reflect the difficulty of the task and the extra load from any heat stress, factored of course by the person's fitness. Heart rate is easy to measure, and few people question the relationship between HR and work load.

Heart rate is an effective measure for two reasons. First, it is an indicator of the relative difficulty of a task, especially when the same person performs several tasks for comparison. Second, it is the best indicator of cardiovascular stress and will indicate whether a person is experiencing excessive stress.

Excessive stress is occurring when the actual heart rate exceeds 85% of the age-predicted maximum, as noted in Table 9-3. Excessive stress is also considered to occur when the person's average heart rate exceeds more than 40% of the difference between the resting HR and the age-predicted maximum HR for an entire work shift.

Using heart rate as an indicator of work load is based on the relationship between oxygen consumption and heart rate. Astrand states: "Since in a given person there is a linear relationship between O_2 uptake and heart rate, the heart rate or pulse rate under certain standardized conditions may be used as a rough

[3]Ibid., p. 280.

TABLE 9-3 Averages of Maximal Heart Rates Published by 10 American and European Investigators

Age	Maximum Heart Rate
20–29	190
30–39	182
40–49	179
50–59	171
60–69	164

Source: *Exercise Testing and Training of Apparently Healthy Individuals: A Handbook for Physicians.* Copyright American Heart Association, 1972. Reproduced with permission.

index of O_2 uptake in a given test. By comparing the pulse rate obtained during work with the pulse rate during different bicycle ergonometer work loads, a rough estimate of the severity of the work can be made."[4]

Brouha also discusses this relationship, as well as the ease of monitoring HR compared to oxygen consumption: "For plant surveys, study of the heart rate reaction seems to be the most direct, simple and often the only method available for evaluating stress 'on the job.' Because of its close relation to cardiac output and oxygen consumption within the range of many industrial occupations, the heart rate can be utilized to gauge the stress imposed by muscular activity and can be obtained with minimal interference in the subject's freedom of motion and performance ability."[5] Table 9-1 illustrated that relationship.

9-2-2 A Practical Method of Using the Assessment Techniques

The uses of work physiology, especially the easier measures, are approximations of more complex situations. There are many factors that can affect the energy required to perform a task and the human's responses in meeting those energy needs. For our purposes, work physiology should be used as a means of determining whether a job is suitable for the work population or whether it needs some improvements. If improvements are necessary, the work physiology approach should help to pinpoint areas needing improvements.

[4]Ibid., p. 437.
[5]Lucien Brouha, *Physiology in Industry* (Oxford England: Pergamon Press Ltd, 1960), p. 95.

In some studies, there will be a gray area between these two, an area where questions about job suitability remain. Then more accurate measurements are needed to determine whether or not the situation is okay. This study should involve medical personnel, a nurse, a physician, or a work physiologist. Of course, it may be more economical to go ahead and correct the situation rather than incurring the cost of additional study and then, later, possibly also incurring the cost of the correction as well.

The evaluation of jobs should begin with a comparison of the work loads shown in Tables 9-1 and 9-2. If these indicate a questionable situation, heart rate assessments can be done. The heart rates may begin with recovery heart rates, and then, as warranted, with actual working heart rates. If that does not indicate a suitable work load, or if there are still questions, oxygen-consumption work can begin.

A few notes of caution are in order. These methods are useful for large-muscle groups working regularly. Work with small-muscle groups will not affect the HR and O_2 significantly. Static work will generally increase the HR dramatically. Mental work will not affect O_2, and usually does not show up on the HR. Emotional and psychological stress can increase the HR.

The values obtained from the tables should be used as estimates of the work load for a group of workers rather than for assessing individual performance. The values of HR and O_2 will be specific for a person, so several people should be tested. Extrapolations to other people are not recommended.

There will be significant individual variation based on fitness, gender, weight, and heredity. The specific muscle groups involved, the work position, environmental temperatures, and emotional stress all affect the heart rate. Later in this chapter some adjustments to these figures are discussed for gender, age, the muscle groups, and environmental temperatures.

Individual fitness and weight have the greatest potential to affect the measurements. Testing multiple workers can prevent a few from causing biased results. For identifying excess cardiovascular stress for these people, the maximum working heart rate of 85% of age-predicted maximum is a reasonable limitation, although a medical examination is desirable.

9-2-3 Methods and Equipment

The equipment used in the measurement of work stress and energy expenditure is outlined below. Direct caloric intake measurement is difficult and not accurate for industrial tasks, so monitoring heart and oxygen consumption is predominate.

Heart monitoring is done in one of three ways:

1. By direct monitoring of the pulse. This is the least expensive and the least accurate. Normally, the pulse is monitored immediately after work (during the recovery phase). Brouha discusses this method extensively in

his book, *Physiology in Industry*. This is a good "quick and dirty technique," but other techniques are now more accurate.
2. With electrocardiogram (EKG) recording equipment. With miniaturized electronics, it is now common to have lightweight, portable units that can record the heart rate (in fact, they record the entire waveform) for a several-hour period. The recording is made on slow-speed magnetic tape, which can then be interpreted later. A log of activities must be kept and matched with the heart rate tape to correlate work activities with periods of high cardiovascular stress. This equipment may be available for loan or rental from local cardiac specialists or a local hospital. A photograph of the equipment is shown in Figure 9-2.
3. With EKG telemetry equipment. With this equipment, the worker wears a miniature transmitter that sends a tone corresponding to each heart beat. Working at a distance to avoid work interference, the ergonomist can monitor the heart rate, the task currently being performed, and the work intensity. This allows a "real-time" evaluation of each task. A photograph of the equipment is shown in Figure 9-3.

Monitoring oxygen consumption is more difficult. Several pieces of equipment are needed. First, a unit is necessary to monitor the volume of air inspired or expired by the worker. The Max Planck respirometer, a common device for this, forces expired air through a meter. At the same time, a small sample of the exhaled air is saved in a rubber bladder. This sample is then tested with an oxygen analyzer for the percentage of oxygen remaining. Room air is also tested to determine the percentage of oxygen inspired. Calculations can then be made using the air pressure, volume of air expired, percent oxygen expired, and percent oxygen inspired to determine the liters of oxygen (at standard temperature and pressure) used on the task. These same types of samples, taken at rest, can also be used to determine the resting, or basal, metabolism of the person. Photographs of this equipment are shown in Figures 9-4 and 9-5.

Using this same equipment, it is possible to determine the "work" at several points on a standardized treadmill, bicycle ergonometer, or step tests. The pamphlet "Ergonomics Guide to Assessment of Physical Work Capacity" is a useful reference on determining physical work capacity by these methods. Figure 9-6 shows a person walking on a treadmill with a full set of equipment.

The oxygen consumption can be plotted against the heart rate to determine a person's state of fitness. The more oxygen consumed per heartbeat, the greater the person's fitness. There will be a range of fitness levels between individuals. In most cases the plot will be linear over the normal work range for each person. A representative plot of this relationship is shown in Figure 9-7.

Chap. 9　The Hard and Hot Jobs　203

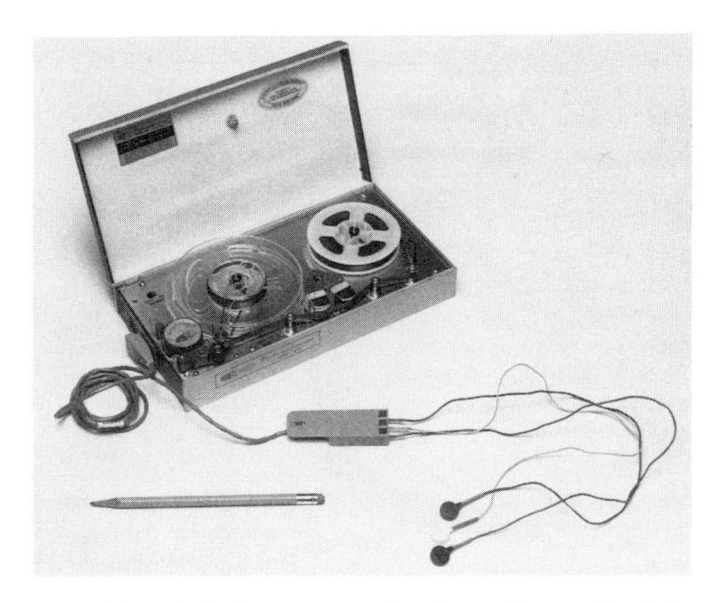

Figure 9-2 Avionics heart rate recorder. *Source:* Eastman Chemicals Division of Eastman Kodak Company.

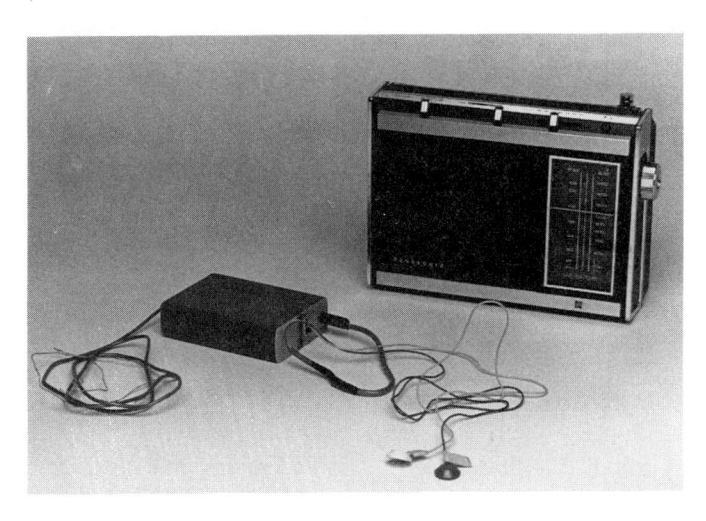

Figure 9-3 Heart rate telemetry device. *Source:* Eastman Chemicals Division of Eastman Kodak Company.

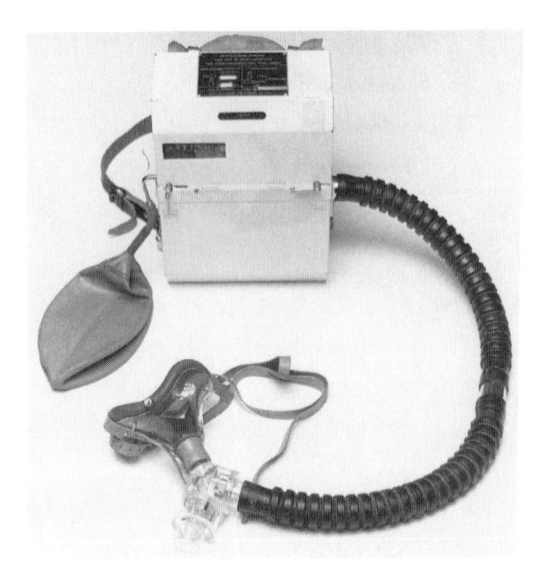

Figure 9-4 Max Planck respirometer. *Source:* Eastman Chemicals Division of Eastman Kodak Company.

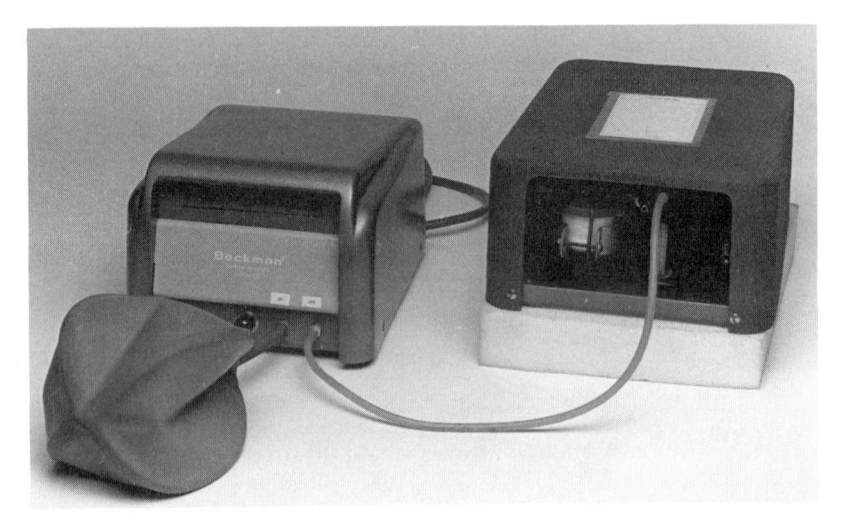

Figure 9-5 Beckman oxygen analyzing instruments. *Source:* Eastman Chemicals Division of Eastman Kodak Company.

9-3 THE HARD JOB

The early parts of this chapter were designed to acquaint the reader with the problems associated with a fair day's work and to introduce the concept of work physiology as a scientific, repeatable method of analyzing those problems. This section covers the "hard job," that is, a job that is physically demanding.

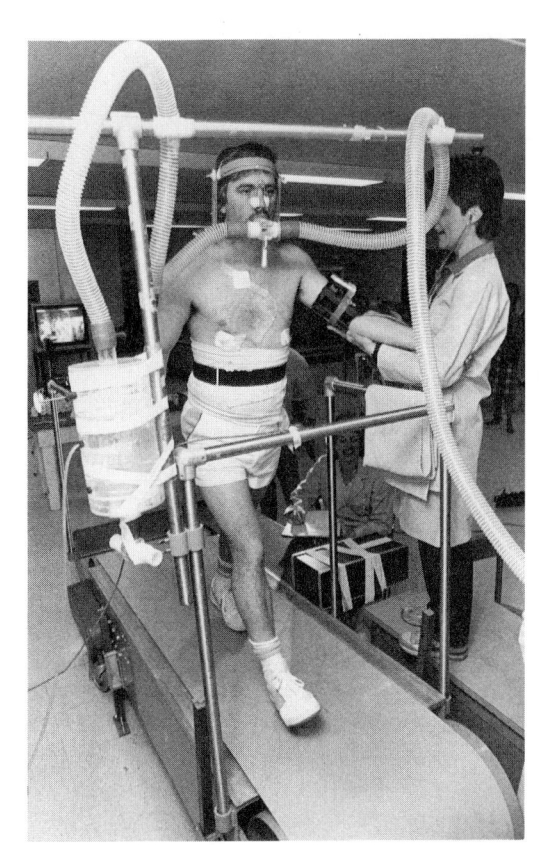

Figure 9-6 Person on a treadmill with a full set of heart rate and oxygen monitoring equipment. *Source:* Wellness Center and Human Performance Lab, Physical Education and Recreation Dept., East Tennessee State University, Johnson City, Tenn.

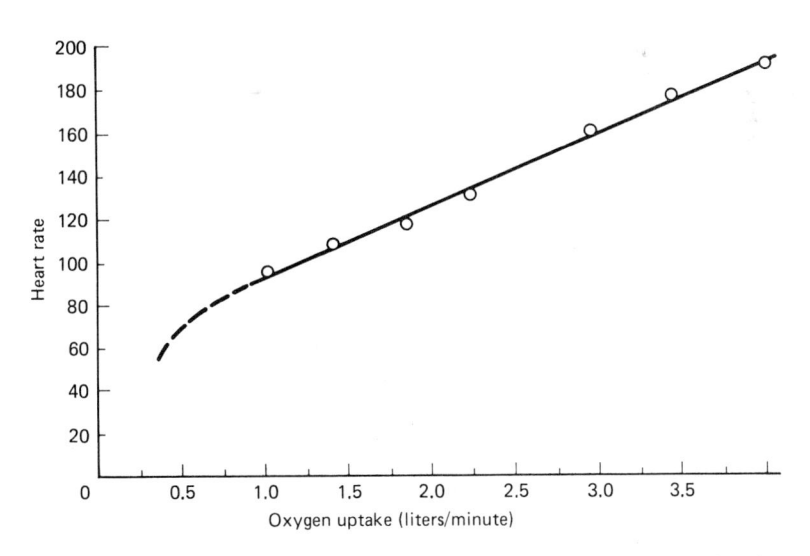

Figure 9-7 Plot of the relationship between heart rate and oxygen consumption during a stress test.

Analysis of "hard jobs" will answer such questions as:

- How difficult is the job?
- Why is it difficult—which element/task is causing most of the problem?
- What possible solutions might work, and which are best?
- How many of these corrections are needed?

The process described earlier in relation to the "problem job" is useful in diagnosing problems and developing practical solutions.

The initiation of the process is the identification of a problem. These characteristics often indicate a need for further study: A fast work pace imposed by the work equipment—conveyors are notoriously bad; imposed by the task—especially emergency tasks such as firefighting, rescue work, or emergency repair work; or imposed by management—through work quotas or difficult financial incentives. Excessive sweating is another sign, both during the work or for periods of time afterward. A job is unusually difficult when people have trouble performing double shifts on the job. Finally, regular self-selection away from the job, such as refusals to accept the job and/or requests for job transfers, also indicates problems.

On the other hand, some favorable characteristics of jobs include self-regulation of the work pace and/or of rest and recovery breaks. Also, a knowledge of the symptoms of heat stress and encouragement to respond to them is another good sign. When people are able to judge their own state of fatigue/stress and respond accordingly, they usually will not exceed their physiological limits and injure themselves.

9-3-1 Major Problem Situations

The hard job is a concern because of the stress on the cardiovascular system resulting from elevated heart rates and blood pressure (which come from the high oxygen demands associated with the excessive energy demands). The problems with high-energy-demand jobs can show up in four general ways:

1. Excessive demands for an entire work shift
2. Peak work loads (high demands for short periods of time)
3. High work loads for a few muscle groups
4. Inappropriate work/recovery cycles

Each of these is discussed below to aid in understanding that type of problem, the significant characteristics, and how possible solutions should be formulated.

Excessive demands for an entire work shift. Excessive work requirements result in fatigue, overstress, long-term health problems, and employee selection problems. Excessive work requirements are defined by high heart rates, or indirectly, by excessive oxygen usage, kilocalorie requirements, or task difficulty. These methods of identifying "hard jobs" are discussed in limited detail.

For the entire work shift (assumed to be 8 hours), the average heart rate should be elevated no more than 40% of the difference between the resting heart rate and the age-predicted maximum. This will avoid excessive stress on the cardiovascular system. General estimates of suitable heart rates for the 8-hour work shift are in the range 110 to 115 beats per minute average for the entire work shift. When calculating the heart rates of the employees is too time consuming, the alternative methods of estimating full-shift work demands become more desirable.

The use of oxygen consumption is helpful in determining excessive rates for a variety of work times. Astrand has researched the acceptable levels of the percent of oxygen uptake for various prolonged work times, a graph of which is reproduced in Figure 9-8.

This is useful in determining "excessive" work rates. A level of one-third of the person's maximum aerobic power is generally considered acceptable for the 8-hour work shift. Garg and Ayoub state: "It is generally accepted that, for a healthy young male, 8-hr average metabolic rate should not exceed approximately 5 Kcal/min or 33% of the individual's maximum aerobic power."[6]

The tasks not only have to be known, but the work population must also be identified, because it will determine what is a "suitable work load." The limitation of 5 Kcal/min average is a useful figure for young healthy males. Adjustments to this figure need to be made for other work populations. Astrand discusses this: "The women's power is on an average 70 to 75% of that of the men. In both sexes there is a peak at eighteen to twenty years of age, followed by a gradual decline in the maximal oxygen uptake. At the age of sixty-five, the mean value is about 70% of what it is for a twenty-five-year-old individual. The maximal oxygen uptake for the sixty-five-year-old man (average) is the same as that typical for a twenty-five-year-old woman."[7]

In many cases, particularly when initial estimates of work load are being developed, estimates of work load from Table 9-2 are suitable. The various activities, including rest, must be summed to determine the total work load in kilocalories. This process will also highlight which tasks are the major users of energy, which is helpful in developing solutions.

---

[6]Arun Garg and M. M. Ayoub, "What Criteria Exist for Determining How Much Load Can Be Lifted Safely?" *Human Factors,* 22(4), 1980, pp. 475-486. Copyright 1980 by The Human Factor Society, Inc., and reproduced by permission.

[7]Astrand and Rodahl, *Textbook of Work Physiology,* p. 305.

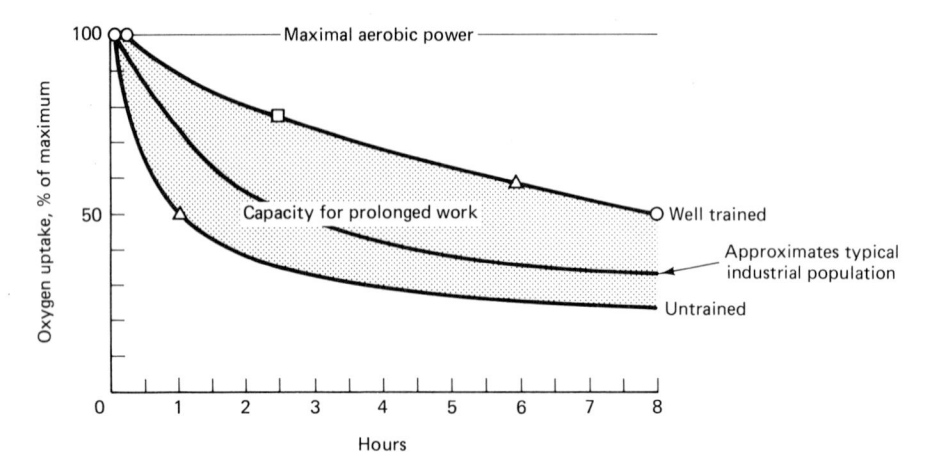

[Graph: Oxygen uptake, % of maximum vs Hours (0-8), showing Maximal aerobic power at 100%, with curves for Well trained, Approximates typical industrial population, and Untrained, bounding a "Capacity for prolonged work" region.]

Figure 9-8 Approximate work times based on the percentage of oxygen uptake utilized. *Source:* Per-Olof Astrand and Kaare Rodahl, *Textbook of Work Physiology* (New York: McGraw-Hill Book Company, 1970), p. 292.

Peak work loads. Peak work loads can overstress a person even though the total work load for the workday is relatively modest. The peak activities can occur as a short burst of energy or as a sustained high level over a longer period.

The diagnosis of peak work problems is made from elevated heart rates (over 85% of the age-predicted maximum), from energy expenditures that exceed those in Astrand's graph (Figure 9-8), "approximate work times based on the percentage of oxygen uptake utilized," or from extended recovery times. These situations are shown in Table 9-4.

During extremely high work loads, a person can develop a very high heart rate. This may occur even before enough fatigue sets in to diminish the work rate.

A person can get energy from two sources, aerobically (with oxygen) or

TABLE 9-4 Peak Work Load Situations

Work Load	Diagnosis	Concern/Comment
Extremely high	HR exceeds 85%	Peak HR and cardiovascular stress
Very high	Rapid fatigue; very high HR	Anaerobic work (limited duration, adds only 25 kcal); energy debt; extended recovery time
High	Percent energy curve vs. time; long recovery	Need recovery breaks

anaerobically (without oxygen). For most industrial tasks the work is aerobic; that is, the person is able to reach a steady state where the oxygen coming into the body meets the energy needs. Anaerobic work is characterized by very short work loads of very high intensity, for example, a short sprint generally requires anaerobic energy. The high intensity causes the body to seek energy from places other than through the normal metabolism in the muscles; the very short work period results from quite limited amounts of this energy, only about 25 Kcal worth. This type of work is not typical of industry except perhaps during emergency work. A graph (Figure 9-9) of aerobic and anaerobic energy yields for a maximal effort indicates the relatively low percentage of energy coming from the anaerobic process.

High work levels are characterized by long recovery periods. Since full recovery is not possible during the typical 10-minute coffee break, a longer rest period is needed or the person needs to be assigned a light-duty task such as logging information, where recovery can take place.

To detect this situation, measure or estimate the energy required, calculate the percent of oxygen uptake required, and then compare it with suitable work times as noted in Astrand's graph, Figure 9-8. Monitoring the

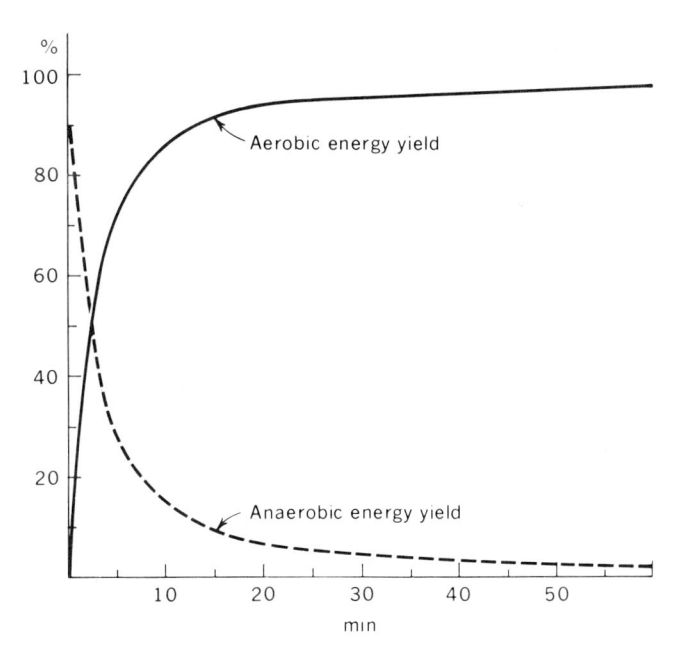

Figure 9-9 Relative contribution in percent of total energy yield from aerobic and anaerobic processes during maximal work. *Source:* Per-Olof Astrand and Kaare Rodahl, *Textbook of Work Physiology* (New York: McGraw-Hill Book Company, 1970), p. 304.

recovery heart rates will provide supporting evidence of this situation. If the recovery rate is long or occurs slowly, there are reasonable concerns calling for further investigation.

A discussion of Brouha's methodologies can reveal some important concepts regarding the evaluation of fatigue induced by peak loads. He uses the heart rate recovery curve, which is measured immediately after work during the initial recovery phase. It is determined by "counting the pulse from 30 sec to 1 min after work stops; from 1½ to 2 min; and again from 2½ to 3 min. From these counts a 'heart rate recovery curve' can be constructed."[8] A graph of several heart rate recovery curves for light, moderate, and heavy work is shown in Figure 9-10.

Brouha goes further to explain: "The onset of fatigue can also be predicted from heart rate recovery curves. When satisfactory recovery does not take place between a series of successive operations, the initial pulse rate after work becomes progressively higher and the heart rate during the recovery period remains at an elevated level for a progressively longer time. A return toward the resting level that is slow points definitely to the the existence of

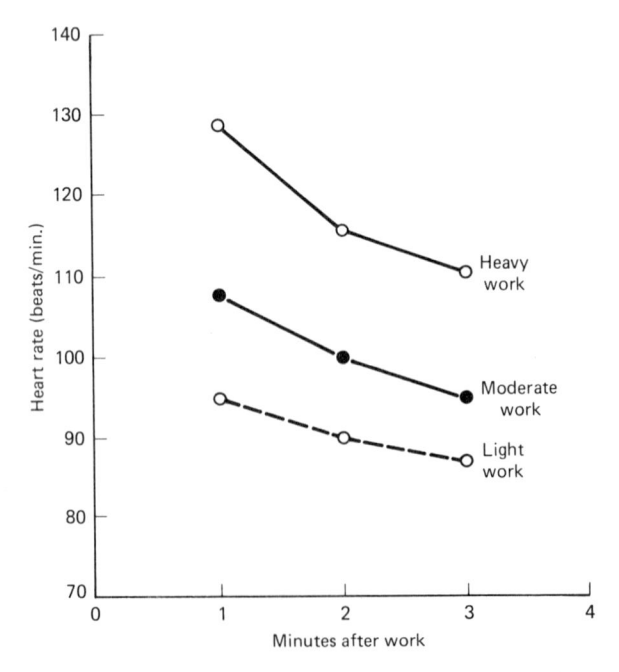

Figure 9-10 Higher level of heart rate recovery curves with increasing work load. *Source:* Lucien Brouha, *Physiology in Industry* (Oxford, England: Pergamon Press Ltd., 1960), p. 96.

[8]Brouha, *Physiology in Industry,* p. 100.

Chap. 9 The Hard and Hot Jobs   211

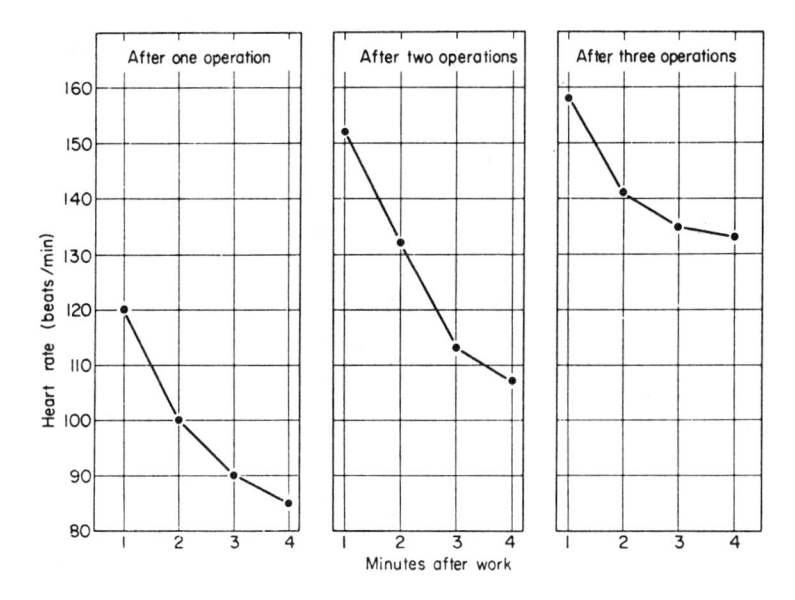

Figure 9-11 Increasing level of heart rate recovery curves and decreasing speed of recovery after successive operations. *Source:* Lucien Brouha, *Physiology in Industry* (Oxford, England: Pergamon Press Ltd., 1960), p. 106.

'physiological fatigue' in a worker."[9] Examples of increasing recovery curves after successive operations are shown in Figure 9-11.

The important point for the application of industrial ergonomics is to be able to use these curves to distinguish suitable work rates from those causing problems. Brouha discusses his diagnostic techniques: "Normal curves include all cases in which the third pulse rate is at least 10 beats slower than the first; and those in which all three pulse rates are below 90 beats/min. No recovery curves are those in which the difference between the first and third pulse rates is less than 10 beats/min and in which the third reading remains above 90 beats/min."[10]

He provides a useful guideline to distinguish work levels which can be sustained for the work shift from those that cannot. "It has been our experience, during the past 15 years and in many industrial operations, that when the average value of the first recovery pulse is maintained at about 110 beats/min or below, and when the deceleration between the first and the third recovery pulse rates is at least 10 beats/min, no increasing cardiac strain occurs as the day progresses. Whether this level is produced by the work load and heat exposure, it appears that such stress can be sustained throughout the shift

[9]Ibid., p. 105
[10]Ibid., p. 108.

in a physiological steady state, provided the sequence of work and rest periods is adequately organized."[11]

High work loads for a few muscle groups. There are situations where the heart rate indicates a higher work load than the actual work load. These situations are important since work load predictions from the heart rate responses will be artificially high. Similarly, if one tries to establish peak heart rates from an estimate of the required work load, the true heart rate will be much higher, with possibly serious consequences.

According to Astrand: "In arm exercise, the maximal oxygen uptake is about 70% of what is attained in leg exercise. The fact that work with arms plus legs does not further increase the oxygen uptake and cardiac output as compared with leg work alone might be interpreted as an indication that the pumping capacity of the heart is the limiting factor during heavy exercise."[12] This means that when a person is using the arms predominantly (a blacksmith or someone positioning material on a conveyor line), the heart rate/oxygen uptake factor should be converted by the 70% change from normal.

Another situation that causes errors in determining heart rate and work load relationships is static work. Static work occurs when a person maintains force, but with no movement. This can occur while holding a heavy part while someone else places a fastener, or while shoving on a heavy object with little movement. Static work requires energy since the muscles are producing force, yet the lack of movement also "locks up" the circulatory system, preventing the flow of blood. Although the muscles continue to signal for more oxygen, causing the heart rate and blood pressure to increase, little extra oxygen actually gets to the muscles. Astrand states the relationship clearly: "Static (isometric) exercise also increases the heart rate above the value expected from work load."[13] The amount of heart rate increase will depend on the muscles involved, the intensity of the exertion, and the duration of the effort, so a clear prediction of the inflated heart rate is not possible.

Work/recovery cycles. In some cases, the work load appears to be acceptable, yet problems exist with job performance. The concern may be the arrangement of the work/recovery cycles. Sometimes work/recovery cycles are referred to as work/rest cycles; recovery is a more appropriate term, provided that recovery from the work cycle is really the reason for breaking from work.

Diagnosis of work/recovery cycle problems is a little more difficult than for other types of work load problems. The diagnosis begins with a calculation of the overall work load, by summing the work periods and the rest periods.

[11]Ibid.
[12]Astrand and Rodahl, *Textbook of Work Physiology*, p. 294.
[13]Ibid., p. 67.

This load is compared to an acceptable level, based on Astrand's work rate curve (Figure 9-8), factored for the expected population (age and gender) and for the muscle groups (whole body or arms). For example, if the work population is young males, an acceptable rate is 5 Kcal/min for an 8-hour shift. If the population involves females, the maximum rate should be factored by 70%. When the overall work load is too high, problems are the work level or the relative work and rest times.

If the overall work load is acceptable, smaller increments of time should be evaluated in the same manner. With this evaluation, be careful to adjust the work load based on the shorter time increment. Natural increments of time are half or quarter shifts, times between work breaks, or for an hour or two of work. In each case, keep looking for periods of time whem the work load exceeds the acceptable load, causing an energy deficit to build. The deficit should not exceed 25 Kcal before recovery is initiated.

Suppose that a worker lifts 20-kg (44-lb) bags from the floor to waist high, for an entire shift. If the rate when lifting is 10 bags per minute, the energy expenditure is 8.2 Kcal/min. This level exceeds the acceptable level of 5 Kcal/min, so recovery must be introduced. The recovery itself requires about 1 to 1.5 Kcal/min for basal metabolism. To determine the recovery required:

$$\text{work } (8.2 \text{ Kcal/min} \times X \text{ min}) + \text{rest } [1.25 \text{ Kcal/min} \times (480 - X) \text{ min}]$$
$$= \text{total } (5 \text{ Kcal/min} \times 480 \text{ min})$$

Therefore, work time equals 259 minutes and recovery time equals 221 minutes.

9-3-2 Solutions to the Hard Job

It is important to diagnose problems before creating solutions for them. The correct diagnosis can save considerable time in searching out solutions as well as ensuring that the correct problem gets solved and stays solved.

Usually, several solutions can be generated for each problem. In choosing among these solutions, the following criteria are useful.

- Effectiveness in meeting the needs diagnosed earlier. If the problem has to do with an excessive work load, the solution must resolve that problem. This criterion alone will eliminate many potential solutions.
- Workability means that the solution must work in the real-life setting. It is one thing to develop a solution on paper and often another to ensure that it works on the job.
- Ease of administration examines the long-term implementation of the solution. For example, if the solution is to control the work pace, administration will probably be difficult. Similarly, if the solution is to install an industrial robot, the administration (after implementation) will be easy.

- Cost remains a useful criterion for ranking various alternative solutions. A better concept might be "value," the combination of cost and the other factors mentioned previously.

There are many solutions to physically demanding tasks and jobs. Many general solutions are listed, both to provide insight into the wide variety of possible solutions and to provide examples of solutions to common problems. There are, obviously, many specific solutions to the unique problems that one may encounter in any given work setting.

Redesign of the physical environment. Redesign of the physical environment remains the favored approach to completely eliminate or greatly reduce physically demanding jobs. Automation completely removes the person through the use of automated equipment (e.g., industrial robots). Materials handling equipment and other innovations also fall into this category. These changes may be expensive, especially in existing facilities.

Mechanization is the use of equipment to assist the person, but without fully removing the person. The use of tools, conveyors, lift trucks, and so on, are examples of mechanization. Normally, mechanization will also reduce the time to perform a task. So with high labor costs, these improvements are often cost justifiable. Static work is a special type of problem and one that readily lends itself to mechanical assistance. Hoists, jacks, and positioning guides are excellent ways to reduce or eliminate static work.

Methods and workplace improvements can often result in reduced work loads. These are most effective when they are permanent, such as moving workstations closer together, aligning several workstations, or changing the heights of workstations.

It may be possible to change or altogether eliminate the task/job/work process. It is worth the time to think about, especially if the problem is serious. For example, it may be possible to purchase material in bulk rather than using many smaller bags. One part of the process may be subcontracted or subassembled to reduce extensive handling in-house.

Changes in the task and job. Changes in the task/job are also effective ways to make improvements. Their major drawback is that the changes may have to be reinstituted periodically if people drift back to using previous methods. The initial costs of most of these changes are low, although there may be increases in operating costs.

The realignment of tasks within jobs can solve some of these problems. If there is one difficult job, it may be possible to break that job up and spread the harder tasks among a number of jobs, where the smaller tasks are not as limiting. Job rotation, particularly on an hourly or 2-hour basis, is a "cheap way" to break up a job. Rotation on a daily basis is often a step in the wrong direction since it requires hard physical work without any "break-in period."

The 8-hour work period will completely fatigue and overstress the unconditioned worker. With daily rotations, the person may only perform the job every week or two. This is adequate time for any training or conditioning of the cardiovascular system to wear off.

Conversely, at times it may be desirable to consolidate many physically demanding tasks into one job, provided of course that job is outside the normal promotion sequence and is filled by natural selection. The job needs to have a reputation of being a "tough job," so that only those with good physical conditioning will accept the job. Fortunately, it is not unusual for some people to want these jobs since they provide the side benefit of physical conditioning.

Changes in the work/recovery cycle are also ways to suitably modify physically demanding tasks. It is possible to add some recovery breaks if the job requirements are excessive or if the peak periods last for several hours. The work schedule can be modified to reduce the period of work so that additional recovery periods are not needed. The work/recovery cycle is also important. The relationship of fatigue to recovery is not linear; increases in fatigue result in increasingly longer recovery periods. Shorter work/recovery cycles may be useful when fatigue buildup or long recovery periods are a symptom.

Changes in the people performing the job. Changes in the people performing the job is a final type of correction. If people have increased capacity, they will be able to perform adequately the more physically demanding jobs.

Selection of people is one possible alternative. Natural selection (e.g., offering the job, and then letting the employee decide whether or not to accept it) is preferable to management selection because it eliminates the validity requirements that are so difficult to establish. Management selection now requires developing tests that adequately predict whether a person can perform the job satisfactorily. Although management selection is occasionally used, it is a difficult, time-consuming process.

Fitness training is another method of increasing the capacity of the work force. It is not used extensively in industry, although it is common in professional sports and is often used for firefighters, security guards, the police, and military personnel.

Rotation of people, mentioned earlier, is also a way to increase work capacity. If people work for only several hours on a job, they can work at a far higher percentage of their aerobic capacity than if they work for a full shift. If some jobs are demanding and others are not, rotation will "increase capacity" by moving people up the worktime/percent aerobic capacity chart.

A summary of these methods for solving the "hard job" is provided in Table 9-5. Also included are the four major diagnoses of the hard job problems and the match between these diagnoses and the solutions.

TABLE 9-5 Solutions to the Hard Job

	Diagnosed Concern			
Solution Strategies	Overload	Peak Work	Muscle Overload	Work/ Recovery
Redesign				
Automation	×	×	×	×
Mechanization	×	×	×	×
Workplace redesign	×	×	×	×
Change task/job				
Task realignment	×	×	×	
W/R cycle	×	×		×
Rotation	×	×	×	×
Change people				
Selection	×	×	×	
Training	×	×	×	
Rotation	×	×	×	×

9-4 THE HOT JOBS

Heat stress can cause the same cardiovascular problems that a heavy work load can cause—fainting, strokes, and heart attacks. These problems may be the result of rises in the core body temperature or the result of increases in the heart rate in an attempt to rid the body of extra heat. Heat stress may be an additional stressor to an already physically demanding job, or it can be a concern by itself.

Some situations where heat stress would be a potential problem are:

Repair of hot equipment such as bakery processing lines

Operation or maintenance of boilers in power plants

Monitoring of furnaces and lines in steel and aluminum processing

High humidity and hot work areas, near water quenches or baths

Paced work in a hot warehouse during the summer

The solution to these problems will depend on the extent that the cardiovascular strain is caused by the work load or by the heat load. Diagnosis in these situations is directed at determining the severity of the problem and at isolating the work load stress from the heat load stress.

The diagnosis is the road map for the solution of these problems. Depending on the diagnosis, a variety of alternative solutions will be posed with different costs and degrees of acceptability and workability.

9-4-1 Background and Concept

As discussed earlier in Section 9-2, increases in work or heat load can result in increases in heart rate. During heat stress, the circulatory system acts like the cooling system on an automobile. The increase in heart rate is due to the circulatory system pumping blood to the skin surface for cooling before returning to pick up more heat inside the body. The red, flushed appearance of someone with heat stress is simply the amount of blood on the skin's surface. The system works this way whether the heat is imposed from external sources, is generated by the working muscles, or is a combination of both.

Heat is generated in people when the muscles give off heat as a by-product of work (although in the case of shivering, the rapid muscle movement is caused by a need for heat). Heat also comes from a radiant heat source such as the sun or heat-producing equipment; heat comes from convective sources such as warm air in a room or work area that is hot; heat also comes from conduction or direct contact with warm objects, such as lying on the floor of a still-warm furnace during repairs.

Heat will move in either direction, toward the person or away from the person. The amount of heat transferred is a function of the temperature difference and other variables. This section will deal with heat stress where the body is trying to get rid of heat rather than where the body is trying to conserve heat in cold stress situations.

Since the human being is a constant heat producer, there is always a need for controlled heat loss. Thermostats in homes are set at 70°F to ensure this heat loss from the 98.6°F human being.

It is common to model temperature regulation in the human being as

$$M \pm R \pm C - E = S$$

Here M is the metabolic heat gained as a by-product of the work done; always a heat gain.

R is the radiant heat gained or lost, depending on the nearby, direct-line sources of heat relative to the person; a person will gain heat from the sun or a furnace and will give heat to a cold wall or window.

C is the convective heat gained or lost from exposure to air; this is a gain if the ambient temperature exceeds the skin temperature of 95°F and a loss if lower than that; the air velocity and temperature differential determine the rate of heat transfer.

E is the evaporative heat lost due to the evaporation of moisture from the skin (sweating does no good until the moisture evaporates from the skin, so wetted skin in high-humidity areas is ineffective and heavy or protective clothing can reduce this benefit substantially); this is always a heat loss.

S is the net heat stored by a person; long term this must be zero, although short term a person can tolerate some heat gain with a resulting temporary rise in the core body temperature.

The most important factors in this equation are the metabolic heat gain and the evaporative heat loss. In most industrial situations, the other factors are smaller. Work can easily produce enough heat to injure a person. Evaporation is the major heat loss for the person, and if evaporation is not possible due to high humidity, clothing, or low airflow, there can be serious problems getting rid of any heat generated. The other factors can be significant, but high radiant and convective heat loads are much easier to identify before they cause serious problems. There are many other factors that also influence the heat gain in people, as noted in Table 9-6.

9-4-2 Analysis

There are many methods of assessing heat stress situations. The method that is the most convenient in the industrial setting is the WBGT (wet-bulb globe temperature), an algebraically determined sum of environmental con-

TABLE 9-6 Factors Influencing Heat Gain

Environmental Factors

Humidity: high humidity retards evaporation, which is the major method of losing heat
Air velocity: faster air movements cause more rapid heat transfer of convective heat; it also enhances evaporation

Individual Factors

Fitness: personal fitness enhances heat tolerance, both by providing a greater margin of heat load acceptable and by providing a more efficient method of dispelling heat generated
Age/gender: generally younger workers and males are more heat tolerant than older workers and females, although there is considerable overlap of heat tolerance among these groups; changes to the work population can enhance previously unnoticed problems
Obesity: high body fat reduces heat tolerance by acting as an insulator and by contributing to the mass that must be cooled

Situational and Job-Based Factors

Clothing: increasing clothing retards the effects of evaporation; impermeable clothing can completely negate the effects of evaporation, thus increasing heat stress substantially; specialized clothing can help to reflect radiant heat loads
Acclimatization: repeated exposure to the hot environment will condition a person for greater heat tolerance; the increased use of air-conditioned homes, automobiles, shopping malls, etc., will generally reduce acclimatization effects
Duration of exposure: the human being can tolerate high temperatures for very short periods of time; determination of safe exposures to high temperatures is beyond the scope of this book
Work pacing: self-paced work and the allowance of worker-initiated recovery breaks can allow better worker control of heat stress situations (these are difficult to establish in industry, but they are not unheard of—maintenance workers in furnaces and foundries usually follow this regimen); the workers should have some training to spot the early signs of heat exhaustion and heat stress

ditions. Other methods include the Heat Stress Index (HSI), the sweat rate, sweat evaporation, core temperature monitoring, and computerized modeling of heat transfer situations.

WBGT involves straightforward calculations, is easy to use and understand, and has good reliability for industrial work. An understanding of the background and purpose of this method of assessing heat stress in exposed workers provides a better appreciation for its utility. "In 1971, the American Conference of Governmental Industrial Hygienists (ACGIH) published a notice of intent to establish threshold limit values (TLVs) for heat stress. These proposed TLVs were based on the assumption that acclimatized, fully clothed workers whose deep body temperatures are maintained at 38°C or less are not subject to excessive heat stress.

Obviously, it is not practical or economical to monitor every worker's heat stress by measuring deep body temperatures. Thus, industrial hygienists have turned to the measurement of certain environmental factors which correlate well with physiological response to heat. Of the several environmental indices of heat stress currently in common use, the wet-bulb globe temperature (WBGT) was, in the opinion of the ACGIH, the most suitable index for assessing the heat stress in exposed workers. WBGT is defined as follows:

1. For use outdoors where a solar load is imposed,

$$WBGT = 0.7NWB + 0.2GT + 0.1DB$$

2. For use indoors, or outdoors with no solar load,

$$WBGT = 0.7NWB + 0.3GT[14]$$

These values of calculated WBGT can be compared with predetermined levels to assess the degree of the health hazard. The levels are shown in Figure 9-12, which shows work rates, work-rest regimes, and WBGT levels.

A description of the WBGT and its use is provided in the same reference.

> WBGT is computed by appropriate weighting of the Vernon globe (GT), dry-bulb (DB), and natural wet-bulb (NWB) temperatures. The natural wet bulb is depressed below air temperature by evaporation resulting only from the natural motion of the ambient air, in contrast to the thermodynamic wet bulb, which is cooled by an artificially produced fast air stream, thus eliminating the air movement as a variable.
>
> Originally the interpretation of the levels of WBGT was for military activities of recruits in the following manner: above 30°C (86°F) WBGT, activities to be curtailed; above 31°C (88°F) WBGT, suspended entirely. For those in the latter stages of training, and hence acclimatized to heat, the levels are 31.0°C and 32.2°C (88 and 90°F), respectively.[15]

[14]*An Improved Method for Monitoring Heat Stress Levels in the Workplace* (Cincinnati, Ohio: HEW Publication No. (NIOSH) 75-161, May 1975), p. 6.

[15]*The Industrial Environment—Its Evaluation and Control* (Washington, D.C.: U.S. Government Printing Office, 1973), pp. 423-424.

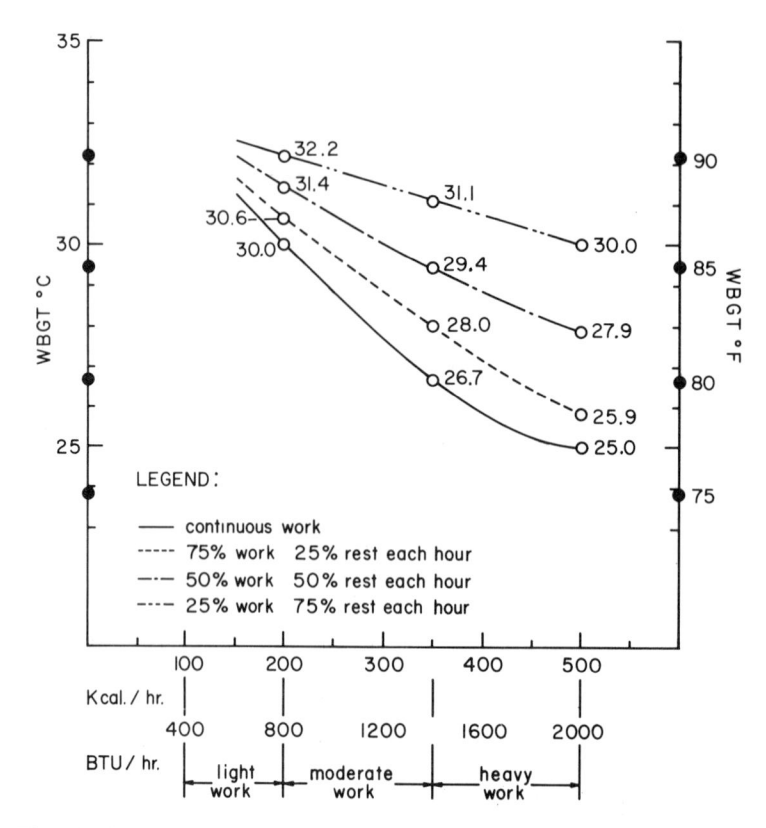

Figure 9-12 Permissible heat exposure threshold limit value. *Source: The Industrial Environment—Its Evaluation and Control* (Washington, D.C.: U.S. Government Printing Office, 1973), p. 424.

Ramsey provides descriptions of the work load levels used in the WBGT evaluations in Table 9-7.

There are a number of environmental, personal, situational, and job factors that influence heat tolerance that are not specifically noted in the WBGT. Ramsey has done considerable work in this area and has determined some modifications to the threshold WBGT. These factors are noted in Table 9-8.

9-4-3 Using the WBGT in Industry

There are four major concerns associated with the practical use of the WBGT in the industrialized setting. These are:

1. An understanding of what is important relative to heat stress situations compared to what factors are easy to measure
2. The equipment for gathering data

Chap. 9 The Hard and Hot Jobs 221

TABLE 9-7 Selected Types of Work Classed according to Work Load Level

Work Load	Energy Expenditure Range
Level 1: Resting	Less than 100 kcal/hr
Level 2: Light	101-199 kcal/hr

Sitting at ease: light hand work (writing, typing, drafting, sewing, bookkeeping); hand and arm work (small bench tools, inspecting, assembly or sorting of light materials); arm and leg work (driving car under average conditions, operating foot switch or pedal)

Standing: drill press (small parts); milling machine (small parts); coil taping; small armature winding; machining with light power tools; casual walking (up to 2 mph)

Lifting: 10 lb less than 8 lifts/min; 25 lb less than 4 lifts/min; 45 lb less than 2 lifts/min

| Level 3: Moderate | 200-299 kcal/hr |

Hand and arm work (nailing, filing); arm and leg work (off-road operation of trucks, tractors, or construction equipment); arm and trunk work (air hammer operation, tractor assembly, plastering, intermittent handling of moderately heavy materials, weeding, hoeing, picking fruits or vegetables); pushing or pulling lightweight carts or wheelbarrows; walking 2-3 mph

Lifting: 10 lb at 10 lifts/min; 25 lb at 6 lifts/min; 45 lb at 4 lifts/min

| Level 4: Heavy | 300-399 kcal/hr |

Heavy arm and trunk work; transferring heavy materials; shoveling; sledgehammer work; sawing, planing, or chiseling hardwood; hand mowing, digging, walking 4 mph, pushing or pulling loaded hand carts or wheelbarrows; chipping castings; concrete block laying

Lifting: 10 lb at 14 lifts/min; 25 lb at 10 lifts/min; 45 lb at 7 lifts/min

| Level 5: Very Heavy | above 400 kcal/hr |

Heavy activity at fast to maximum pace; ax work; heavy shoveling or digging; climbing stairs, ramps, or ladders; jogging, running, walking faster than 4 mph

Lifting: 10 lb greater than 18 lifts/min; 25 lb greater than 13 lifts/min; 45 lb greater than 9 lifts/min

Source: James L. Smith and Jerry D. Ramsey, "Designing Physically Demanding Tasks to Minimize Levels of Worker Stress," *Industrial Engineering,* May 1982, p. 50, Table 5. Reprinted with permission from *Industrial Engineering* magazine, May 1982. Copyright Institute of Industrial Engineers, 25 Technology Park/Atlanta, Norcross, GA 30092.

3. A method of estimating temperatures indoors during hot periods and of relating indoor and outdoor temperatures realistically
4. Using the WBGT to quickly flag potential problem areas and then having a strategy that allows a full analysis of serious problems

A discussion of each of these four items completes this section.

Heat stress measures. The evaluation of heat stress is one of those areas where it is possible to gather multitudes of temperature data with little effort. Unfortunately, the more significant factors, such as work load and evaporative cooling, are usually estimated. This leads to a situation with false precision and perhaps, erroneous conclusions. Without accurate measures of metabolic work, one must realize that there is substantial range in the answers

TABLE 9-8 Modifications of Threshold WBGT

Factors	Modification WBGT C	WBGT F
1. Unacclimatized, not physically fit	−2	−4
2. Air Velocity:		
Velocity above 1.5 mps (300 fpm) and air temperature below 35°C (95°F)	+2	+4
3. Clothing:		
Shorts, seminude	+2	+4
Impermeable jacket or body armor[1]	−2	−4
Raincoats, firefighter's coat[1]	−4	−7
Completely enclosed suits[1]	−5	−9
4. Obese, elderly	−1 to −2	−2 to −4
5. Female	−1	−2

[1] Modification for increased air velocity not appropriate with impervious clothing.

Source: James L. Smith and Jerry D. Ramsey, "Designing Physically Demanding Tasks to Minimize Levels of Worker Stress," *Industrial Engineering,* May 1982, p. 48, Table 4. Reprinted with permission from *Industrial Engineering* magazine, May 1982. Copyright Institute of Industrial Engineers, 25 Technology Park/Atlanta, Norcross, GA 30092.

developed. The calculations should be repeated several times, first for the estimated work level, then for the maximum and minimum work loads possible in this situation. When the work load plays such an important part in the heat stress situation, its range of values must be fully explored.

Equipment for gathering data. The equipment for the measurement of the temperatures needed for the WBGT calculation is relatively straightforward. That is one of the strengths of the WBGT. Instrument specifications are noted in *The Industrial Environment — Its Evaluation and Control,* page 426. A unit can be fabricated that will monitor these temperatures for extended periods of time to determine the overall seriousness of the situation, as shown in Figure 9-13. Commerical units, similar to that shown in Figure 9-14, are also available, are becoming more compact, and have faster response times.

Estimating temperatures. The WBGT is based on a stable estimate of the work situation for a 2-hour period (the approximate time between rest breaks in industry). In reality, though, even during that short a period, the factors are changing. The work load may be variable and the heat load normally changes as the day progresses. Except for cases of very high temperatures such as in smelting, or of environmentally controlled areas such as offices,

Chap. 9 The Hard and Hot Jobs 223

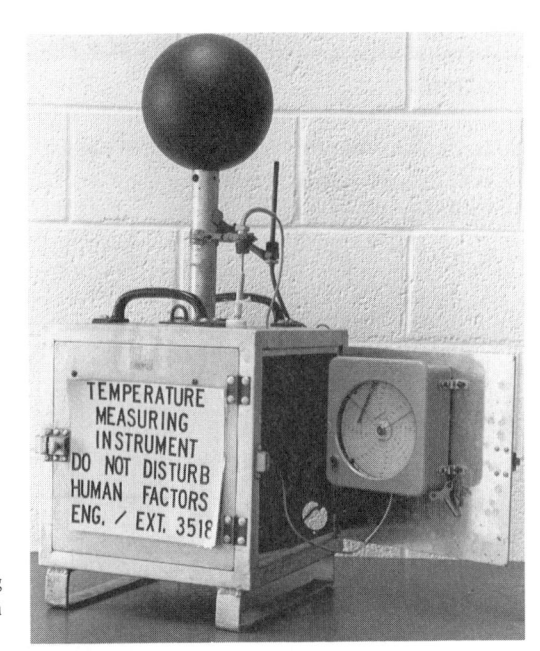

Figure 9-13 24-hour WBGT monitoring unit. *Source:* Eastman Chemicals Division of Eastman Kodak Company.

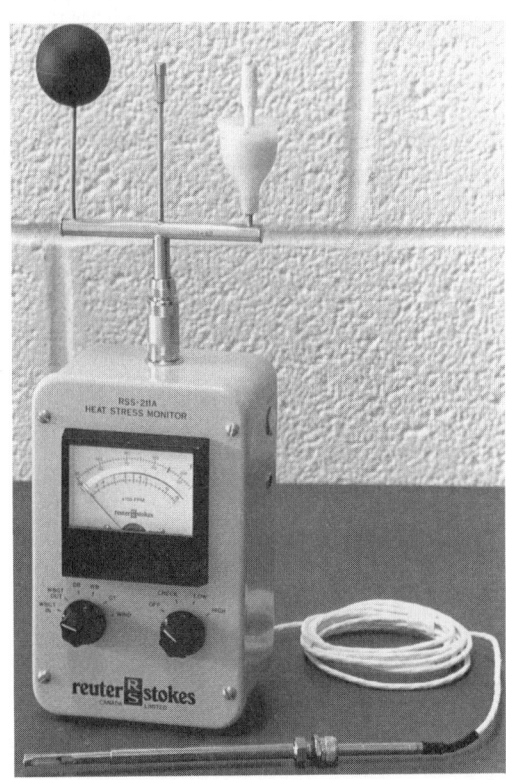

Figure 9-14 Commercial WBGT monitoring unit. *Source:* Eastman Chemicals Division of Eastman Kodak Company.

there will be a change in the ambient temperature over the day and with the seasons.

This variability for a 24-hour day is shown in Figure 9-15 for representative climatological data for Kingsport, Tennessee. The important point is that severe heat stress may be a problem only for part of the day and then only for a few days each year.

It is not practical to develop extensive preventive measures for situations

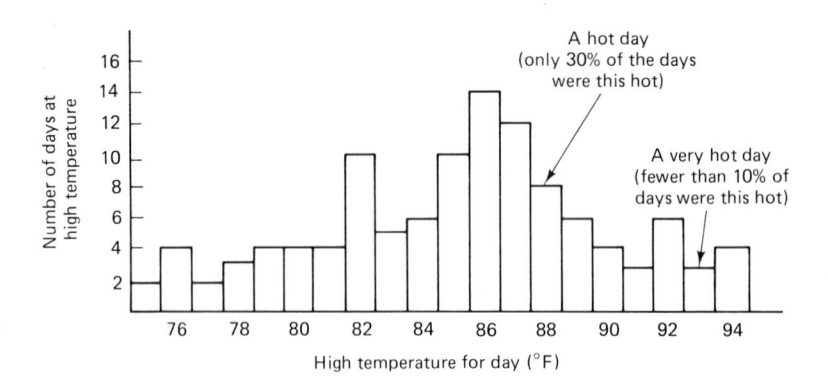

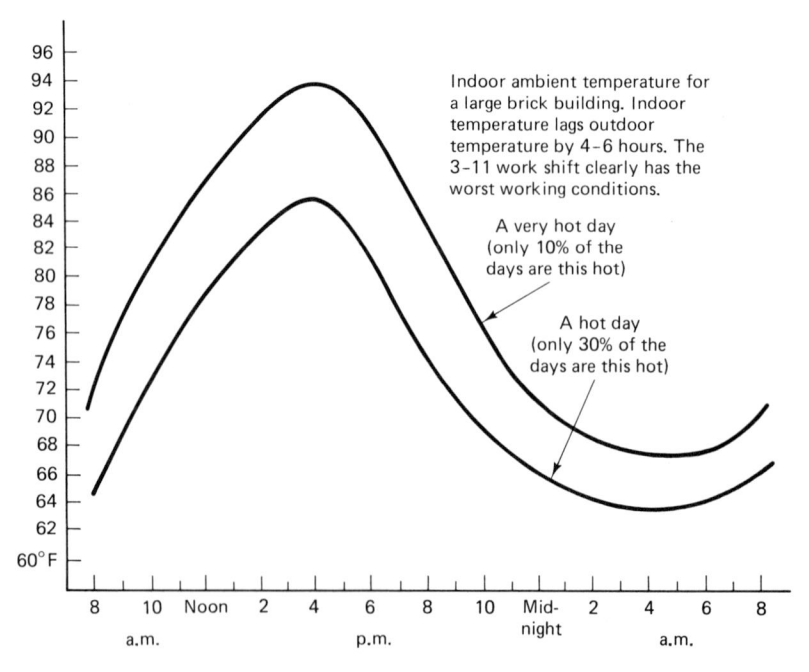

Figure 9-15 Representative daily climatological data for Kingsport, Tennessee, during the summer months.

that occur infrequently. It may be more economical and practical to provide extra rest pauses or to rearrange work schedules to other parts of the day. One situation often quoted occurred when a roofing crew requested to work from 5 a.m. until 1 p.m., to avoid the heat of the day. Their request was approved and they finished the job a full week early, saving considerable amounts of money. Even following the pattern set by the military, simply slowing or stopping the pace of work above certain WBGT levels may be less expensive than investing in expensive air-conditioning systems.

A tiered strategy for diagnosing heat stress problems. In the diagnosis and analysis of a heat stress study, it is possible to develop enough information to answer virtually any question that line supervisors may ask about the situation. Usually, however, they do not have many questions; the one that they want answered is usually: "Should I buy the fan the workers have asked for?"

With heat stress situations it is easy to spend more to analyze the problem than it costs to implement the solution. The power of detailed analysis of heat stress is that it can answer questions thoroughly, yet often that power is not needed. It is far more important to initially separate potential problems into three areas. The first type of problem is one where everyone agrees that there is a problem and generally what the solution should be. Another situation is where people agree that there isn't a problem and that study probably is not needed. The remaining situations are those where study is warranted, but only if the study will provide useful information. The study should, for example, further clarify the work risks involved or provide additional alternatives for resolving the situation.

A tiered approach works well for these needs. Certain criteria are established to go from the first stage to the second stage. The first stage is the awareness of a potential problem, and the second stage is a detailed diagnostic of that situation.

Suitable criteria to move past the first stage involve both work load and the WBGT measures:

Workload (kcal/hr)	WBGT Limit (°F)
Less than 200	87
200–300	84
Over 300	79

When any of these stages are exceeded, a second-stage analysis is initiated. Figure 9-16 shows the first-stage limits superimposed on the earlier WBGT threshold limits. The advantage is that the first stage requires only estimates of work load and simple measures of WBGT. The WBGT threshold limits require more data, which may not be needed to make an elementary analysis.

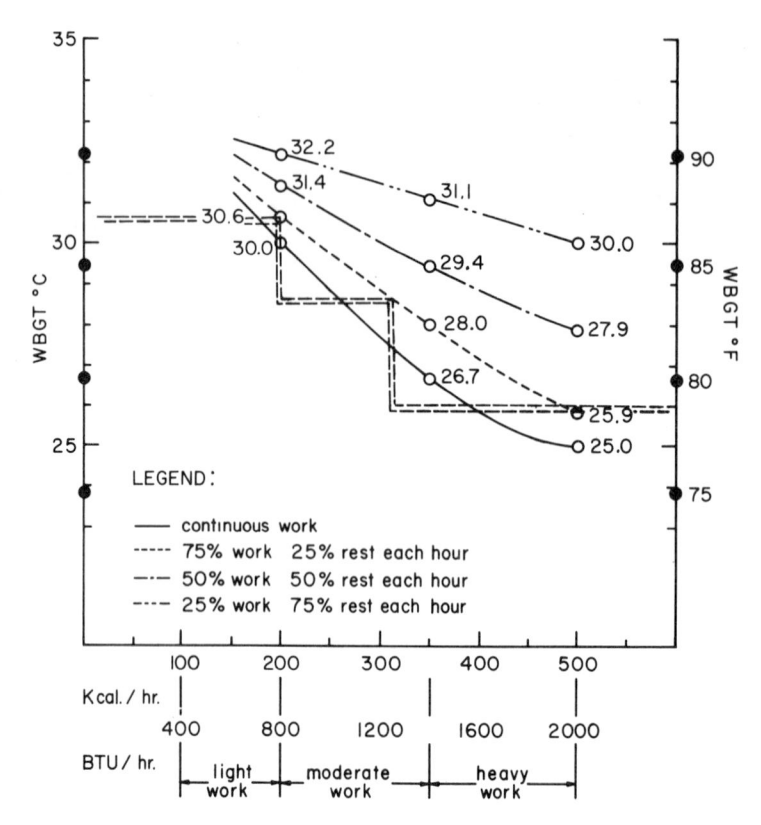

Figure 9-16 Tiered-approach first-stage limits superimposed on WBGT threshold limits.

While the first stage is essentially a "snapshot" of the worst conditions during the work day, the second stage is more realistic. It involves looking at the work load throughout the day, determining the work population, determining adjustments to use, interviewing workers and supervisors, and detailed checks of health situations. Other measures (HSI, etc.) may be initiated for another look at the situation. The stage two analysis should go a long way toward clarifying the situation and determining what solutions, if any are needed, are possible. The stage two analysis will probably (but may not definitively) determine whether or not the situation is suitable. If there are questions, a stage three analysis is warranted.

Stage three analysis requires the physiological monitoring of workers on the job for evidence of heat stress. Normally, this will involve monitoring of the heart rates and oxygen consumptions of people at work and during recovery to assess the degree and intensity of the stress. This stage, although expensive and time consuming, will be able to answer detailed questions about the relationships of the workers, the tasks, and the working environment. It

Chap. 9 The Hard and Hot Jobs   227

is very detailed and time consuming, requiring between 4 and 10 weeks for the full analysis. Only when the question is very specific, the implementation costs very high, or the number of people involved very large is this level of detail undertaken. Even then, it should utilize the unique skills of the industrial engineer for work and methods analyses, medical personnel for health problems, and heating, ventilation, and air-conditioning (HVAC) engineers for possible design problems.

The tiered approach to the analysis of heat stress situations should make effective use of limited resources by ensuring that a problem is not overstudied, that solutions are effective, and that solutions are needed to correct the problem (and that, indeed, there is a problem). Figure 9-17 illustrates the effectiveness of the tiered approach to solving heat stress problems.

9-4-4 Prevention of Heat-Related Problems

The analysis of heat stress can be beneficial for the prevention of potential heat stress situations. A few typical situations are discussed below.

Facilities design. It is possible to predict what kind and how much air conditioning is required in new facilities. Air conditioning can lower the air temperature, control humidity, or change airflows. Depending on the environmentally imposed heat load and on the internally generated heat load, the responses to these different changes can be estimated.

For example, when the work is hard, a cool temperature will not be the most effective way to remove heat. Evidence of this can be seen by the way lumberjacks work in a cold environment and still remove clothing to keep from overheating. When the job is hard, it would take a tremendous cooling effort to get rid of the excess heat. A far better method is to remove the heat

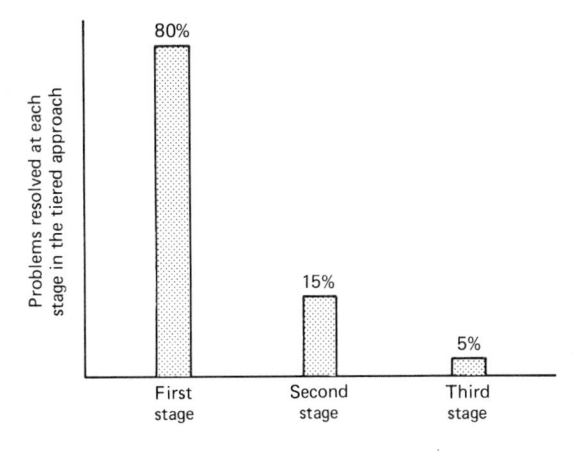

Figure 9-17 Solving problems using the tiered approach to heat stress problems.

through evaporation by increasing the air velocity and providing cool (but not cold) dry air.

In a facility, it may not be desirable or even necessary to provide high air velocities throughout the entire work area. This may be far more expensive than it is worth. Localized "temperate zones" or microenvironments are often adequate to meet the human needs and are far more economical. The ergonomist can provide critical design information to the HVAC engineer on where the temperate zones should be and on the thermal conditions required in these areas. These zones should be located to keep the operator on the job rather than away from the job, as a cool isolated break area might do. For example, in a process industry the temperate zone should be by the control panel.

An example of a temperate zone is shown in Figure 9-18, where an appropriate environment is contained by clear plastic strips. Another example,

Figure 9-18 Microenvironment in the OB-1 area. *Source:* Eastman Chemicals Division of Eastman Kodak Company.

Figure 9-19, shows localized air conditioning washing the front of the control panel area from special ductwork. These same concepts are possible with flexible pieces of ductwork, called elephant snouts, and with portable heat pumps.

Changing worker population. Heat stress analysis can be used to predict facilities and job changes necessary to accommodate a changing worker population. As trends in the worker population are noted, one can change modifiers and determine whether there is a need to redesign the HVAC facilities or the job (pace, W/R cycles, etc.) to safely accommodate the new group.

Figure 9-19 Ductwork cooling system in the B-12 control area, second floor. *Source:* Eastman Chemicals Division of Eastman Kodak Company.

For example, a group of temporary summer workers may not be heat acclimatized and may need to phase in to the job to avoid heat problems. This would be a temporary job pace accommodation. Another situation may involve the introduction of more female and/or older workers in jobs previously held by young males. This situation may require HVAC changes or permanent pace accommodation to avoid health problems.

Changing work duties. When the jobs in the work area are redesigned, it is now possible to predict whether there may be heat stress problems before the changes are actually implemented. Although there is an overall trend toward less physically demanding jobs, changes are made in consolidating jobs that may require more effort, either all day long or for short periods of time. These changes can be evaluated early in the proposed design process based on estimated work loads and known heat loads. The proposed jobs may be entirely suitable for the worker population, but if not, alternative work task designs can be considered or HVAC facilities redesigns can be considered. It is better to know about this ahead of time, when prevention can still be done, rather than finding out about a hazardous situation from a case of heat illness.

Changes to the work/recovery (W/R) cycles and work pace can be evaluated in a similar manner. Prediction and prevention of health problems is important and worthwhile.

There are a set of variables that come together in heat stress evaluation. Changes in one or more of these variables can be evaluated relative to the stress on the person or as changes to the other variables to avoid stress on the person. The variables are:

The job: its component tasks, pace, W/R cycles, clothing
The person: acclimatization, fitness, gender, age
The environment: temperature, radiant heat, humidity, air flow

9-4-5 Solutions to Heat Stress Problems

The effective resolution of heat stress problems is a four-step process: diagnosis, developing possible solutions, choosing a solution, and implementation and follow-up. These steps are discussed below.

Diagnosis. The early parts of this section provide an understanding of heat stress and its components. Proper diagnosis will determine which factors are excessive (either alone or in combination) and will provide an idea of how much they contribute to the heat stress (sensitivity analysis).

It is common to find one or two main contributors, such as temperature and work load. Other factors, such as humidity, the W/R cycle, and air flow, may play a secondary role. By knowing which variables are critical and how significant these variables are, looking for solutions becomes a more directed approach.

Solutions. There are appropriate solutions to heat stress problems based on the problem diagnosed. A list is provided in Table 9-9.
There are some general solutions; obviously, there will be other unique and customized solutions available for any situation.

Choosing a solution. The third step is the choice of a solution to the heat stress problem. Normally, there are several solutions that will resolve the problem, as discussed below.

TABLE 9-9 Checklist for Controlling Heat Stress and Strain

Item	Actions for Consideration
Components of heat stress	
M, body heat production of task	Reduce physical demands of the work; powered assistance for heavy tasks
R, radiative load	Interpose line-of-sight barrier Furnace wall insulation Metallic reflecting screen Heat-reflective clothing Cover exposed parts of body
C, convective load	If air temperature above 35°C (95°F) Reduce air temperature Reduce airspeed across skin Wear clothing
E_{max}, maximum evaporative cooling by sweating	Increase by Decreasing humidity Increasing airspeed
Acute heat exposures	
R, C, and E_{max}	Air- or fluid-conditioned clothing; vortex tube
Duration and timing	Shorten duration of each exposure; more frequent work periods are better than long work periods to exhaustion
Exposure limit	Self-limited, based on formal indoctrination of workers and foremen on signs and symptoms of overstrain
Recovery	Air-conditioned space nearby
Individual fitness for work in heat	Determine by medical evaluation, primarily of cardiovascular status Careful break-in of unacclimatized workers Water intake at frequent intervals Non-job-related fatigue or mild illness may temporarily contraindicate exposure (e.g., low-grade infection, diarrhea, sleepless nights)

Source: The Industrial Environment—Its Evaluation and Control (Washington, D.C.: U.S. Government Printing Office, 1973), p. 569.

Cost: This is the overall cost to solve the problem, including additional study time, equipment/facility modification costs, operating costs for HVAC equipment, and labor costs for recovery. The cost may easily vary by a factor of 2 to 5.

Administrative controls versus design solution: Some solutions are permanent one-time changes (HVAC equipment), whereas others are administrative changes (W/R cycles) that require ongoing monitoring to be effective. The design solutions are normally the most desirable.

Workability: The overall ease of workability of the solution is important. Has it been used before? Will the workers accept it? Is it perceived to be effective?

Robustness/utility: To what degree will the solution resolve both the diagnosed problems and related problems? A robust solution will handle a wider variety of situations and will allow more workers to do the job with less attention to their work pace. This may be confused with overdesigning a solution, but it involves seeking the broadest solution.

Implementation and follow-up. The ergonomist should be involved with implementation of the solution. If discussions are held concerning work pace or symptoms of heat stress, it is important to ensure that workers understand the importance of the guidelines. Similarly, designers of HVAC systems may encounter system design limitations previously unknown. The ergonomist must be available to help determine which factors are essential and which can be modified.

After implementation, follow-up is necessary to ensure that the solution is working as planned and has solved the original problem without creating others. This is a small step in the overall resolution of a heat stress problem, yet it may be the most essential. Without an effective, working solution, all the previous efforts have been wasted.

Follow-up is also valuable to the ergonomist, since it provides an opportunity to determine the effectiveness of the solution relative to its design. It also allows the ergonomist to see the effectiveness of any modifications made to the original design, thus providing the ergonomist with one more "design trick" to be used in new design areas.

9-5 SUMMARY

Chapter 9 provides the industrial ergonomist with the knowledge to evaluate the "tough jobs." The chapter provides the basic material on work physiology, which is the technology used to measure stress on the cardiovascular system.

Both the hard physical job and heat stress were discussed and examples

Chap. 9   The Hard and Hot Jobs   233

of analyses were shown. The diagnosis of problems and the creation of workable solutions are keys to the successful resolution of problems.

There are clearly situations in industry where the industrial ergonomist should seek the assistance of medical personnel or trained physiologists. Yet with the background material in this chapter, the ergonomist will understand the techniques of the study and will be able to clearly explain the problem and the types of analyses needed.

QUESTIONS

1. Why are jobs that overstress the cardiovascular system potentially more dangerous than jobs that overstress the musculoskeletal system?
2. What are the two major characteristics of jobs that can result in overstressing the cardiovascular system?
3. What three methods are available to assess the difficulty required in performing a job? Which one is the easiest to use?
4. What is a typical energy cost for resting? For walking at 3 mph? For performing light arm and leg work?
5. When using oxygen consumption measures to determine the energy requirements for a task, what are the precautions that should be observed?
6. Why is the heart rate an effective measure of the energy required on the job?
7. What are several cues that a job is difficult enough to warrant further study using the work physiology techniques?
8. What are four ways that high energy demand jobs can show up on the job?
9. What percentage of a person's maximum aerobic power should not be exceeded for an 8-hour work shift? For 1 hour?
10. For work of 30 minutes' duration, what percent of the energy is supplied through anaerobic process?
11. When monitoring heart rates on the job, what is the conclusion to be drawn when the initial pulse rate after successive units of work becomes progressively higher?
12. For 8-hour work shifts, the average metabolic rate is 5 Kcal/min for young healthy male workers. What adjustments should be made for female workers? For older workers? For predominately arm work?
13. What is "static work"? Give an example. Why is it dangerous?
14. If a task performed by young healthy males requires a metabolic level of 6.5 Kcal/min, what is the amount of recovery time required during the 8-hour work shift? If the same task is performed by young healthy females, what is the amount of recovery time required?
15. What criterion are useful in the evaluation of solutions to the hard jobs?
16. What general types of solutions are effective in dealing with hard jobs?
17. Show the model of temperature regulation in human beings. Describe the elements.
18. List some general factors which influence heat gain in a person.

19. Why is "false precision" a danger with heat stress measurement in industry?
20. Why is a "tiered approach to heat stress" an effective industrial tool?
21. What are the typical solutions to heat stress problems?
22. What are the primary methods of controlling heat stress if the metabolic rate is high? If the radiant load is high? If the convective load is high? If the evaporative rate is low?

Chapter 10

THE ENVIRONMENT

This chapter discusses the effect of the work environment on the person and on task performance. There are several common environmental influences that the industrial ergonomist must be familiar with—thermal concerns, lighting, noise, and vibration. In each area, the ergonomist should understand why that factor affects the person or performance, how to diagnose the problem, and what solutions will be effective in different settings.

10-1 INTRODUCTION

"My performance is way down," said the caller, "and nothing I do seems to improve it." The caller supervised a warehouse and had been experiencing lower productivity for about a month. After a brief investigation, the situation became obvious. The stacker drivers worked in an unheated warehouse, and since it was January, they were getting cold. The lost productivity was caused by the time spent in front of some heaters in the loading dock area. There were two choices: provide more heat on the stackers or create a microenvironment to protect the workers from the cold. A cab was installed for the stacker (to create the microenvironment) and productivity returned to the previous levels.

Chapter 2 discussed the six categories of problems normally dealt with in industrial ergonomics. Those six are anthropometric, endurance, strength, manipulative, environmental, and cognitive. This chapter deals with the environmental problems.

Environmental problems are those that involve the surroundings of the worker. Heat stress, cold stress, lighting, noise, and vibration are the typical problems that one will experience in the practice of industrial ergonomics. There is a relationship between the person, the task, and the environment, as shown below.

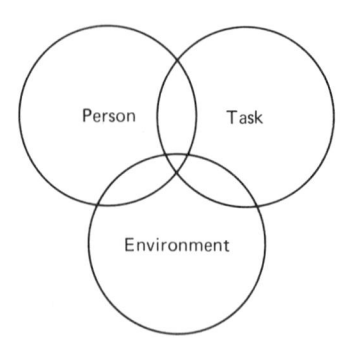

[Venn diagram showing three overlapping circles labeled "Person", "Task", and "Environment"]

The environment alone may directly affect the person, for example being outside on a cold day. However, it is usually a combination of the task and the environment that seriously affects the person, such as hard work on a hot day. Neither alone is really harmful, but the combination can be deadly.

Solutions to some environmentally imposed problems, such as excessive noise, can be compensated directly by removing the noise source or by modifying the job to reduce exposure time. The same holds true for exposure to glare or vibration; the solution can be to remove or control the source of the problem or to redesign the job to reduce the exposure.

Some environmental conditions affect most directly the sensory systems: light and vision, noise and hearing, vibration and feeling. These have been studied most thoroughly and are reasonably well understood. Other conditions cause problems with the physical systems: for example, heat and cold affect the endurance (cardiovascular) system, while vibration affects manipulation.

As we noted in Chapter 9 the environment does make a difference. In the absence of heat, more work can be done with less risk. The environment must be considered because it can cause or contribute to damage to the person's health or safety. The environment can also affect job performance and reduce personal comfort.

The three systems, environment, task, and person, usually must be studied together. Most problems result from the interaction of two or three of these systems, and consequently, good solutions result from changes from all systems. It is important to avoid the mindset that says that heat problems can be solved only with air conditioning; changes to the task or person are equally effective.

As was mentioned in Chapter 9, this is an area where detailed quantitative measurements are possible. It is important to keep the need for a solution in mind and avoid the common problem of overmeasurement. Besides, without reasonably strong correlations of the impact on the person or performance, it is easy to develop a false sense of precision from the extensive data. It seems as if we know a lot when, in fact, we really do not.

A general strategy for dealing with the environmental factors is first to be aware that they are present. The burden of environmental factors is not necessarily measured directly, but often affects the person and job performance. Second, design within known limits if at all possible. Fortunately, with environmental factors it is possible easily to quantify design limits and incorporate them into design standards. Most problems can be avoided through this method. Third, when ergonomic problems occur on the job, remember that the environmental factors are contributors and consider them in the diagnosis.

The more common environmental problems of cold stress, lighting, noise, and vibration are presented in this chapter. (Heat stress is so intimately tied to the work load that it was important to deal with those factors together.) These environmental concerns adjoin other industrial hygiene concerns, such as contaminations and exposures through the air or skin contact. The classical industrial hygiene concerns are beyond the scope of this book and should be studied elsewhere.

10-2 THERMAL CONCERNS

The thermal concerns are heat stress and cold stress. Both result from specific combinations of work load, temperature, humidity, airflow, clothing, and work surroundings.

10-2-1 Heat Stress

Heat stress is such a major area of concern and it ties so closely with the work load that it was dealt with as a major factor in Chapter 9. The heat load and work load do combine to cause health and safety concerns for the human being that neither could do alone. Solutions to heat stress problems may result from changes in the environment (temperature, air velocity, humidity), the task (clothes, W/R cycle), or the person (gender, age, fitness, acclimatization). The interrelationship of the person-task-environment for contributing to problems and for variety in problem solutions is clearer with heat stress than for any other situation.

10-2-2 Cold Stress

The effects of cold stress are not as dramatic as those of heat stress. Cold seems to be easier for workers to cope with through their normal methods (e.g., by wearing more clothes). The ill effects of cold stress are seen locally by frostbite, and system-wide by the lowering of the core body temperature.

Cold stress is determined by the task and the temperature. High work rates increase the metabolic rate, resulting in an internally generated heat load. In a cold environment, the internal heat load offsets the cold and makes the person more comfortable. In fact, a person's natural response to cold is this very situation—shivering produces the heat that the person needs for warmth.

Diagnosis of cold stress. The dynamics of cold stress are similar to those involved in heat stress. A conceptual understanding of either heat stress or cold stress greatly aids in understanding the other. The major differences come in the actual solutions designed and implemented.

The formula

$$M \pm R \pm C - E = S$$

applies for cold stress. Here M is the metabolic component, the heat generated internally from work. It occurs at the rate of 4 Btu per kilocalorie of energy used.

R is the radiant heat transfer; in this case the heat is radiated from the warm person to the cold external surfaces. It is most noticeable when the person is motionless in a cold environment. Outdoors with a bright sun, R gained by the person can still be significant. Similarly, radiant heaters at work can be effective in providing warmth.

C is the convective heat transfer and it can be quite large when the temperature is low and the air velocity is high. The wind-chill index illustrates this factor. The absolute temperature differential between the skin and the air determines the rate of heat exchange. However, in still air, a layer of insulating air will cover the skin reducing that differential. As the air velocity increases, the rate of heat exchange goes up. Increasing the insulation of the skin from the air is the preferred method of controlling heat loss.

E is the evaporative heat loss, which includes sweating and the loss from heating and moisturizing air entering the lungs. E is assumed to be about 25% of M,[1] thus reducing the effectiveness of the metabolic heat generated through work.

S is the heat storage (in this case, negative heat storage or heat loss), which can be up to 40 to 80 kcal/m^2 before undue discomfort develops.[2] For the 1.8 m^2 of skin found on the typical human being, the value is 72 to 144 kcal of heat loss before discomfort begins. Therefore, there is some tolerance

---

[1] *The Industrial Environment—Its Evaluation and Control* (Washington, D.C: U.S. Government Printing Office, 1973), p. 569.

[2] Ibid, p. 571.

for error in calculating a heat balance, particularly if regular work breaks are provided for warming or if there is self-pacing of the W/R cycles.

The major factors in cold stress are the work load and convective heat transfer. (This parallels the situation with heat stress, where the major factors normally are the work load and evaporative cooling.) Therefore, control of cold stress is largely a matter of controlling the heat loss from convection or of increasing the heat load by increasing work.

Increases in work load may be limited by other factors, such as an already high work load, limited energy availability, a belief that the job is too hard already, or a lack of extra work to do. While *M* affects the thermal equation, it is difficult to use as a control mechanism in ergonomic solutions to cold stress problems.

The other major factor, convective heat loss, is controlled through the insulation of the warm skin from the cold air and by reducing the effect of air velocity. The unit of measure for the insulating value of clothing is the "clo." Representative clo values are as follows:

Clothing	Approximate Clo Units
Shorts	0.55
Shorts, T-shirt	0.75
Coveralls	1.00
Coveralls plus over garments	1.50

The relationship between the work load, the ambient temperature, and the clo level is shown in Figure 10-1.

It was noted earlier that the air velocity combines with the ambient temperature to increase the heat loss by convection. The relationship between the ambient air temperature and the air velocity (wind speed) is called the wind chill and is shown in Figure 10-2. Figure 10-3 shows the perceived level of chill for still air and for slow moving (9 ft/sec) air.

Solutions to cold stress problems. There are many solutions to the problems of cold stress. The concept of a microenvironment is probably the most widespread method of dealing with cold, whether it is a heated automobile or an enclosed office. Some of the more universal solutions are listed in Table 10-1 on p. 242.

There are many other solutions, of course, depending on the particular situation. Avoiding extreme cold and "taking the environment with you" are useful concepts to remember in developing solutions to problem situations. The selection of the best solution from among those available involves the same criteria as in a heat stress solution: the cost, administrative controls versus design solutions, workability, and the robustness/utility of the solution.

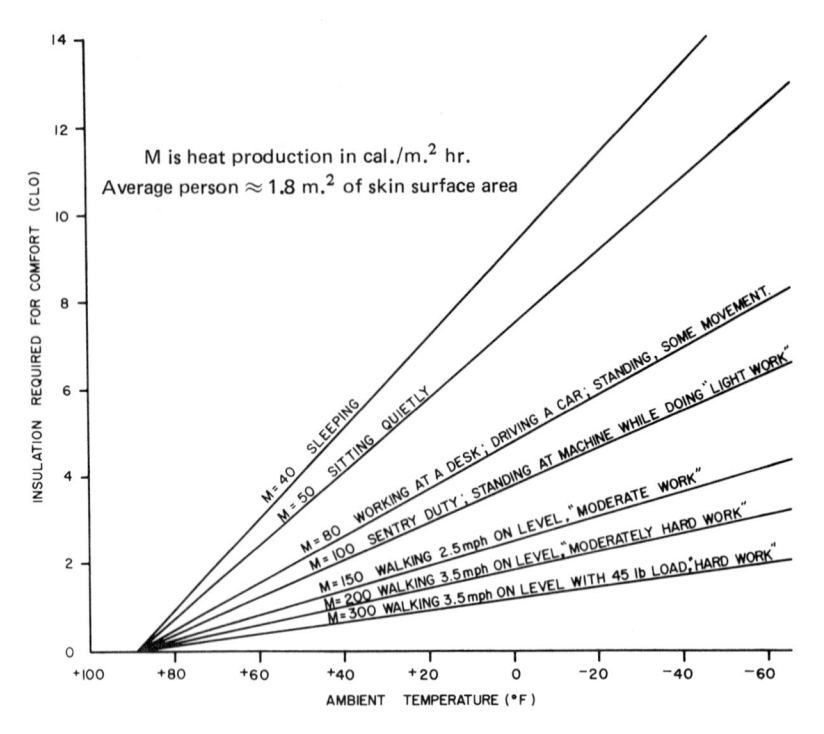

Figure 10-1 Prediction of total insulation required for prolonged comfort at various activities in the shade as a function of environmental temperature. *Source: The Industrial Environment—Its Evaluation and Control* (Washington, D.C.: U.S. Government Printing Office, 1973), p. 570.

An example can illustrate these points. Operating a stacker in a metal warehouse in the winter can be a very cold job. The temperature is low, the area is shaded from any of the sun's radiant energy, and the movement of the stacker creates a high airflow. In one case, productivity was declining, there was a high level of work breaks as workers went to stand by the radiant heaters, energy bills were increasing, and the workers were still miserable. A close examination of the situation revealed severe cold stress with several alternative solutions. The use of heaters on the stackers was not possible since the stackers were electric and the heaters required a high energy load. Raising M was not possible. Creating a microenvironment was the chosen solution, and cabs were installed for the stackers. The reduction in air velocity was enough, as shown in Figure 10-3, to improve the job substantially.

10-3 LIGHTING

Within most industrial and service areas lighting is not a widespread or serious problem. Usually, it affects only a few tasks or a few areas. In the past many

Chap. 10   The Environment   241

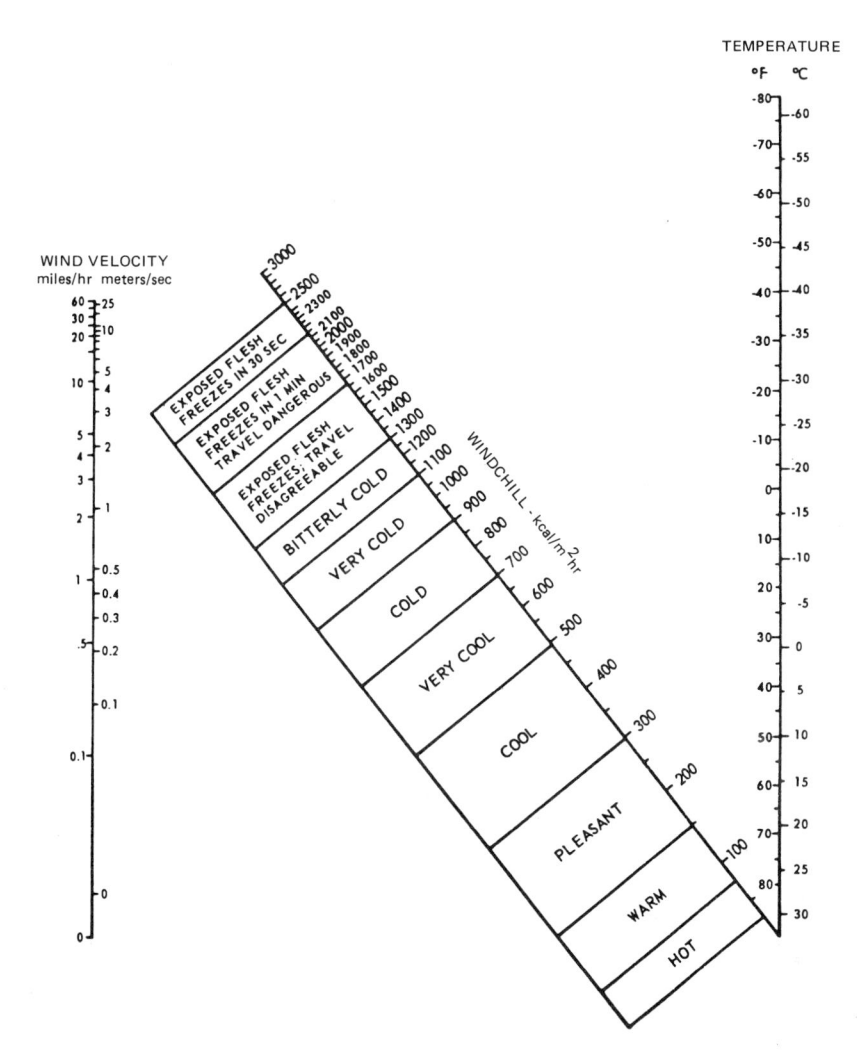

Figure 10-2 Nomograph for determining wind chill. *Source: Bioastronautics Data Book* (Washington, D.C.: U.S. Government Printing Office, 1964), p. 123.

lighting concerns were handled simply by increasing the illumination level. Now, however lighting problems are recognized as being more complex than that.

There are several ways to classify lighting:

- General or specialized lighting
- Natural or artificial lighting
- Primary or secondary lighting

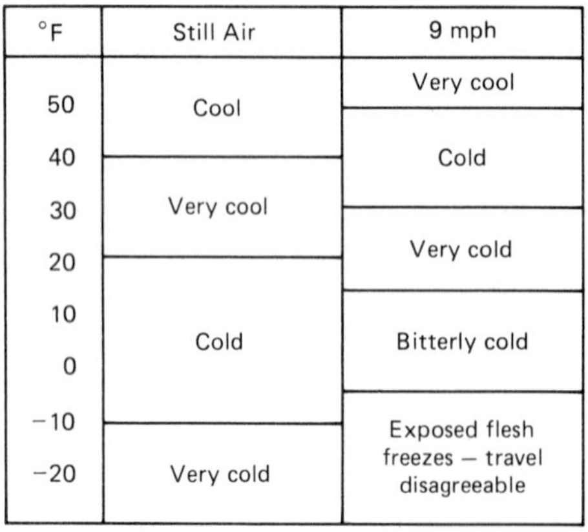

°F	Still Air	9 mph
50	Cool	Very cool
40		Cold
30	Very cool	
20		Very cold
10		
0	Cold	Bitterly cold
−10		
−20	Very cold	Exposed flesh freezes — travel disagreeable

Figure 10-3 Effect of air velocity on subjective cold stress.

TABLE 10-1   Solutions to Diagnosed Cold Stress Problems

Item	Possible Actions
Heat components	
M	Increase by:
	Increasing work load
	Increasing work pace
	Reducing recovery time
	Shivering
R	Increase by:
	Providing radiant heaters
	Utilize sun's radiant heat
	Remove clothing if heat source nearby
	Removing walls, barriers to any nearby heat source
C	Increase by:
	Wearing more clothes
	Raising air temperature
	Lowering air velocity
	Creating microenvironment
	Wind-breaking clothes
E	Reduce by:
	Decreasing air velocity
	Layering clothing
Work schedule	
Duration	Faster work/recovery cycle, with warm recovery area
Recovery	Warm recovery area
Other	Self-limit exposures
Clothing	Increase clo value
	Reduce airflow through clothes

Lighting from overhead sources over a broad area is general lighting. Most tasks can be accomplished with this type of lighting, although additional specialized lighting may be required. Specialized lighting is good for highlighting quality defects, or for other tasks that are largely visually dependent. In some cases, specialized lighting is a better choice than increasing the general level of illumination because through contrast it can highlight the visual target.

Naturally occurring light is used widely because of its low cost, but most job sites have artificial lighting as a supplementary source or for specialized lighting. Artificial lighting is usually better than natural lighting for specialized lighting. The contrasts of these two methods is shown in Table 10-2.

Primary lighting is the lighting needed to perform a task. It should be designed with a knowledge of the task and the nearby work area. Secondary lighting is the light needed to move about and for safety needs. Secondary lighting is usually adequate and if not, generally can be corrected by increasing the illumination level.

Lighting problems may also be task dependent or independent. Task-independent problems (tripping in a darkened corridor) tend to be with secondary lighting and are usually resolved with increases in the general illumination levels.

Task-dependent problems are more complex, since lighting may be one of the primary ergonomic concerns. For example, visual display terminals (VDTs) found at computer workplaces are significantly affected by lighting. Some other highly visual tasks are detail work, such as drafting, and jobs involving color discrimination, such as color sample matching. A good tool for assessing the importance of lighting is task analysis. Task analysis can determine the percentage of task performance that requires vision and can also determine the importance of these tasks in the success of the job.

The analysis of these tasks is very difficult and the solutions are highly varied. This book will provide an overview for dealing with visual tasks, but for specialized situations, the reader should consult additional references.

10-3-1 Illumination Levels in Industry

Illumination consists of both the quantity of light and the quality of light. Either of these may contribute to lighting problems. The footcandle is the normally accepted measure of the quantity of illumination falling on a surface. The quality of light is the distribution of light within the visual field. Quality problems are commonly seen as glare, contrast, or shadows. These problems are discussed more fully in Section 10-3-2.

The task itself has the most impact on the level and type of light needed. The visual requirements of a task increase as the size of the objects (or details) being viewed decrease in size. The visual requirements also increase as the contrast between the objects (or details) and those of the immediately surrounding areas decreases.

TABLE 10-2 Natural versus Artificial Lighting

Natural Lighting	Artificial Lighting
Low cost	Easier to control
Highly variable (dusk, clouds, direction of windows)	Repeatable
Very bright at times	Can be uniform over wide areas

Sometimes, the task can be altered to enhance either the size or the contrast, if they are a problem. For example, magnification can be used to enhance size, and the contrast on metal surfaces can be improved with a dye to highlight fine scratches. If the task is difficult and the size or the contrast cannot be altered, then either the time alotted for the task or the illumination level can be increased for proper task performance.

A list of common tasks and the necessary illumination levels are shown in Table 10-3. Although the minimum level of illumination shown is 30 footcandles, levels of 5 footcandles are acceptable for aisleways.

TABLE 10-3 Illumination Levels for Tasks

Area: Clothing Manufacture (Men's)	Footcandles on Tasks[1]
Receiving, opening, storing, shipping	30
Examining	2000[2]
Sponging, decanting, winding, measuring	30
Piling up and marking	100
Cutting	300[2]
Pattern making, preparation of trimming, piping, canvas, and shoulder pads	50
Fitting, bundling, shading, stitching	30
Shops	100
Inspection	500[2]
Pressing	300[2]
Sewing	500[2]

[1] Minimum on the task at any time.

[2] Can be obtained with a combination of general lighting plus specialized supplementary lighting. Care should be taken to keep within the recommended luminance ratios. These seeing tasks generally involve the discrimination of fine detail for long periods of time and under conditions of poor contrast. The design and installation of the combination system must not only provide a sufficient amount of light, but also the proper direction of light, diffusion, color, and eye protection. As far as possible it should eliminate direct and reflected glare as well as objectionable shadows.

Source: *The Industrial Environment—Its Evaluation and Control* (Washington, D.C.: U.S. Government Printing Office, 1973), p. 350.

Measurement of light levels is not difficult with the proper instrumentation. Light levels are very difficult to estimate subjectively due to the human being's remarkable ability to adapt to widely different light levels. Any light level almost always appears fully adequate if a person has long enough to adapt.

The specification of light levels should be based on estimates for adjustments to the light levels. However, in situations such as darkroom work, adjustment times may be so long that special accommodations such as early entry times and low-light-level break areas are necessary.

10-3-2 Problems with Lighting

Generally, problems with lighting result from the quality of the light rather than from the level of the illumination. Glare, contrast, or shadow can cause serious decrements in performance. These problems need to be identified and corrected.

Glare occurs when a bright light shines in the person's eyes and reduces the visibility of the object to be seen. The glare can be from a light source in the field of vision near the object, or may be reflected from the object itself. Corrections include shielding light sources or relocating sources of reflected glare. The layout of the VDT workplace calls for the display screen to be perpendicular to nearby windows, thus eliminating sources of direct or reflected glare without further measures.

Contrast is caused when the illumination level of the object is significantly higher or lower than the visual surroundings. The ratio of the luminance of the object and the nearby surroundings should be controlled to about $\pm 1:3$, for farther surroundings the acceptable ratio increases to $\pm 1:10$, and it goes to $\pm 1:20$ to $\pm 1:40$ for the farthest surroundings, including windows. Luminance is a measure of the light reflected at each surface. When nearby surfaces have high luminance, this affects the eye's normal adaptation of light from the object/task, both during the task and after any brief glances about.

Shadows can cause areas where it is difficult to see, especially when the eye is adjusted for higher light levels. Nonuniform light levels are normally controlled through diffuse lighting such as fluorescent lights. Shadows may be useful for specialized tasks, since they can actually enhance task performance. In inspection tasks, use of directed light to graze the surface will highlight surface irregularities. Inspection tasks are covered in more detail in Chapter 11.

Service and maintenance tasks offer their own special concerns for proper lighting. For normal tasks in operations, proper lighting is usually provided as a matter of course. Lighting for proper "visual access" during maintenance and service work is seldom considered and rarely implemented. Working inside cabinets and under equipment is difficult enough without the additional burden of not being able to see the task properly. Trouble lights and flashlights are compensations for problems that should be eliminated during equipment and facilities design.

10-3-3 Solving Lighting Problems

The process of solving lighting problems begins with the identification of a problem area, from actual complaints on the job, low task performance levels, or with concerns from a proposed task.

Diagnosis of the situation follows that. Quantitative information about light levels and reflectances is necessary. A clear understanding of the task and the importance of visual information must also be known. A simple task analysis can be beneficial to compare the task requirements and the normal human capabilities on the job. Mismatches in the requirements and capabilities point out the specific problem areas that require solutions.

The solutions are rather specific to the task, but common solutions include:

- Controlling the light source by shielding or relocation
- Changing the task to require less fine visual perception
- Changing the object to control reflectance
- Changing the pace for better visual adaptation
- Changing the illumination level

Obviously, there are many other specific solutions, but they are usually developed from a detailed task analysis and the specific needs identified.

10-3-4 Additional Reading

American National Standard Practice for Industrial Lighting, A 11.1-1965 (R 1970). New York: Illuminating Engineering Society, 1970.

American National Standard Practice for Office Lighting, A 132.1-1966, New York: Illuminating Engineering Society, 1966.

IES Lighting Handbook, 5th ed. New York: Illuminating Engineering Society, 1972.

Lighting Survey Committee of the IES, "How to Make a Lighting Survey," *Illuminating Engineering,* 57, 1963, p. 87.

10-4 NOISE

Noise and sound are two concepts for the same phenomenon; *noise* is used for unwanted sound, whereas *sound* means communication of pleasant sounds.

Noise can cause hearing damage, which may be either temporary or permanent. The same noise will not cause uniform hearing damage in all people. Like most ergonomic concerns, it is difficult to establish unique limits that absolutely predict damage to the human being. Noise also interferes with communications on the job, it can cause annoyance, and it may reduce performance on mental (e.g., concentration) or physical tasks.

Too little noise or nonuniform noise can also cause problems. It can be "too quiet" to work when noise levels are very low. When it is very quiet, transient noises such as passing automobiles, nearby conversations, or closing doors become annoying. Sound-masking systems or noise generators may be required to alleviate these problems.

10-4-1 Noise Limits

Noise can have three detrimental effects on a person: hearing loss, interference with communications, and annoyance resulting in performance losses. Noise limits are set by the two more quantitative criteria: hearing loss and speech interference.

Hearing loss. Permanent hearing loss results from higher sound pressures (or "decibel levels") during high noise exposure. The frequency of the noise is important, since higher-frequency sound is more harmful than low-frequency sound. The duration of the exposure also contributes to hearing loss. These factors have been incorporated into a set of limits of permissible noise exposures for industrial jobs. The standards are time based and weighted for intensity (decibel level) and frequency (the weighting factor for dBA). Table 10-4 provides the standard.

Since the susceptibility of people varies, there is some controversy over these standards. The classical trade-off for ergonomic limits is present here—

TABLE 10-4 Noise Exposure Limits

Time Duration per Day (hr)	Sound Level (dBA)
8	90
6	92
4	95
3	97
2	100
1.5	102
1	105
0.75	107
0.50	110
0.25	115 max.

Source: Federal Register, 36(105), May 29, 1971.

the percentage of people who sustain damage at different exposures is weighed against the overall costs to industry associated with lower exposure levels. The levels are a compromise for both.

The decibel scale is logarithmic, with significant increases in the intensity of the noise only slightly changing the decibel reading at higher levels. Some representative values of the A-weighted frequency, dBA, are shown in Figure 10-4.

Intermittent noise causes less hearing loss than does continuous noise. The hearing mechanisms have a chance to recover before each new exposure, and the likelihood of permanent damage is reduced. Similarly, impulsive noise has a different (usually reduced) effect on hearing loss at the same dBA level. Limits are available for both of these types of noise but they are complicated to apply and exceed the scope of this book. Several references for additional reading are provided.

Speech interference. The detrimental effect of unwanted background noise on oral communication is well known. Research has been done to determine the effect that various noise levels have on speech communications.

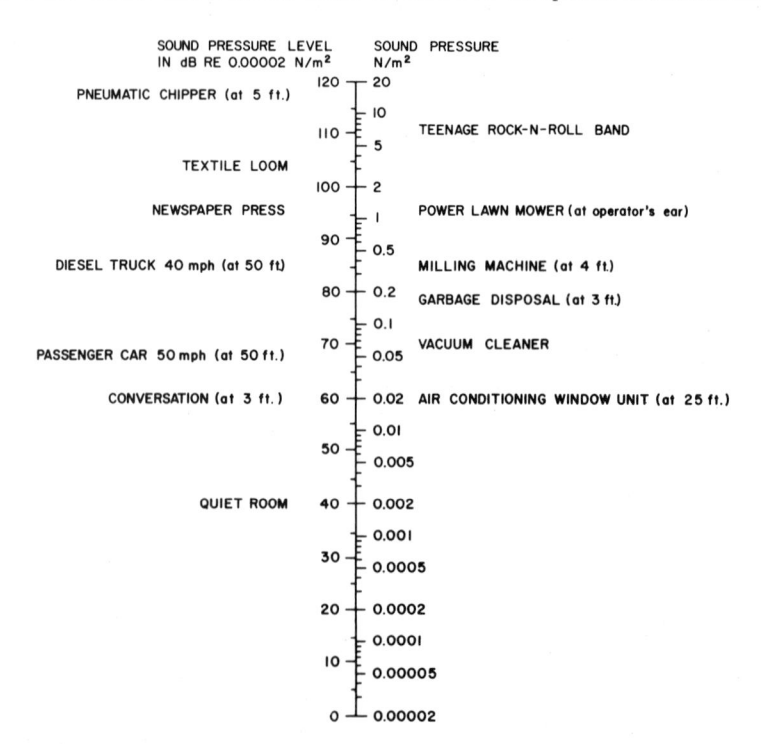

Source	Sound Pressure Level in dB RE 0.00002 N/m²	Sound Pressure N/m²	Source
PNEUMATIC CHIPPER (at 5 ft.)	120	20	
		10	
	110	5	TEENAGE ROCK-N-ROLL BAND
TEXTILE LOOM			
	100	2	
NEWSPAPER PRESS		1	POWER LAWN MOWER (at operator's ear)
	90	0.5	
DIESEL TRUCK 40 mph (at 50 ft)			MILLING MACHINE (at 4 ft.)
	80	0.2	GARBAGE DISPOSAL (at 3 ft.)
		0.1	
	70	0.05	VACUUM CLEANER
PASSENGER CAR 50 mph (at 50 ft.)			
CONVERSATION (at 3 ft.)	60	0.02	AIR CONDITIONING WINDOW UNIT (at 25 ft.)
		0.01	
	50	0.005	
QUIET ROOM	40	0.002	
		0.001	
	30	0.0005	
	20	0.0002	
		0.0001	
	10	0.00005	
	0	0.00002	

Figure 10-4 Examples of noise at various dBA levels. *Source:* Adapted from *The Industrial Environment—Its Evaluation and Control* (Washington, D.C.: U.S. Government Printing Office, 1973), p. 301.

TABLE 10-5 Noise Level (dB) versus Ease of Communication

Noise Level (dB)	Ease of Communication
30–40	Communication in normal voice satisfactory
40–50	Communication satisfactory in normal voice at 3 to 6 ft; and raised voice from 6 to 12 ft; telephone use satisfactory to slightly difficult
50–60	Communication satisfactory in normal voice at 1 to 2 ft; raised voice from 3 to 6 ft; telephone use slightly difficult
60–70	Communications with raised voice satisfactory at 1 to 2 ft; slightly difficult from 3 to 6 ft; telephone use difficult; earplugs and/or earmuffs can be worn with no adverse effects on communications
70–80	Communication slightly difficult with raised voice at 1 to 2 ft; slightly difficult with shouting from 3 to 6 ft; telephone use very difficult; earplugs and/or earmuffs can be worn with no adverse effects on communications
80–85	Communication slightly difficult with shouting from 1 to 2 ft; telephone use unsatisfactory; earplugs and/or earmuffs can be worn with no adverse effects on communications

Source: Adapted from *Standard—Human Engineering Design Criteria; MSFC-STD-267A* (Huntsville, Ala.: George C. Marshall Space Flight Center, National Aeronautics and Space Administration, 1966), p. 279.

Table 10-5 presents that information for interpersonal and telephone conversations.

Annoyance resulting in performance losses. This is the most difficult area for the ergonomist. In these circumstances, there is an alleged interference with work performance, and usually there is a cost associated with its resolution. This is akin to the comfort situations discussed in earlier chapters.

Changes to reduce annoyance are difficult to quantify and to sell. If work performance is degraded, what improvements will result with different reductions in the noise level? It is difficult to predict specific levels of improvement and therefore very hard to indicate what the organization will get for its money. At the same time, these problems tend to occur when the noise levels are in the range 65 to 80 dBA, where corrections are increasingly difficult and expensive to implement.

The ergonomist is thus faced with justifying high costs when performance improvements are difficult to predict. Without tangible evidence of a favorable cost/benefit ratio and in the absence of legislated standards, many projects are simply left unapproved.

10-4-2 Problem Identification, Diagnosis, and Resolution

There are three levels of noise problems—hearing loss, speech interference, and annoyance—which result in performance losses. The processes used for problem identification, diagnosis, and resolution are similar, although not identical, for the three areas.

Identification. Measurements of noise levels and exposure times will reveal whether the hearing loss standards are exceeded or whether speech interference is expected. Work performance changes are difficult enough to identify without the added burden of specifying that the cause is noise and not other factors. Annoyance problems may be the easiest to identify since they are usually brought to the ergonomist's attention by other people who have determined that noise is causing them work problems.

Diagnosis. Diagnosis of hearing-loss problems is done by gathering dBA noise levels and exposure times, which are compared with the standards. Since the problem is a combination of noise level and exposure time, changes in either or both can solve the problem.

The diagnosis of speech interference problems is done through the measurement of noise levels. The time of exposure is unimportant if the high noise levels coincide with the need to communicate.

The diagnosis of work performance problems is much more difficult because there are many causal factors for performance problems. At the same time, there are no stated noise levels where performance deteriorates. Although higher noise levels and exposure times are reasonable indicators that noise is a contributor to performance problems, they may not be enough to justify action. Finding work units with similar tasks but with reduced noise exposure and comparing performance may help determine noise as the real cause.

Resolution. The resolution of these problems again depends on the type of problem experienced. For hearing-loss concerns, the exposure time or the noise source can be modified. For speech interference and annoyance problems, the noise source is modified or isolated from the person. Table 10-6 shows the types of solutions suitable for these three types of problems. A list of engineering solutions for controlling noise at the source is provided in Table 10-7.

Isolation of the person and the noise source can be accomplished by distance and/or barriers. Increasing the distance between the person and the noise source is effective, but the distances required may be prohibitive in an industrial environment. Barriers are effective to either contain or redirect the noise. Care must be taken to ensure the barriers insulate rather than conduct the noise. Controlling the exposure time is effective only for hearing-loss problems. Job rotation and job redesign are other ways to control exposure time.

Earmuffs and earplugs are additional administrative methods for controlling noise problems. These seem to be most effective for the higher noise levels associated with hearing loss, although they can also reduce speech in-

Chap. 10 The Environment

TABLE 10-6 Types of Solutions Suitable for Types of Noise Problems[1]

	Control Noise Source	Isolate Person and Noise Source	Control Exposure Time	Earmuffs or Earplugs
Hearing loss	Engr.	Engr.	Admin.	Admin.
Speech interference	Engr.	Engr.	×	Admin.
Annoyance	Engr.	Engr.	×	Admin.

[1] Engr., engineering solution to the problem, usually a permanent one-time solution; Admin., administrative solution to the problem, usually inexpensive, but must be monitored regularly to ensure effectiveness; ×, not appropriate solution.

TABLE 10-7 Control of Noise through Engineering Solutions

1. Planning
   a. Specify quieter equipment
   b. Design to isolate noise sources
2. Substitution
   a. Use quieter equipment
   b. Use quieter process
   c. Use quieter material
3. Modification of the noise source
   a. Reduce the driving force on vibrating surface
      (1) Maintain dynamic balance
      (2) Minimize rotational speed
      (3) Increase duration of work cycle
      (4) Decouple the driving force
   b. Reduce response of vibrating surface
      (1) Add damping
      (2) Improve bracing
      (3) Increase stiffness
      (4) Increase mass
      (5) Shift resonant frequencies
   c. Reduce area of vibrating surface
      (1) Reduce overall dimensions
      (2) Perforate surface
   d. Use directionality of source
   e. Reduce velocity of source
   f. Reduce turbulance
4. Modification of the sound wave
   a. Confine the sound wave
   b. Absorb the sound wave
      (1) Absorb sound within the room
      (2) Absorb sound along with its transmission path

Source: Adapted from *The Industrial Environment—Its Evaluation and Control* (Washington, D.C.: U.S. Government Printing Office, 1973), pp. 533-535.

terference problems. Plugs and muffs could be used to reduce annoyance from noise, but the annoyance of wearing them would probably offset any work performance improvements. Monitoring to ensure that personnel wear the plugs and muffs must be done often, and this is a significant administrative drawback for these items.

In extreme conditions, communications may need to be handled with lights or with very loud piercing noises such as sirens. Sirens are usually reserved for emergency situations, but lights can be effective for other types of communications as well.

10-4-3 Additional Reading

Noise is a complex area and a science in its own right. The treatment of noise in this book is intended to provide an overview to common industrial concerns and solutions. For more complex problems and additional information, consult the following works.

BERANEK, L. L., *Noise and Vibration Control.* New York: McGraw-Hill Book Company, 1971.

HARRIS, C. M., *Handbook of Noise Control.* New York: McGraw-Hill Book Company, 1957.

KRYTER, K. D., *The Effects of Noise on Man.* New York: Academic Press, Inc., 1970.

PETERSON, A. P. G. and E. E. GROSS, *Handbook of Noise Measurement.* West Concord, Mass.: General Radio Company, 1967.

10-5 VIBRATION

Exposure to vibration can result in Reynaud's syndrome or in other damage to nerves, muscle, or bone. Vibration can be whole body or localized. Exposures to whole-body vibration come from construction equipment, tractors, lift trucks, and other vehicles. Localized vibration results from the use of power tools such as chain saws, drills, and jackhammers, or from the operation of high-speed or powerful equipment such as textile looms or metal stamping machines.

There is no standard on vibration because the incidence of problems is highly varied and not thoroughly reported. Reynaud's syndrome is the most common symptom of localized vibration. It occurs most often with repeated exposure to vibration in the frequency range 40 to 125 hertz. Reynaud's syndrome is often called "dead fingers" because one of the symptoms is a lack of sensation in the hands. It is also known as "white fingers," from the impaired circulation in the hands. The lack of sensation is most pronounced when the hands are exposed to the cold, which further reduces blood circulation. Preventive measures include handles with damping, handles of suitable size for the hands (especially handles that allow a looser grip while maintaining control), and exhaust of air away from the hands.

Like noise, vibration is a complex phenomenon, well beyond the scope of this text. Additional information is appropriate if complex problems are present. A good place to begin looking for more information is Robert D. Soule, "Vibration," Chapter 26 in *The Industrial Environment—Its Evaluation and Control,* U.S. Department of Health, Education and Welfare, Public Health Service, Center for Disease Control, National Institute for Occupation Safety and Health, (Washington, D.C.: U.S. Government Printing Office, 1973).

10-6 SUMMARY

The work environment can play a major role in the health and safety of the person at work. The environment can also enhance or degrade work performance. The industrial ergonomist must understand the environment and its relationship to both the person and the task. The ergonomist must be able to recognize environmentally related problems, diagnose the major contributors to its cause, and develop solutions that resolve that problem without creating others.

QUESTIONS

1. Explain how the three systems—the person, the task, and the environment—must be studied together to determine problems by using heat stress as an example.
2. What is the model of temperature regulation used for cold stress situations?
3. What are the primary methods of controlling cold stress if the metabolic rate is low? If the radiant load is low? If the convective load is low? If the evaporative rate is high?
4. What determines the type of lighting needed?
5. Illumination has both quantity and quality aspects. If the illumination level is the primary quantity concern, what are the quality concerns?
6. What are five common solutions to lighting problems?
7. What is the difference between noise and sound?
8. What are three detrimental effects that noise can have on the human being?
9. Why is annoyance so much harder to deal with than hearing loss problems?
10. What are the common methods of controlling noise problems?
11. What are the primary sources of vibration in the industrial environment? What is the most common symptom of localized vibration?

Chapter 11

THE HUMAN ERROR

This chapter on human error provides a look at the types of errors made in the industrial and service sectors. The ergonomist should be familiar with the types of errors that can occur and should have a model to understand how those errors can occur. Diagnosis and correction/prevention of errors in seven different situations completes the chapter.

11-1 INTRODUCTION

My friend was telling me his problems: "The cost of materials was over $10,000, and we will have another late shipment." An operator had added the wrong amount of a critical ingredient, and the batch wouldn't set up. "It might be reworked, but more than likely, we will have to start over again. This is the second time this month—I can't afford for this to happen again." An investigation revealed a string of similar errors, some reported and some not. A loading sheet was later found to the culprit, instead of an errant operator. With a few simple corrections to the loading sheet, that particular "mistake" was not to happen again.

There are many situations where human errors have resulted in unnecessary problems. All too familiar are airline and automobile tragedies attributable to human error. The story of Three Mile Island demonstrates both the huge economic losses resulting from "human error" and the huge costs for the rest of the nuclear power industry to correct all those bad designs. It becomes quite apparent that the magnitude of the error is not necessarily related to the cost of the consequence of the error.

Of course, these are dramatic examples, but what about other less dramatic situations? Errors occur every day, many times around the world, and although individually they may not amount to that much trouble or expense, collectively they represent a huge cost. Some typical situations are a lost order, misplaced shipment of material, the wrong order in a restaurant, a form that requests ambiguous information, the form that's filled out wrong, a misoperation in a control room, a part that's left out in an assembly operation, a bad product passed by an inspector, or the misuse of a written procedure.

An error can be defined as a situation where an action other than the one desired takes place. This definition assumes a desire to do the correct thing, that is, that the error is not intentionally produced. This chapter is designed to discuss error-producing situations, to point out the huge potential losses associated with those situations, and to describe corrective actions.

11-1-1 Error Rates and Costs

With their frequency and potentially high cost, human error situations are a very lucrative area for the ergonomist. The problem is simply to identify the error-producing situation and to convince others that substantial improvements are possible.

Error rates vary greatly with different situations. Error rates range from as low as a fraction of a percent to rates as high as 50% (or higher with bad designs that run counter to established human expectations or stereotypes). Alan Funt, in "Candid Camera," made a career of placing unsuspecting people in situations where expected actions produced unexpected results. Unfortunately, with improper design work, similar human errors occur in industry, but with much less humorous results.

An error rate of 1% is considered acceptable (or even good) for human beings, yet an error rate of 1% for a machine is usually intolerable. Error rates are predictable even though the error rate varies with the particular task, with the frequency with which that situation occurs, with the training that people have received, and with their own past experiences. Error rates for many tasks are well known from historical data.

The cost of individual errors will also vary considerably. Although very expensive errors get the most attention, it is often the low-cost but higher-frequency situations that are the most expensive to industry.

11-1-2 Error Detection Is Difficult

The detection of "problems" (i.e., of error-causing situations) is more difficult than with other ergonomic areas. Identifying error-producing situations is difficult for several reasons. First, there is a strong tendency to expect errors. One hears, "They're only human," "People make mistakes," and "You can't make it idiot-proof" as evidence of the futility of trying to avoid human errors. These lead us to believe that errors should be expected and are usual, that errors are unavoidable, and that an error is not a bad situation.

People tend to be quick to correct their own errors. Often "mini-errors" occur, only to be corrected before they are reported or noticed by others. The correction of these mini-errors, although desirable from an operations standpoint, will prevent the identification of causes of errors and may lead to failure to correct bad situations.

Of course, many tasks do have very low error rates and trends are hard to establish. This inhibits analysis since each error is seen as a random occurrence and common causes do not become apparent.

Trends may be overlooked because it is not always easy to report errors. There is a strong cultural tradition to blame the operator when errors do occur. Just because a task is performed correctly part of the time, that is frequently seen as evidence that the task is well designed, and therefore the only cause of errors is the operator. Yet the error may be induced from a design flaw or it may be the result of a malfunctioning instrument. Punishment does not lead to prevention as much as it leads to nonreporting of errors.

It is common to find that many people make similar mistakes. However, when everyone makes the same mistake, it is really not a mistake. This is more likely a system error rather than an "operator error." The person has been set up to make the error.

The investigation of errors is very important in order to pinpoint the cause and correct that cause. Like most ergonomic situations, it is more profitable first to determine the cause and then to determine solutions for that cause. The proper understanding of error-producing situations and their analysis is covered in the next section.

11-2 ERROR-PRODUCING SITUATIONS: A MODEL PLUS DISCUSSION

Prevention and correction of error-producing situations is often easier when the ergonomist understands when and how errors occur. In this section a model of error-producing situations is described. Following that, several analysis techniques for identifying errors and/or trends in errors are discussed.

11-2-1 The Model

There are three separate stages where errors can occur, as shown in Figure 11-1. These are errors in perception, errors in decision (logic), and errors

Chap. 11   The Human Error   257

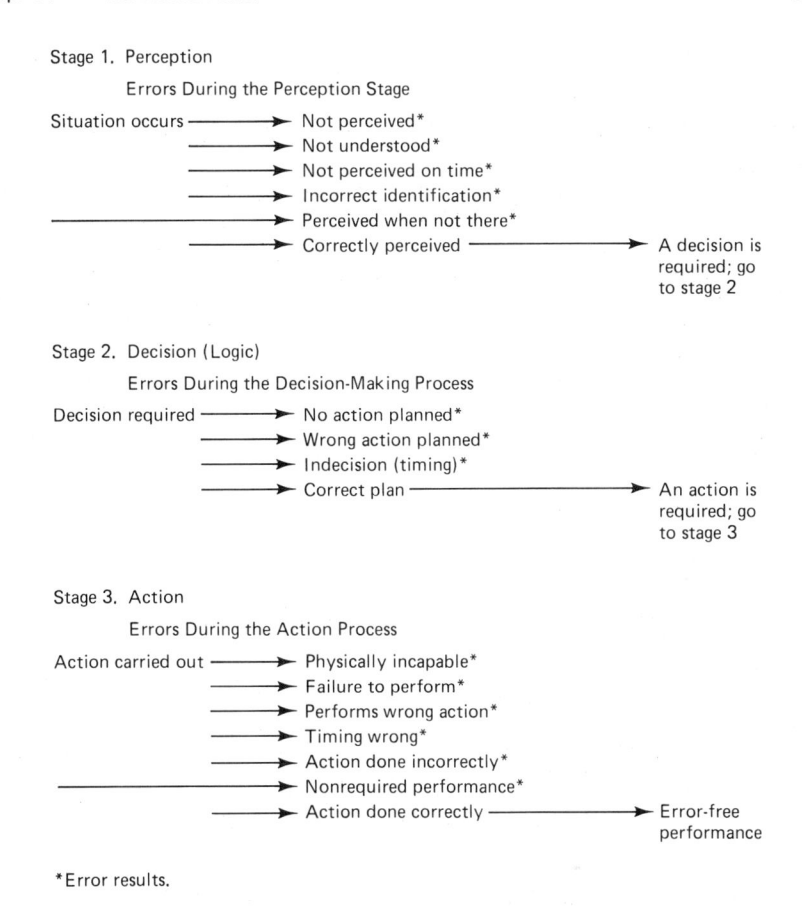

Stage 1. Perception
    Errors During the Perception Stage
Situation occurs ─────► Not perceived*
                ─────► Not understood*
                ─────► Not perceived on time*
                ─────► Incorrect identification*
                ─────► Perceived when not there*
                ─────► Correctly perceived ─────► A decision is
                                                   required; go
                                                   to stage 2

Stage 2. Decision (Logic)
    Errors During the Decision-Making Process
Decision required ─────► No action planned*
                  ─────► Wrong action planned*
                  ─────► Indecision (timing)*
                  ─────► Correct plan ─────► An action is
                                              required; go
                                              to stage 3

Stage 3. Action
    Errors During the Action Process
Action carried out ─────► Physically incapable*
                   ─────► Failure to perform*
                   ─────► Performs wrong action*
                   ─────► Timing wrong*
                   ─────► Action done incorrectly*
                   ─────► Nonrequired performance*
                   ─────► Action done correctly ─────► Error-free
                                                         performance

*Error results.

Figure 11-1 Model of human error situations.

in action. An example with this model for a person performing word processing requires that (1) a misspelled word is seen and known to be incorrect, (2) a decision is made to correct the spelling and that the correct spelling is determined, and (3) that the correct actions are taken to correct the word. If there are any flaws along the way, such as not recognizing the incorrectly spelled word, not knowing the correct spelling, or not understanding the word-processing software enough to correct the word, an error will result.

Operational errors will occur at one of these three general areas. Different tasks are more prone to errors at different stages, so determining the cause of errors should begin with this model. The next section expands this model by listing and discussing the types of errors that occur at these three general stages. There are many causes for the errors that occur, and they are discussed specifically in Section 11-3.

Perception. The first stage is where the situation that requires action is perceived by the person. There are five errors that can occur during the perception stage.

The first error is to fail to perceive or discriminate (see, hear, feel) a correctly presented signal, for example, overlooking a radar blip on a screen.

The second type of error is to fail to understand the signal or that it is important. A misunderstood signal on a control panel is an example of this. The signal is clearly seen, but it is not understood to be important or to require action.

The third type of error is one of timing. The signal is perceived, but not in a timely manner. Temporarily overlooking an emergency light is an example of this.

The fourth type of perception error is to identify the signal incorrectly. A signal is seen, but the wrong one is detected. This might occur when, as an emergency light flashes, the one next to it is perceived to require action.

The fifth type of perception error is a false signal, that is, to falsely perceive a signal that is not actually there. The false signal often causes incorrect action. Inspection tasks are loaded with false signals, where the inspector falsely believes a part is defective and rejects a good part.

The only correct result of the perception stage is to perceive and understand a situation where an action is required. Then one can go on the next stage, which requires a decision to be made.

Decision. The second stage of this model requires the person to make a decision about the situation. There are three types of errors that can occur during the decision-making process.

The first error is to plan no action, when action is indeed required. Not responding to a high-pressure light illustrates this.

The second error is to plan the wrong action from among an array of possible actions. Many situations have several possible actions, with complex decision rules for choosing among them. Following the wrong logic or not understanding the choices typify this error.

The third type of error is one of timing. Making the correct decision, but making it too late, is an operational error. Deciding on the correct part after the assembly has gone by illustrates this.

The only correct result of the decision stage is a timely and correct plan for action. If the decision stage (and the perception stage) is handled correctly, an error can occur only at the action stage.

Action. The third stage of the model is the action stage. Here the person must correctly carry out the action planned in the second stage. There are six types of errors that occur during action processes.

The first type of error is the physical inability to take the correct action.

The person might not be able to reach the correct control switch. Or the control valve may require such high forces that the person is not able to turn it.

The second type of error occurs when the person fails to perform the required action, although not due to physical inability or to an intentional desire to fail. Interruptions, forgetting, and other task requirements can cause this error.

The third type of error occurs when the person takes the wrong action. For example, the person turns the wrong control valve or pushes the wrong typewriter key.

The fourth type of error is one of timing. The correct action is taken, but the action is either too early or too late to be effective. Video games are particularly exciting because the element of timing is required for correct action.

The fifth error type occurs when the action is done incorrectly. There are several subsets for this error. The error of omission occurs frequently, where an operational step is omitted or a part is left out during assembly. The error of addition involves adding an extra part or step when it is unnecessary. Substitution is the use of a new step in place of the correct step. The error of sequence involves using all the correct steps, but taking them in an incorrect order. A transposition of adjacent steps or numbers in series is a classical sequence error. One last action stage error is the performance of a nonrequired step, such as changing a control setting when it is not required. The correct end to the action stage is the correct performance of the planned action.

11-2-2 Analysis of Error-Producing Situations

The analysis of error-prone situations is required to determine the causes of problems prior to developing solutions. Causes are easier to see and solutions are easier to develop when there are abundant error data.

Often, however, error data are incomplete or inaccurate. Errors may be ignored, or they may be corrected before anyone knows about them. Systems need to be established to catch the errors so that the proper analysis can take place. These systems need to take the fear out of reporting an error. Blaming the person who reports the error often results only in failure to have any more errors reported, rather than in their elimination.

When is a situation bad enough to warrant extensive attention? Error rates vary widely and so do the costs of errors. Examining the costs of rework should provide a starting point in the determination of fruitful areas.

There are four data sources that can help provide the information needed to analyze for cause. The first source is customer complaints. When a customer complains, the problem is real. These complaints will not reveal all the errors that are occurring, but they will reveal the more serious ones. The information may provide a starting point to search for errors on the job.

The second data source is empirical data. These gross measures will reveal the total number or cost of errors without tying them down to specific tasks or action. One good empirical measure is the "yield" rates for a product, where the total output is factored by the total input. Comparing yields for parallel lines, or with competing products, can identify weaker areas. Trends in production rates or operating costs can also help to pinpoint weak areas. Whenever there are weak areas, data can be investigated further to help clarify causes.

The third data source is existing error data. While dependent on the quality of the data, analysis of past data can pinpoint weak areas. Accident data are frequently used in this way. The larger volume of data is usually helpful in identifying trends. Some common areas of problems surface at the same point in the operating process, at certain times in the work shift, with particular crews or operators, and so on.

If the data are inaccurate, better investigative techniques need to be established and implemented. Although not helpful now, they will help in the future.

The fourth source of data is called "near-miss investigation." It involves asking: "When did an error almost occur?" The responses to this reveal weak points in the operational system, as well as potential solutions to problems. People are much more open to this type of question, because there was no actual loss involved, and they can report what they did to save a bad situation. In some cases, especially where accidents are usually fatal, near-miss investigations are the only way to gather firsthand data about the error. Many early human factors design flaws in aircraft were discovered with this tool.

11-3 ERROR-PRODUCING TASKS

Seven general tasks that often result in errors in industry are discussed below. Table 11-1 presents a matrix showing the more likely errors for each of these tasks. A description of the seven general tasks, some examples, classic causes of errors, and some standard solutions and preventive measures are presented in this section.

11-3-1 Control and Display Tasks

In control and display tasks, the operator receives information from displays on an instrument panel. Based on information from these signals, a diagnosis is performed and actions are planned. The actions are usually carried out by adjusting controls on the instrument panel. The information is normally in real time; the response to the action may be in real time or it may lag the action taken.

TABLE 11-1 Common Errors for Given Tasks[1]

	Control and Display Tasks	Labelling and Sign Usage	Inspection Monitoring	Information Processing	Forms Usage	Instruction	Assembly
Stage 1: Perception							
Not perceived	1	2	1	3	2	3	3
Not understood	1	1	2	1	1	3	3
Not perceived on time	1	2	1	2	3	3	1
Incorrect identification	1	2	1	2	3	3	3
Perceived when not there	2	3	1	2	3	3	3
Stage 2: Decision (logic)							
No action planned	2	1	2	2	1	2	1
Wrong action planned	1	2	1	1	1	1	2
Indecision (timing)	1	3	1	2	3	2	1
Stage 3: Action							
Physically incapable	3	3	3	3	3	3	2
Failure to perform	1	2	2	1	3	2	2
Performs wrong action	1	2	2	2	2	2	2
Timing wrong	2	3	3	3	3	3	1
Action done incorrectly (omission, addition, substitution, sequence)	1	1	2	1	1	1	1
Nonrequired performance	2	3	1	2	3	3	2

[1] 1, Primary concern; 2, secondary concern or consideration; 3, not typical.

Examples of use. Control and display tasks are common in control rooms for the petrochemical, chemical processing, and electrical power industries. With more sophisticated manufacturing techniques, however, the use of control and display tasks will become more common in traditional industry as well.

The classical situation involves monitoring a number of displays for unusual conditions. Although normally inactive, the job is punctuated with bursts of high activity. Normal operations are routine to the point of being boring, resulting in vigilance decrement. When an upset does occur, it may not be immediately perceived unless it is large enough to trigger an alarm. Upsets often come in clusters and the conflicting demands for action may overstress a person. Changeovers occur irregularly, and provide only occasional bursts of activity.

Causes of errors and solutions. Classical causes of errors are poorly designed displays, confusing controls, inappropriate control–display relationships, and questionable layouts.

Display design has been the subject of much research. Some of the more common problems are illegible or small displays, using the wrong type of display for the information needed, using look-alike displays for widely varying functions, and having displays obscured by glare or hidden from the line of sight. An evaluation of several displays is shown in Figure 11-2.

Controls may be hard to grasp or handle, controls may have ill-defined or excessive set points, and controls may have confusing or inconsistent motions (stereotypes) to achieve desired actions. Controls include rotary knobs, pushbuttons, foot pedals, thumbwheels, toggle switches, hand wheels, and cranks. Rarely will the practitioner be forced to actually design these controls, and if that is needed, there is ample information on the design of these controls from other sources. More often, the practitioner will be asked to specify the appropriate control for a given situation. Table 11-2 compares several controls, and Figure 11-3 and Table 11-3 describe operational stereotypes.

Controls and displays are usually related, yet the relationship between the two can be confusing. Appropriate grouping procedures should be used to keep them linked together. Grouping procedures include placing controls under displays, and marking off logical groups of controls and displays.

Layouts can help prevent errors. For example, displays must be readable from where the operator normally sits and in the operator's normal line of sight. Controls should be usable in the normal standing (or sitting) position and near the hands. Controls certainly should be located near the instrument display.

Other problems include the monitoring of too many displays, and a similarity of alarms, so that one sounds like another. These both cause problems in the perception of the correct problem. When major problems occur in con-

Chap. 11 The Human Error   263

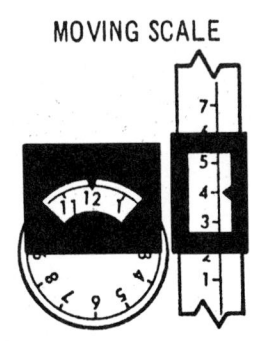

COUNTER   **MOVING POINTER**   **MOVING SCALE**

Display	Quantitative Reading	Qualitative and Check Reading	Setting Number	General Use
Counter	Excellent	Fair (changes not detected)	Good	Fair
Moving Pointer	Fair	Good	Good	Good
Moving Scale	Fair	Poor	Fair	Fair

Figure 11-2 Evaluation of types of displays. *Source:* Adapted from Harold P. Van-Cott and Robert G. Kinkade, eds., *Human Engineering Guide to Equipment Design* (Washington, D.C.: U.S. Government Printing Office, 1972), p. 82.

TABLE 11-2 Comparison of Control Types

Attribute	Rotary Selector Switch	Thumb-Wheel	Push-Button	Pedal
Number of control settings	Up to 24	Up to 24	2	2
Space requirements	Medium	Small	Large	Large
Chance of accidental activation	Low	Low	High	Medium
Ease of visually identifying position	Good	Good	Poor	Poor
Speed	Medium	Medium	Excellent	Excellent

Source: Adapted from Harold P. VanCott and Robert G. Kinkade, eds., *Human Engineering Guide to Equipment Design* (Washington, D.C.: U.S. Government Printing Office, 1972), p. 579, Table 8-3.

Figure 11-3 Associated direction of motion for increases.

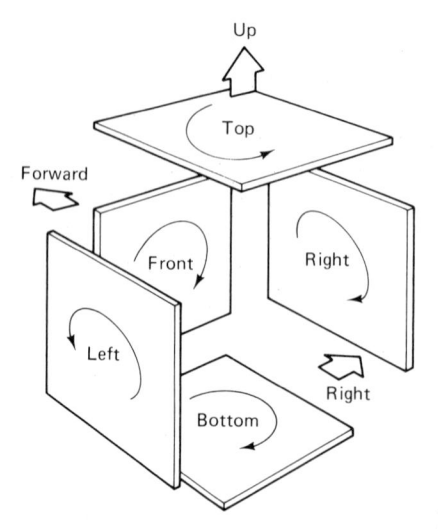

Curved arrows point to direction of increase

trol rooms, the displays all compete for attention. This contributes to errors in making decisions and in carrying out necessary actions correctly.

The solutions to these problems are to purchase only instruments that are well designed from the ergonomic standpoint. When complex instrument panels are assembled, the control-display relationships and layouts should be carefully planned. Computerized control systems can be used to display only important information and allow background information useful for diagnosis to be called up as needed. They can also create an order for major problems by displaying critical items first and blocking out less important information until the major items are corrected. Also, computerized workplaces can reduce a roomful of displays to a more convenient console.

TABLE 11-3 Operational Stereotypes

Following is a list of associated up-down expectations for controls. The first position is the up position and the second position is the down position.

On-off	Start-stop
High-low	Up-down
In-out	Fast-slow
Raise-lower	Increase-decrease
Open-close	Accelerate-decelerate
Forward-reverse	

Additional reading

EDWARDS, ELWYN, and FRANK P. LEES, eds., *The Human Operator in Process Control.* London: Taylor & Francis Ltd., 1974.

Military Standard 1472B, *Human Engineering Design Criteria for Military Systems, Equipment and Facilities.* Washington, D.C.; U.S. Department of Defense.

ODOM, JAMES A., *Applying Manual Controls and Displays—A Practical Guide to Panel Design.* Minneapolis, Minn.: Micro Switch—a Honeywell Division, 1984.

VAN COTT, HAROLD P., and ROBERT G. KINKADE, eds., *Human Engineering Guide to Equipment Design.* Washington, D.C.: U.S. Government Printing Office, 1972.

WOODSON, WESLEY E., and DONALD W. CONOVER, *Human Engineering Guide for Equipment Designers,* 2nd rev. ed. Berkley, Calif.: University of California Press, 1965.

11-3-2 Labels and Signs

Information presented on lables (and signs) is read, processed, and acted on. Labels and signs are used almost everywhere, and their use is assumed to be effective. Yet they can confuse or provide ambiguous information. Format and size also cause user problems. An additional concern is that they need to be effective for unfamiliar users as well as providing information for experienced users.

Examples of use. A label may highlight a situation requiring action, may outline decision steps or alternative actions, may give instructions for action, and may identify a container. Labels can provide handling information or directions for the use of a product.

Signs are also widely used to provide instructions and directions. They provide "labels" where there is a large distance involved.

Causes of errors and solutions. Labels and signs are used almost everywhere, yet they are not often checked for effectiveness. They may not be readable, their message not understandable, and their directions confusing or ambiguous.

Some common problems with labels and signs are the size of the letters and their readability, the contrast between the letters and background, the understandability of the message, and the use of error-prone coding systems such as long strings of digits and confusing alphanumeric codes.

The size of letters is a function of the viewing distance of the label or sign. Table 11-4 provides information on the letter height relative to viewing conditions and distance.

Readability of the message is enhanced with contrasting colors between the message and its background. Table 11-5 provides appropriate contrasts.

Message understandability is a complex issue, especially as international trade expands. The use of graphics to convey certain messages is increasing,

TABLE 11-4 Height of Letters and Numbers (inches)[1]

	Low Light Levels	Adequate Light Levels
Critical labels (information, data, emergency labels)	0.25	0.15
Noncritical labels (nonemergency labels, instructions, identification labels)	0.10	0.10

[1] These values are for a standard 28-in. viewing distance. Increasing the viewing distance will proportionately increase the letter height.

Source: Adapted from Harold P. VanCott and Robert G. Kinkade, eds., *Human Engineering Guide to Equipment Design* (Washington, D.C.: U.S. Government Printing Office, 1972), p. 89, Table 3-11.

although all graphics are not universally recognized. Testing for understandability with the user population is essential. Symbols should also be provided with a readable message.

Additional reading

Military Standard 1472B, *Human Engineering Design Criteria for Military Systems, Equipment and Facilities.* Washington, D.C.: U.S. Department of Defense, 1974.

VAN COTT, HAROLD P., and ROBERT G. KINKADE, eds., *Human Engineering Guide to Equipment Design.* Washington, D.C.: U.S. Government Printing Office, 1972.

WOODSON, WESLEY E., and DONALD W. CONOVER, *Human Engineering Guide for Equipment Designers.* 2nd rev. ed. Berkley, Calif.: University of California Press, 1965.

11-3-3 Inspection and Monitoring

In both inspection and monitoring situations, the human being is used to detect unusual situations, often from among confusing signals or highly

TABLE 11-5 Appropriate Color Contrasts for Labels and Signs[1]

	Letters					
Background	*Black*	*Blue/Green*	*Red*	*Orange*	*Yellow*	*White*
White	5	4	3	2	×	×
Yellow	4	3	2	1	×	×
Orange	2	1	1	×	1	1
Red	2	1	×	×	1	2
Green/blue	×	×	1	1	2	3
Black	×	×	1	2	3	4

[1] 5, Very good readability; 4, good; 3, adequate; 2, poor; 1, very poor; ×, avoid.

irregular data. The perception of the proper signals from among the "noise" is critical. Low-level signals, ambiguous signals, and vigilance decrement make this type of task difficult.

Numerous studies in the literature show that the percentage of defects detected can be as low as 30 to 40% of all defects. At the same time, the number of good items rejected (or false signals) can be as high as 10% of the total defects discovered.

Examples of use. Inspection and monitoring tasks all involve long periods of observation, during which time few defects or signals are found. Examples of monitoring include operators in control rooms waiting for signals or security guards watching doors or video cameras. Inspection tasks include both the inspection of continuous processes, such as a plate glass line, and the inspection of discrete products, such as circuit boards. Proofreading is also an inspection task.

For these tasks, true performance levels are hard to establish and monitor. Secondary audits are not always effective (since the second inspector is subject to the same errors that the first inspector made), and hidden defective samples may not be "hidden" well enough to be reliable. Therefore, feedback on performance is infrequent and often inaccurate.

Changes in the rate or number of defectives found may be due to changes in the product quality or to changes of the inspector's level of sensitivity. Fatigue and boredom certainly reduce performance, but quantitative reductions in performance are not predictable.

Causes of errors and solutions. One classical problem is vigilance decrement, where detection performance deteriorates with increasingly better levels of products. In other words, the fewer the actual defects, the lower the percentage of defects that are found. Since the inspector has little physical activity (and therefore is assumed not to be too busy), the inspector may be called on to monitor additional product lines, thus diluting an already ineffective performance.

Training of inspectors is hard to evaluate, especially when the training was completed some time ago. Often there are judgment calls to determine whether or not something is a defect. Standards of judgment may drift over time. Variability among inspectors is often present. Using "standard" defectives so that all inspectors can calibrate themselves or check questionable decisions is good practice. Rotation of inspectors to offset fatigue and vigilance decrement may be useful, provided that there are enough fully trained inspectors.

Specialized lighting is useful for highlighting defective parts. Grazing a surface will highlight surface irregularities. Glass inspection is done by observing shadows from a light shown through the glass rather than by looking directly at the glass.

Designing inspection jobs so that inspectors search for a few defects is better than having them inspect for a large number of defects. People tend to look only for a few defects anyway, so asking someone to inspect for too many defects will not produce an effective inspection.

Visual acuity for both static and dynamic vision should be assessed. Dynamic visual testing is critical for inspection of moving parts. Proper work site design and visual conditions are very important to good inspection performance. Distractions from a fatiguing work site will quickly affect inspection performance.

Additional reading

DRURY, C. G., and J. G. FOX, eds. *Human Reliability in Quality Control.* London: Taylor & Francis Ltd., 1975.

HARRIS, DOUGLAS H., and FREDERICK B. CHANEY, *Human Factors in Quality Assurance.* New York: John Wiley & Sons, Inc., 1969.

11-3-4 Information Processing

Number and word processing requires the handling of large volumes of data. Although the processes are largely routine and the data are normally handled with a very low error rate, the extremely high volume of data results in enough errors to be troublesome. Action stage errors predominate here.

Examples of use. The world of number and word processing is increasing every day. Common number- and word-processing tasks are keypunching, data entry, coding information, and typing. Characteristically, experienced people involved in handling large volumes of data have a very low error rate. Yet with the high volume, some errors occur and can be difficult to find and correct.

Causes of errors and solutions. Gallagher[1] discusses short-term memory and the rapid deterioration of this information with time. He states that this information is vulnerable to displacement by new information, although it can be retained as long as full attention is given to it by rehearsal.

Errors in information processing are primarily stage three errors. Long strings of consecutive digits and alphanumeric codes especially cause transposition, omission, addition, and sequence errors. Gallagher also covers many design guidelines, as noted in the following paragraphs.

Letters are easier to recall when they are pronounceable, or approximate the English language. When possible, have vowels between consonants to ease

---

[1] C. C. Gallagher, "The Human Use of Numbering Systems," *Applied Ergonomics,* December 1974, p. 220.

pronunciation. Random letters are more difficult to recall than random numbers.

The use of both alphabetic and numeric characters will increase the quantity of symbols and provide a wider range for the code. For example, numeric symbols provide 10 symbols and the alphanumeric system provides 34 symbols, even after the letters I and O are removed. An advantage of using both numbers and letters is that they can be arranged in a pattern to mark off certain fields of information without having to resort to hyphens or spaces.

Most people will break up long strings of digits into natural subgroups of three or four numbers. Those who do not use these natural subgroups have higher error rates. Providing long numbers already broken up into these natural groups will enhance their usability. Good grouping of data can increase the data-entry speed by as much as 20% without causing an increase in the number of errors created. Caplan has also worked extensively in code error reduction, and his guidelines for the design of codes are given in Table 11-6.

It has been common practice to verify certain information as a means to improve accuracy. This expensive practice acknowledges the weakness of human beings in this area, yet is usually ineffective. However, some recent innovations are augmenting human capabilities in these areas. Bar-coding systems are now used to identify products faster and more accurately than people can. Other vision recognition systems are becoming more practical.

Additional reading

CAPLAN, STANLEY H., "Guidelines for Reducing Human Errors in the Use of Coded Information," *Proceedings of the Human Factors Society, 19th Annual Meeting* (Santa Monica, Calif.: Human Factors Society, 1975), pp. 154-158.

TABLE 11-6 Coding Error Considerations

General structure
  All codes should be the same length
  Numeric codes are preferable to alphanumeric codes

Numeric code structure
  Length should not exceed four or five digits
  Longer digits should be grouped in threes and fours, and separated by a hyphen
  For handwritten codes, avoid using either 0 or 6

Alphanumeric code structure
  Use a specific position for numbers and letters to avoid look-alike substitutions
  The letters B, D, I, O, Q, and Z and the numbers 0 and 8 should be avoided if they can be mixed up
  Where possible, use pronounceable words and syllables instead of random letters

Source: Adapted from Stanley H. Caplan, "Guidelines for Reducing Human Errors in the Use of Coded Information," *Proceedings of the Human Factors Society 19th Annual Meeting, 1975* (Santa Monica, Calif.: Human Factors Society, 1975), pp. 154-158. Copyright 1975 by the Human Factors Society, Inc. and reproduced by permission.

GALLAGHER, C. C., "The Human Use of Numbering Systems," *Applied Ergonomics,* December 1974.

11-3-5 Forms and Paperwork

These tasks involve the collection of information, often from a variety of people. Decision stage errors predominate, especially when the form or the requested information is unfamiliar to the user.

Examples of use. Forms are used to assemble specific information in a specific format. Questionnaires and surveys are often "one-time use" documents used to gather information from a wide variety of people.

Order forms are used routinely to collect and transmit information. They must ensure that specific data are provided in a uniform format. This form collects information from many people, but for rapid processing, the information should be complete and uniform.

Log sheets and check sheets are used to gather routine data on a frequent basis. They need little explanation since the users are familiar with the form and there is maximum space for the information.

Causes of errors and solutions. The causes of errors come from misunderstandings about the information desired and how it is to be entered. Questions are often unclear to the user.

Forms should have a title at the top of the page. If there is not one, the users will soon make one up and use it informally. Unfortunately, they may make up several different titles, resulting in confusion. The author should use words familiar to the form user.

Instructions should be provided on the forms, if possible, especially for order forms, where specific information in uniform formats is required. Instructions and examples can be printed on the back of the form, or the instructions can be interspersed among the requested information.

The information should be planned and sectionalized to group similar information. For example, on purchase orders, vendor information such as name, address, and billing information should be grouped together. Each section should be defined and separated from the other sections by heavy lines. The groups should be placed in logical sequences on the form.

Each section should have a short title, for example "Vendor Information." The information requested should be captioned when possible. For single pieces of information, a box design with a built-in caption is best, and for repeated data, column headings are best. These are illustrated in Figure 11-4.

Ensure that there is adequate space for the answer requested. The space should be suitable for handwritten or typewritten responses. Do not automatically extend the information space to the edge of the page, since that may

NAME	DATE
STREET ADDRESS	
CITY AND ZONE	STATE

QUANTITY	DESCRIPTION

Figure 11-4 Box caption and column heading caption.

imply more information than is actually needed. Boxes, as shown in Figure 11-5, for character information are better for both the reader and the writer. If the information is to be read at a distance, increasing the size of the boxes is a suitable way to increase the size of the writing.

Like all designs, the form should be tested and revised before it goes into widespread use. The testing should be done with representatives of the user population. Colored forms have little value if the information is to be photocopied. If they are required for normal transactions, and may infrequently be copied, the pink form should be labeled somewhere "This is the pink copy."

Shealy[2] noted an interesting effect in the design of a form to capture accident information. He listed over 300 possible agents of accidents and/or injuries. On the finished reports, however, he discovered that the closer an item was to the beginning of the list, the more likely it was to be chosen. People were unwilling to search the entire list for the best agent and were willing to use the first agent that came close. He solved this "by giving the major categories in an organization chart format so that the person filling out the form would first locate the general category and then search for the specific."

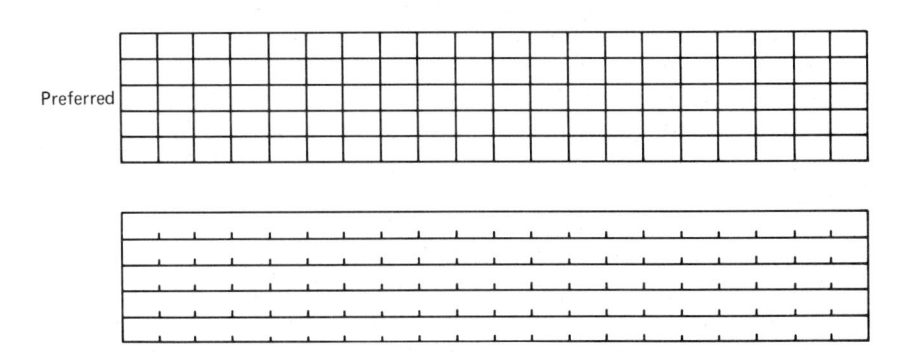

Figure 11-5 Examples of character information boxes.

[2]Jasper E. Shealy, "Impact of Theory of Accident Causation on Intervention Strategies," *Proceedings of the Human Factors Society, 23rd Annual Meeting, 1979* (Santa Monica, Calif.: The Human Factors Society, 1979).

11-3-6 Instruction

Instructions may be presented orally or in writing. The clarity of instruction is often overestimated by the presenter, especially if the presenter is well experienced at the task. Decision stage errors predominate from not knowing the alternatives available and/or how to choose from among them. Action stage errors may also occur when actions are unintentionally performed incorrectly.

Examples of use. Instructions are found in everything from operations training to assembly of purchased toys. Instructions are used for self-assembly of products, for references during operation of equipment, and for the repair and service of equipment.

After initial exposure and experience, the instructional information is often kept as reference documentation when problems crop up. The material is used frequently at first, and then is used less and less as time passes.

The users of instructional information vary widely in their knowledge and experience. Some are novices with no background, whereas others have extensive experience. Writing for this broad an audience is difficult at best.

Causes of errors and solutions. Design the information for the naive user. The experienced user can pick over the unneeded information, but the novice will not be able to add the information that he or she needs.

In the instructions, use words familiar to the user population. Oral information should be preplanned and documented to ensure that the best words are used.

The instructions should be sectioned into natural groups. For example, all instructions about equipment setup should be grouped together. The groups of information should be sequenced for presentation. The physical process provides a natural and understandable sequence. Overview all the major sections at the beginning, to provide a framework for the specific instructions.

The sections should have information to tell the user when that step of the process or section of the instruction is complete. A comment such as "When the bulb lights, go on to section 14" provides both a criterion for completion and directions on what to do next.

The instructions should provide shortcuts for experienced users. It may list steps that can be omitted. The operating instructions may have comments for alternative sequences if certain conditions are met. A good table of contents and/or index will help the infrequent user find the needed information quickly. The instructions must provide alternative actions that might be taken and the decision rules for choosing among them.

Provide diagnostic information to lead people to discover their own mistakes. Tell them what they will see when something is going wrong, and how to correct that problem.

Pictures or drawings of the actual work is very helpful, especially for assembly work. Showing the tools and equipment needed to do the task will also help the person prepare and perform the task better. With adequate preparation, there will be less need to use the wrong tools. Of course, the instructions should be tested on the user population, both for clarity and for completeness.

11-3-7 Assembly

These tasks require a set of specific steps be performed in a sequential order. Action stage errors predominate, especially omission and transposition errors. Some decision stage errors occur, especially when many alternatives for assembly are possible. Timing can be a problem, especially in a paced environment.

Examples of use. Assembly operations can take several forms. Some assembly operations are done on lines where a series of people each perform one step before passing the component on to the next person. Other assembly operations are self-paced, with materials coming from one bin and finished goods being placed in another bin. In other operations, teams of people share groups of tasks and complete whole subassemblies.

Causes of errors and solutions. The problems that arise are often timing (especially when the pace is critical) or errors involving the transposition, omission, or addition of component parts.

Too many choices among component parts can lead to errors and also to increased operating times. Gatchell[3] noted that operators with 10 parts made 46% more errors and needed 13% more decision time than operators with only 4 parts to choose from.

Omission and transposition errors can be corrected sooner if there is a way to make the error obvious to the casual observer, or if there is a way to make the next step difficult or impossible when a mistake has been made. Designs that allow a mistake to go unnoticed should be avoided.

The choice of paced or unpaced work comes up regularly. The evidence seems to favor self-paced rather than machine-paced work. Efficiency will drop off as one works either faster or slower than at his or her own "natural pace."

Relief operators must be considered in assembly work. Gatchell found that relief operators made from three to five times more errors than the regular operators. The training and skill development for relief operators should be given careful attention.

---

[3]Susanne M. Gatchell, "The Effect of Part Proliferation on Assembly Line Operators' Decision Making Capabilities," *Proceedings of the Human Factors Society, 23rd Annual Meeting, 1979* (Santa Monica, Calif.: The Human Factors Society, 1979).

New operators will experience a learning curve as they gain skills on the job. The pace, final inspection, and error feedback all need to be planned for the new operator until satisfactory perform levels are achieved.

Automation is making a clear impact in the assembly area. Automated machines can provide more uniform results and can provide reliable checks of their own work before further assembly. Automatic tests can be used to ensure that all parts are correctly in place, are in the correct assembly, and so on. Bar-coding devices and other vision systems continue to expand the use of automated inspection systems.

11-4 SUMMARY

Human error reduction is a critical element for a successful industrial ergonomics effort. Although many ergonomists concentrate on the physical side of the human being, the reduction of errors is a valuable way to enhance a person's performance in the organization.

One must begin with a belief that people do not have to make mistakes, and that errors can be reduced. With a proper understanding and a suitable diagnosis of the error-producing situations, the problems can be found and corrected.

The situations described in this chapter are representative of what one will find in industry. There are many others, and the ergonomist will find a need for creative solutions to all the problems that will arise.

QUESTIONS

1. Give five examples of human error in industry.
2. Human error situations cannot be corrected until they are detected. Why are human error situations so difficult to detect?
3. What are the three stages of an operation where an error can occur?
4. List the perception stage errors and give examples.
5. List the decision stage errors and give examples.
6. List the action stage errors and give examples.
7. List and discuss the methods of identifying error-prone situations for further study.
8. In control rooms, what task situations make errors difficult to identify and correct?
9. Discuss the causes of errors from displays and controls.
10. For the following situations, what would be the result of a control movement in the up direction:
    Turn on or off?
    Lower or raise?
    Open or close?

Chap. 11   The Human Error   275

Stop or start?
Increase or decrease?

11. What are major causes of errors in the use of labels and signs?
12. A sign will be mounted on a wall 20 feet from an operator. It is a critical sign, and the lighting is normally adequate. How high should the letters be? What are suitable colors for the sign?
13. List four ways to enhance inspection performance.
14. In alphanumeric coding systems, which letters should be omitted because of confusion with numbers?
15. What is the best subgrouping of numbers to ensure accuracy? Give an example.
16. List five ways to enhance the design of forms.
17. List seven ways to enhance the design of instructions.
18. From an assembly standpoint, why is it useful to limit the number of optional parts for an assembly?

Part 4

ADDITIONAL APPLICATIONS OF ERGONOMICS

Chapter 12

THE PURCHASE OF ERGONOMICALLY DESIGNED PRODUCTS

Part 4, which consists of Chapters 12 through 15, goes beyond the classic ergonomic problems. It explains how the ergonomist can have even more influence on the organization by seeking these applications and by integrating ergonomics with other concerns of the organization's management.

Most of an organization's equipment is designed and built outside that organization. So even if an organization has a strong internal ergonomics program, it may find frequent need to redesign equipment to enhance human usability.

This chapter discusses how to purchase ergonomically designed equipment. Vendor assessment, product assessment, product specifications, and vendor relations are included. Specific examples of poor equipment design from the ergonomics standpoint are presented and discussed.

12-1 INTRODUCTION

"This looks like a better way to go," I explained, in my call to the computer department. I had read another article on the ergonomic advantages of using an orange phosphor screen rather than the more traditional green phosphor. I wanted to share what I had found, so that our computer operators could benefit from the company's expertise in ergonomics.

Later, when I checked again, I found that we were continuing to purchase green phosphor CRTs. "Our current vendors don't offer this screen yet" was the explanation, "and other vendors' equipment either isn't compatible with our current equipment, or it is more expensive." Finally, we were able to get the orange CRT, but the effort was far greater than I had ever anticipated. It did not involve any questions about the benefits of the orange screen (everyone agreed as to the advantages); it was our inability to place these advantages high enough on the criteria list so that they made a difference in our purchasing decisions.

Purchased equipment makes up a large part of any organization's resources. So even with a strong emphasis on ergonomic design inside the organization, people may be using poorly designed equipment.

It is appropriate for organizations to buy outside, since self-design is far too expensive and time consuming unless the design is proprietary or not available elsewhere. Organizations normally buy equipment from outside and use it as is or modify it to fit their needs. This effectively limits the number of optional designs to a handful, most of which have poor ergonomic designs. Therefore, people outside your organization may have a substantial influence on the productivity, safety, and health of people in your organization.

One major office furniture manufacturer responded positively to requests about the use of ergonomic design criteria for their office chairs. The basis for their statement was that the chair was adjustable throughout the range of dimensions required for most people. Unfortunately, the chair was very difficult to adjust—it required both a screwdriver and a wrench to make adjustments—and no office personnel had ever adjusted their own chair.

Size and adjustability are common problems with purchased equipment. Or the equipment may require excessive force to operate. In some equipment, replacement parts are too heavy. Maintainability is a common ergonomic concern that is usually overlooked. Compatibility with current operating stereotypes is often overlooked until operating errors begin to show up. The purchase of equipment that meets your ergonomic criteria is just as important as the use of those same criteria for in-house designs.

12-2 DIFFICULTIES IN BUYING ERGONOMICALLY DESIGNED EQUIPMENT

The first step in buying ergonomically designed equipment is to ask for it. Unfortunately, ergonomics may be such a "low criterion" that it just does not affect purchasing decisions. Or purchasing personnel or vendors may not know what you are requesting when you come down asking them to "buy ergonomically designed equipment."

12-2-1 One of Many Criteria

Let's face it, ergonomics is only one of many criteria used in the purchase of equipment. There are many other criteria, such as purchase price, delivery time, production rate, and capacity, that play a stronger role in purchasing decisions. All of these are easier to quantify than "ergonomically designed," and all are certainly easier to understand. When one realizes that only the top five to eight criteria actually influence the decision, it is easy to see why ergonomic considerations are often left out.

To purchase "ergonomically designed" equipment, one must focus on the benefits that ergonomics provides—higher performance, fewer errors, safer operations—and use them to justify one piece of equipment over another. This will move the ergonomic criterion to a higher level, where it will be considered.

Cost is often a major criterion, and few decisions are made without its consideration. Normally, however, only the purchase price is used; the operating costs are excluded. Ergonomics is reflected in the operating costs. The formula

$$\text{total costs} = \text{purchase price} + \text{operating costs}$$

must be used in economic considerations.

It is the ergonomist's job to ensure that increases in operating costs due to ergonomic problems are fully considered. Since people have to operate the equipment, performance rates will affect costs. Maintenance and service costs will be much higher for equipment with poor accessibility. Error rates can vary by a magnitude of 10, so poorer designs will have those higher costs as well. Possible increases in injury rates, problems with labor flexibility, and higher fatigue rates can easily offset minor differences in purchase prices. Yet since few people actively consider these factors, poorly designed equipment continues to come into the plant.

Ergonomics is uniquely a user concern, and therefore the user must make it important to the vendor, the manufacturer, and ultimately, to the designer. Vendors are concerned primarily with the traditional criteria of purchase price, so you have to help them see the influence of total costs and to design with total costs in mind.

12-2-2 Vendor Knowledge

Since ergonomics is a relatively new discipline, many vendors do not have strong knowledge in this area. There are five levels of knowledge that vendors can exhibit. An assessment of their expertise will help you know how to approach each one.

For the following discussion, these definitions will be used. Vendors supply a range of products—equipment, machine tools, hand tools, software, workstations, and so on. The word "product" is used to cover the full range of purchases. The product is normally expected to be a commodity item or at least available from several sources. While the word "plant" is used often, the worksite or office is included in this generic description. The "vendor" is used to represent product design, manufacturing, and supply, even though these may actually be different companies.

Naive. This vendor does not even know what ergonomics is. Consideration of the user tends to be lacking. Approach this vendor with care, explaining that user considerations are important in your organization. Examples of products or equipment with good ergonomic design are helpful to explain your needs. You might also suggest some reading sources or technical courses to develop their base of knowledge.

Aware of ergonomics and learning. This vendor has heard of ergonomics, and knows that it is a new area for consideration. The vendor is still trying to learn if it is important and how it will affect the business world. Usually, the vendor can point to some aspects of the product that reflect ergonomics, but these are at a very elementary level.

Show this vendor that ergonomics is important and how it will affect your purchasing decisions in the future. Suggest possible user considerations for the vendor's products. Discussions with in-house designers and/or users of the product could be useful.

Talks ergonomics. This vendor is aware of ergonomics and will point out those design features that reflect ergonomic design. Without a deep understanding, however, the designs may be superficial, like the ergonomic chair that was not easily adjustable, mentioned earlier.

Normally, the design features were already built in and are just being cited by the vendor. Little extra design has been done to include ergonomics as a design criterion beyond what occurs normally. Ask demanding questions and be careful to examine the ergonomic advantages. They may sound good, yet prove to be illusive.

Evolutionary. This vendor is well aware of ergonomics and how it can help the equipment user. Usually, designs are of an evolutionary nature, where each generation of the product is better than the previous generation. An ergonomic design may mean that the current product is easier to use than the last version.

The focus is on improved designs, with user input to document and verify weaknesses. This vendor will welcome design input and critical comments. Feel free to discuss the use of the product and how it performs in your plant. Inquire about design decisions and particular features of the product. Interactive discussions with users and/or in-house designers are productive.

Best. This vendor is also well aware of ergonomics and the benefits to the user. This company will probably initiate discussions of ergonomics as a selling point for the product. This vendor is proactive and will include ergonomic characteristics before other vendors do.

The design focus is on major improvements, rather than on evolutionary design improvements. This vendor seeks design input and wants to see the product in operation, and to talk to the users. Interactive discussions are valuable, since the vendor actively seeks to upgrade the product. This vendor's designs will eventually become the design standard for the industry, so your input is valuable in both the short and long term.

12-2-3 A Long-Term Commitment

The procurement of well-designed products is a partnership. It does not rest singularly with the vendor or with the purchaser. Vendors normally are not willing to put great amounts of time and effort into this area without knowing that this is what the buyer wants and is willing to pay for. Poor buyer response to ergonomically designed equipment sends a clear message to the vendor—the purchase price is the most important, so focus only on it. The buyer has no obligation to buy appropriately designed products.

Great changes in this area do not occur overnight. But beginning an emphasis on ergonomic design is an important first step. During the early 1970s, when noise criteria became important, there were many pieces of equipment that exceeded the 90-dBA noise limits. Gradually, with increased demand from the purchasers, equipment was designed to operate more quietly and vendors actually began to promote the reduced noise levels as an important feature of their equipment. With continued emphasis, ergonomically designed products will enjoy this same success.

12-3 PURCHASED EQUIPMENT

Purchasing, like Monday morning quarterbacking, is easy to critique. Of course, with use, it is easy to find the flaws in the products purchased. Only by evaluating the product before purchasing and/or developing specifications can those problems be avoided. The role of the in-house designer/user is critical and it is this person who has the responsibility to ensure that the evaluations are done and that the specs are developed. The ergonomist is available to assist in this process.

All too often, purchasing is based on a few criteria such as initial cost and delivery when the actual needs are more complex. Including human usability, maintainability, and serviceability criteria are critical to operations.

12-3-1 Operational Cost Increases

The major areas of losses from purchased equipment with poor ergonomic designs are discussed below.

Problems with initial use of the equipment can slow the payoff. Excessive training time, unusual setup requirements, or adjustments to a new method of operation are situations that slow the startup process.

Difficulty providing service or maintenance for the equipment can prove very expensive over the life of the equipment. Accessibility for maintenance and service are often overlooked, special tools may be required, or standard parts are not used, all contributing to expensive operations.

Less than full utilization of the equipment can also hurt payoff rates. Equipment that has a very long time to develop proficiency, equipment that requires special skills, or equipment that develops fatigue when used for long periods of time can all substantially reduce the efficiency rates.

A hidden cost occurs when less than the full work population can use the equipment. For example, heavy equipment may be hard to use by females or older workers, some equipment repairs require fine dexterity, and tool handles may be sized so that small-handed people cannot use the tools well. These problems require selection of employees over the long term.

Error rates may increase with some equipment or decline with others. This can be a significant cost factor in deciding what equipment to purchase. All these operational costs can quickly add up, and offset differences in purchase prices. The cost formula,

$$\text{total costs} = \text{purchase price} + \text{operating costs}$$

is applicable for use in evaluating alternatives and in determining the "best buy." It is up to the user to specify the needs for the purchased equipment, in terms of both operational needs (capacity, rates, etc.) and human needs.

12-3-2 Scouting for Problems

Specifically, problems with purchased equipment can be related with the six types of ergonomic problems discussed in Chapter 2. Those problems with the design of jobs (endurance and environmental factors) are less influenced by purchased equipment since most job design is done in-house. These six areas are discussed below, based on the magnitude of problems that occur with purchased equipment.

Physical size (anthropometry) problems. These are the most common problems with purchased equipment. Designs that do not seem to fit the user population well are quite common. There are complete misfits, as well as less than fully efficient operations. A difference in fit between the size of the equipment and the anthropometry of the user population is the typical cause.

Some examples of problems:

Use of a computer workstation by a variety of people, which requires easy adjustability.

A maintenance access port that limits the size of person who can reach in to work on the equipment.

A control knob or switch is placed so high on a piece of equipment that the reach required limits the population who can perform the task.

Tool handles too large to fit smaller female hands, resulting in reduced work performance.

Often the problems can be avoided by having easily adjustable equipment throughout the size range for the user population. The size of the user population must be known and then used as part of the equipment specifications. Do not assume that the designer knows the size of your work population or has even considered a wide range of user sizes in the design of the equipment.

Human error (cognitive) problems. Many errors result from the equipment that people use. These are "system-induced errors" and result from the operators having different stereotypes (expectations of operation different from the actual method of operating the equipment).

Some examples of this follow.

A control valve that turns the wrong way

Complex or unusual control-display relationships

Computer software with so many "menus" that people cannot remember the correct choice

A control on a forklift truck that raises the mast, when the same control on other manufacturers' trucks lowers the mast or performs another operation

Purchase of poorly marked parts, resulting in equipment assembled incorrectly

Poor equipment assembly or repair instructions

Another kind of cognitive problem occurs when the new equipment operates differently from existing equipment. The operator must learn new methods of operating while still remembering the methods to use with the existing equipment. Mixtures of operating methods in control room situations are common, especially where expansions or replacements have resulted in mixing brands or generations of equipment. Look for the availability and clarity of repair and instruction manuals for equipment and products.

Strength (biomechanical) problems. These problems are often caused by the operational requirements of the equipment in combination with the capabilities of the population, or task requirements, such as work pace. Fortunately, with the increasing use of powered equipment, this situation is becoming less common.

A few examples:

Purchased bags of manually handled material weighing in excess of 25 kg (the bag weight should be less if the total range of the population is expected to handle the bags frequently)

Replacement parts for equipment that weigh too much to handle and have no provisions for mechanical lifting

Heavy step ladders that one person will carry

Push carts that require too much force to push

Heavy hand tools (e.g., 45-lb-impact wrenches)

The task may have certain requirements, such as torquing-down bolts when rebuilding a pump. The required torque is a function of the person's strength plus the lever arm of the wrench (assuming a field repair without powered tools). The length of the wrenches purchased can determine success or failure for people with lower arm strength.

Manipulation (kinesology) problems. These types of situations show up most during repair and assembly problems. Assembly operations are largely a function of product design, not of purchasing. Equipment repair is very difficult for the purchasing department to assess, and normally will be determined only by using the equipment or from the experiences of other users.

A few situations:

Excessively fine and/or frequent adjustments for heavy tools or parts

Parts that can be inserted wrong during assembly

Static work in awkward positions during service or repair

Repair work required in unusual postures

Frequent repairs or service in inaccessible locations

Look for access ports for repair and service. Self-guiding parts (with pointed rather than blunt ends) and parts that are prepositioned are desirable.

Environmental (external) problems. These are problems with light, noise, and vibration. Lighting is task dependent unless it is part of a total system (e.g., display screen of a computer system). Only as part of the system is it easy to specify.

Noise is commonly specified, especially to be less than 90 dBA. Other noise levels can be specified, but usually with less successful results. One large computer manufacturer responded to requests for printers with lower noise levels (the noise was over 60 dBA in a quiet office environment) with a statement that 60 dBA was under the legally required limits.

Vibration is subject to operating conditions. It will occasionally be specified, usually with mixed results.

A few examples:

Lighting on CRT screens

Noise levels for equipment at normal operating locations and positions

Air conditioning to hold certain temperature and humidity levels (this is done for computers, why not for people?)

Stacker operations in cold environments

Endurance (cardiovascular) problems. These problems are rarely the result of the design of purchased equipment, but more often the result of poor design of jobs, the work pace, and environmental (heat) problems.

Some tasks may require work in unusual postures, thus increasing the blood pressure and the load on the heart. For example, working in bent or stooped postures while pulling maintenance, working on a hot piece of processing equipment such as a boiler, or lifting from a conveyor belt that is too high or too low may each contribute to cardiovascular problems.

12-3-3 Evaluation of Purchased Equipment for Proper Ergonomic Design

Evaluations are the best preventive measures, although they may be difficult to carry out. This section describes the evaluation process.

Although it is the role of the in-house designer to request the purchase of ergonomically designed items, there are two backup systems that should also be used to help ensure proper ergonomic designs. First, operations and maintenance management, when approving the blueprints and specifications for the whole design, must ask questions about usability and maintainability. Management is concerned about operating costs, so this is a natural area for them to ask about. Rather than looking on this as an audit function, these people are simply helping to identify operating problems from their perspectives and experiences.

The purchasing personnel can also be a strong ally. If they support the need to consider the person in the design/purchase of equipment (via your selling of ergonomic concepts), they can initiate questions for the vendor about the usability by people. They are also in a good position to ask in-house designers to provide specifications that deal with human operability.

Although all this sounds good on paper, the problems with it working well are (1) the evaluations of equipment and/or vendors for proper ergonomic design, and (2) the specifications on human performance that make the evaluation possible. Evaluation of equipment for proper ergonomic design is difficult to do, especially before the product is in use. For that matter, evaluation is often difficult even when the equipment is at hand.

Difficulties with evaluations. For a proper evaluation, one must assume that the designer understands proper ergonomic design and has specific expectations for human performance. The designer must also know what aspects of the design require the use of ergonomic concepts in design. If the in-house designer were looking for a computer workstation to purchase instead of build, the designer would, for example, be aware of the correct dimensions and the need to control external glare.

The easiest way to do an evaluation is to see the equipment in use under a variety of conditions. However, before purchasing, some difficulties crop up. It is usually not possible to see the equipment in use under a variety of conditions with different users. In fact, often it is difficult just to see the equipment in use. For that matter, sometimes it may be difficult to see the equipment at all. These are real problems in the evaluation process for a potential piece of equipment.

Even discussing ergonomic design with the vendor may be difficult. To one accustomed to discussing price, delivery, and optional features, the usability issues may be a new topic. The marketing representative may be unfamiliar with ergonomics or may not be aware of ergonomic features in the design. These problems simply serve to mask any ergonomics aspects that the manufacturer may have built into the design. The difficulties with dialogue with the marketing representative make it difficult for the purchaser to determine whether or not the equipment is good from the ergonomics standpoint. If the designer is unsure of what ergonomics aspects of the design are critical for the best operation, then all of the issues noted above cause even more difficulties.

How to evaluate. Evaluation of the equipment is often limited to the vendor brochures and product literature. Do they discuss the use of the equipment and its maintenance? Or do they just discuss equipment design and features? What about discussions of the potential users and the range of people the equipment is designed for? Are there photographs of different people using the equipment—short and tall people, males as well as females, anyone with a disability? One strong feature to look for is adjustability to accommodate a wider range of users. If adjustability is discussed, it should mention the range of adjustability and how easy it is to make adjustments. If the equipment is not adjustable, the working dimensions should be provided so that you can assess the usability for your working population.

Prior experience with that manufacturer will often help you evaluate ergonomic design considerations. Favorable experiences in the past are likely to be repeated. Once a manufacturer begins to utilize ergonomics in the design of equipment, the value of ergonomics is established and its application is hard to stop.

If you have not had prior experience with that vendor, you may be able to talk with someone who has used the equipment under consideration. Often, vendors will make arrangements for you to talk with other users. These users can provide insight into the equipment usability.

The reputation of the vendor can tell you something as well. Vendors with a reputation for first-quality design work tend to consider the human usability of their designs.

Obviously, the best way to evaluate equipment prior to purchase is to have clear specifications for that equipment. With quantitative specifications, it is not only easier to communicate your design needs to the vendor, but it will make it easier for different people to evaluate the product consistently.

Determining those specifications is up to the in-house designer (probably with support from the ergonomist) and it is a difficult task. The development of specifications begins with a determination of who the user will be, what percentages of the possible population to design for, and determining actual user needs. This is detailed in the next section.

12-3-4 Specifications

Specifications are hard to develop but invaluable once they are developed. Good specifications should be quantitative, so that an easy assessment can be made.

Good specifications must also be based on realistic needs of the user. Overspecifying may result in increased first costs, while underspecifying may result in increased operating costs.

Examples of specifications. The specifications can take on a wide range of forms. Some of the more general specifications (and less usable) are "ergonomically designed" and "user friendly." No one knows what these terms actually mean, and it is wise to assume that this design is not functionally better than any others on the market.

Some specifications are dimensional specifications. For example, chair specifications should include such things as seat width, seat depth, and seat height, as shown in Figure 12-1. Similar specifications can be provided for desks and tables, standup work benches, and similar equipment. Specifying bag weights not to exceed 50 lb is also a dimensional specification. These are easy to assess by the vendor and the purchaser. However, they must be interpreted in order to understand their impact on the user population. Who can lift 50 lb? Who fits the dimensions for the chair?

Even better than dimensional specifications are functional specifications, since they actually state what it is that the purchaser ultimately wants. Examples of functional specifications are:

"Usable by 95% of the population," for general tasks

"Reachable by the 5 percentile female," for anthropometric concerns

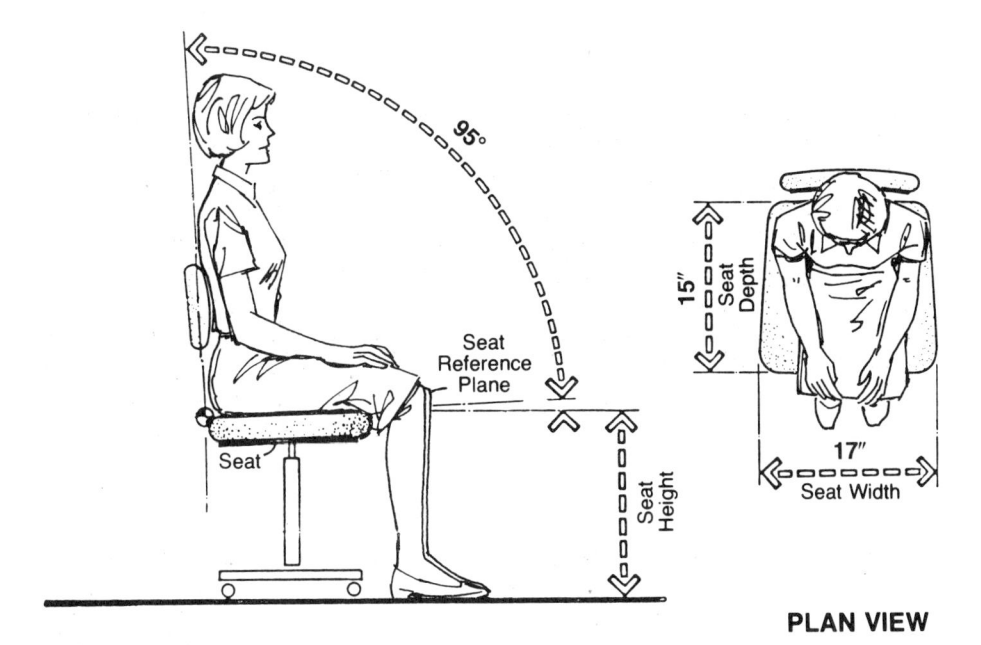

Chap. 12   The Purchase of Ergonomically Designed Products   287

SECRETARIAL CHAIR

PLAN VIEW

Figure 12-1   Drawing of seating specifications.

"Decision time within 2 seconds after alarm sounds," for control room situations

For some equipment, the functional specification might replace another specification. For micrometer calipers, the purchaser is interested in "readability within ±0.0001 in." rather than the previous specification of "accuracy within ±0.0001 in." The functional specification indicates that the person is part of the system and that either the equipment itself or the human user may affect performance.

Functional specifications are difficult to assess even though they are quantitative. Major problems come from the determination of the user population and the required testing to ensure that the proper population percentage can perform the task.

Determining specifications. The process for determining specifications for purchased equipment is the same as determining specifications for in-house designs. Section 7-5-3 discusses the development of design specifications. The steps in that process are shown in Table 12-1.

TABLE 12-1 Use of Anthropometric Data[1]

1. Determine the user population and plan where to get the necessary data about that population.
2. Determine the population requirements, e.g., for anthropometry, to accommodate a reach or a clearance or the middle band of the population. Requirements for strength or cognitive needs can also be determined.
3. Determine the percentage of the population to be accommodated—99%, 98%, 95%, or 90%.
4. Gather the design specifications (e.g., the actual reach height required).
5. Repeat for other design parameters as necessary.
6. As needed, plan trade-offs to meet the needs of the users and the needs of the design.
7. Submit the design specifications to possible vendors.

[1] Detailed descriptions of these steps are given in Chapter 7.

12-4 THE DESIGN OF THE PRODUCT

This section focuses on the processes used by original-equipment manufacturers in the ergonomic design of equipment. The whole area of product design is not covered here; the focus of this section is on the design of products for the industrial/commercial user rather than for higher-volume consumer designs.

The chain of influence now extends past the in-house designer and the purchasing personnel, as noted in Figure 12-2. The chain begins with the users' needs, which are conveyed directly to the in-house designer, or through the ergonomist to the in-house designer. The in-house designer merges the ergonomic design requirements with other design needs and produces equipment specifications. The purchaser utilizes the equipment specifications to locate equipment and to influence the vendor to supply material within the stated specifications (and with the appropriate ergonomic considerations). The vendor utilizes the specifications to influence internal product and equipment de-

```
User ( ⟶ Ergonomist) ⟶
           ⟶ In-house designer
                  ⟶ Purchaser
                       ⟶ Vendor
                            ⟶ Designer/manufacturer
Emphasis on usability and performance ...
```

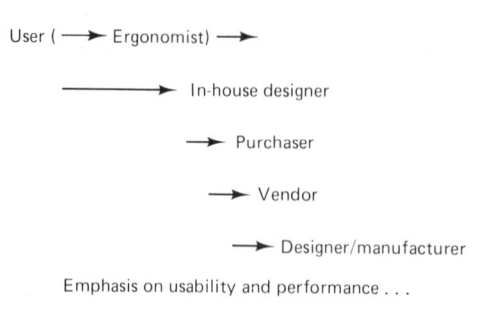

Figure 12-2 From the user to the designer.

sign and manufacture. The product should have an emphasis on human usability and performance.

This is a long trail to follow, and one that should be shortened. There are too many opportunities for the chain to break, resulting in the designer not fully understanding the user's needs. Many of the techniques utilized to test designed products are good for putting the designer in closer communication with the user.

12-4-1 Testing Designed Products

In product design, evaluation of the design is a critical step. The types of evaluation are presented. The easiest method of evaluating a design is to compare it with the design specifications. This is primarily a paper evaluation, and either the design will meet those specifications or it will not.

Another method of evaluation is to mock up the design and test its acceptance with subjects representative of the user population. The mock-up and test provides a real-size model, which helps assess three-dimensional anthropometric fits, as well as giving insight into dynamic movement in the model. The mock-up of a design involves some expense, and so will the testing of a number of subjects. It may not need to be done for each design.

Field trials are the next level of sophistication for evaluation of product designs. Several products are made and sent to customers for use. Evaluations of the ergonomic aspects are conducted by trained personnel while actual users use the equipment. The building of working prototypes is usually expensive. Design changes at this point tend to more expensive than similar changes made earlier.

Ongoing evaluation involves contact with users after the sale of products. Conducted by trained personnel, the evaluation can be expensive, yet it does provide detailed information. This information is useful for the next generation of designs and for critiques of the current design process. These evaluations do not affect the current design unless serious design flaws are uncovered.

The benefits of evaluation are far reaching. Evaluation of product designs and products helps provide better designs. The reputation of those who consistently practice these forms of evaluation is usually higher than that of those who do not.

Evaluation often provides a way to learn things about the product, its use, and even the design process that one could not learn otherwise. It is an effective tool to help create better and better designs.

Finally, through evaluations, one can better predict whether the design will do the job it is intended to do. A good evaluation may also help develop ideas on marketing aspects of the product. Particularly strong usability features should be called to the attention of the potential customer.

12-4-2 Contrasting Product Design with In-House Design

Since the industrial ergonomist may be more familiar with the design processes for designing equipment and facilities for internal use, the differences discussed below are informative for the product design process.

This design process is similar to the process used for developing specifications. One must determine the user population and determine the percentage of that population to design for. The appropriate needs (for reach, clearance, etc.) must be determined.

If the product designer can get many of the specifications from the customer, it will save the time and trouble of developing this material. This is the best way to get this material, but it is not often available.

The differences between in-house design of equipment and the design of industrial/commercial products is marked. A few differences:

- There is usually little feedback on the design. The customer and designer are not in contact with one another. What feedback there is usually is a long time in arriving to the designer. The in-house designer can evaluate the design right after it is built and can easily solicit critiques from the real users.
- The product designer has little chance to redesign, revise, and correct the design once it has started being manufactured. The feedback from the user is usually not available to adjust the design before higher-volume production.
- The product designer has to be more concerned than the in-house designer about product liability lawsuits.
- The designs from the product designer are made faster, with higher volumes, and the time from initial concept to wide distribution is done faster. This requires good forethought to anticipate and correct design problems before they begin to show up in many products.

12-4-3 Relationship to the Customer

Many of the problems between the user and the outside designer can be overcome by close working relationships. This is a win–win situation, and both parties are better off with good communications. There is a lot of useful information that needs to pass between these two. The information begins with the design specifications and continues through product evaluation, including the gathering of information from ongoing evaluations. This information is not available elsewhere, so the designer must be close to the customer.

A philosophy should develop in which the product designer immediately asks: "Who is the customer? Who will use this design?" Following that, the designer must focus on the analysis to determine which aspects of human performance are critical. Finally, the designer should always be looking for ways to increase the percentage of people who can use each design.

This philosophy is similar to that of architects, who readily consult specifications early in each design. Their form of design specifications (door and aisle widths, ceiling heights, etc.) already include the aspect of human usability, so the human being enters the design process very early.

12-5 SUMMARY

The purchase of well-designed equipment from outside vendors is as important as the proper design of equipment and facilities inside the organization. The process of purchasing the best design begins, as usual, with the needs of the users. Those are translated into design needs, which ultimately help determine which products and equipment to buy.

Vendor and product evaluations are important to the success of this effort. Yet these evaluations are difficult since the product may not be available. A collaborative working relationship with the vendors will help develop better designs. The vendor must know how the product functions in the real world, and the buyer wants to have the current product improved in the future. Educating the vendor is a key way to gain support in this effort.

QUESTIONS

1. Why should an industrial ergonomics group with design capabilities be concerned about purchased equipment?
2. Why might a piece of equipment with excellent ergonomic characteristics not be purchased?
3. How can the ergonomist make the consideration of ergonomics more important?
4. When speaking with equipment vendors, what are the levels of knowledge of ergonomics that they might exhibit?
5. List and discuss the types of ergonomic problems that should be anticipated in the purchase of equipment. Give examples.
6. Why is it difficult to evaluate equipment from the ergonomics standpoint before purchase?
7. With the above limitations, how should an ergonomic evaluation of equipment be conducted?
8. Give an example of dimensional specifications and of functional specifications. Which is more appropriate for ergonomic concerns?
9. What methods does the product designer have to evalute a design?

Chapter 13

JOB DESIGN

The use of job design techniques can enhance both job performance and worker well-being. Proper job design must include ergonomics so that people can perform the job. But job design does not stop there. A proper job will also be well designed from the motivational standpoint so that the worker will enjoy the job. The job must also provide value to the organization; that is, the benefits of the job should exceed the costs of the job. These three aspects of job design are presented and integrated in this chapter. The use of people complements the use of equipment to perform tasks and the job designer must determine how the task is to be performed.

13-1 INTRODUCTION

"Three back injuries in two months seems awfully high," said my new client. "We have only had four injuries in the six years since that line was started, and two of those were during the startup phase." An examination of the situation revealed that a new, but particularly difficult task had been added two months ago. The request had seemed simple enough at first. A major customer now required that the cartons be stacked so that the lot numbers were visible on all sides of the pallet. The people at the end of the line were now required to rotate each box before palletizing.

Unfortunately, with box weights of 25 lb, the manipulation to rotate the box was awkward. Working in a stooped position during the first pallet layer and at shoulder height for the final pallet layer was taking its toll on the work-

Chap. 13 Job Design 293

ers. In this case, the added task was "the straw that broke the camel's back." Lifting was not a problem, and rotation by itself was not a problem, yet in combination, the two tasks made the job hazardous.

A job is a collection of tasks formed by a single person. Job design is the process of determining and assembling those tasks. A job may consist of one task repeated indefinitely, or it may consist of a large variety of different tasks.

An effective job is one that a person can do, one that a person wants to do, and one with an output that has value to the organization. Therefore, job design includes knowledge from the fields of ergonomics, motivation, and industrial engineering, as noted in Figure 13-1.

The elements contribute these benefits:

Ergonomics: emphasizes the capability to perform the job, with a focus on injuries and Equal Employment Opportunity possibilities.

Motivation: emphasizes the desire to perform the job, with a focus on continued good performance and a desire to perform the job

Industrial engineering: emphasizes the value of the job for the organization, with a focus on the value of outputs relative to the costs of an effective operation

The inclusion of these three elements is a big job itself and one that won't be completely tackled in this chapter. There is a brief discussion of all three areas, yet the focus will clearly remain in the field of ergonomics. The complete topic of job design is easily a book by itself.

The ergonomist must focus on both the job and the component tasks since reduced job performance or injuries can arise from:

1. A person's inability to perform a required task
2. A person's inability to perform a repeated task many times over
3. A person's inability to perform several tasks in combination

A single task may be so difficult that a person cannot safely perform it even one time. These are the easiest problems to identify. A person may be able to

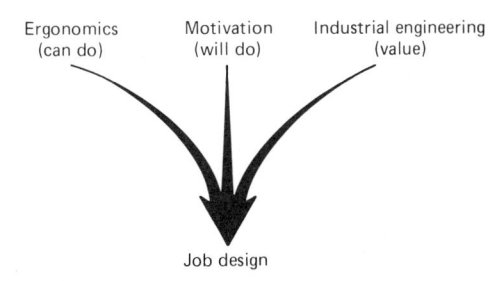

Figure 13-1 Depiction of the three elements of job design.

perform the task once, yet may sustain an injury after repeated performance. For example, putting in one screw is usually no problem, but inserting 100 screws can result in a very sore forearm. Multiple tasks performed in combination may result in problems that each task alone would not cause. This is especially true of endurance situations, tasks requiring monitoring or information processing, and the materials handling job mentioned in the introduction.

The design of the job can have an impact in several general problem areas.

Endurance: This area is strongly affected by the pace and duration of the tasks performed.

Environment: Increasing the time spent in a hostile environment makes it more hazardous to human beings. Task and environmental considerations often combine to cause more serious problems than either one alone could, especially with heat stress.

Strength: The frequency and duration of lifting tasks can turn a reasonable task into an excessive job.

Manipulative: Like strength problems, a task repeated many times can become difficult for the person performing the job.

Cognitive: Over- or underutilization of the person causes problems. Overutilization results in overload errors, and underutilization leads to boredom and inattention and then to errors.

Anthropometric: This is not heavily affected by job design except when a person is confined and works with a bad posture, when the person must repeatedly use access ports that are too small, or when frequent stretching and reaches are required.

Although by definition, jobs are performed by people, tasks can be done by people or machines. In fact, if a task is to be performed, it must be done by either a person or a machine. These two areas complement each other—If a person can't do that task, then a machine must, and vice versa. When designing jobs, it is important to look for machines that can perform some of the difficult, hazardous, repetitive, ornery, and time-consuming tasks that people now perform. These tasks can and should be automated or mechanized to reduce the undesirable aspects of job performance for people.

13-2 UTILIZATION OF PEOPLE

Job design is the process of determining the tasks performed by each person. The utilization of each person is a key ergonomic issue since it affects job performance and injury rates. Overutilization and underutilization of people both cause performance problems.

13-2-1 Optimum Utilization of People

Steve Johnson states the issue clearly: "The goal of ergonomics is *not* to minimize effort; rather, it is to maximize the output with a level of effort that is not harmful to the operator."[1]

Designing a job for optimum utilization, that is, high output without injuries, is difficult. The output may be difficult to measure, effort and injuries may have a complex relationship, and people's capabilities vary widely. Time is often used to determine utilization, even though measures of strength, endurance, and information processing are better predictors of performance for specific jobs.

Variability among people was discussed in the anthropometry section. Proportionally, there are even wider ranges in strength and endurance than there are in anthropometric measures, especially when gender and age are considered. Designing jobs with this range of capabilities is difficult indeed. Jobs that some people find satisfactory for a lifetime cannot be performed for even a single shift by others.

13-2-2 Overutilization of People

Overutilization of people is the more common occurrence. Overutilization comes from efforts to increase productivity, to get an emergency job done, or simply to meet pressing deadlines. Although some temporary increases in performance may not prove harmful, extensive requirements to perform above one's capabilities can be very detrimental.

Exceeding a fair day's work is a good working definition of overutilization. Conceptually, there are three ways to establish a fair day's work. After defining a fair day's work, a discussion of overutilization of the person and some methods of dealing with it is possible. The most common method of determining a fair day's work is time based. It involves measuring the work and adding up work time, break time, and other activities to equal 480 minutes (for an 8-hour work shift). This method of determining a fair day's work is usually adequate, except under conditions of high physical work or where high mental loads are required. With high physical work loads, a person exceeds endurance or strength capabilities before the time-based standard is reached. With high mental loads, a person begins to make mistakes or mental errors regardless of time standards. (It is difficult to measure mentally demanding tasks because it is impossible to observe the thought process.) Figure 13-2 illustrates this concept.

People vary widely on all three of these methods of assessing work loads.

---

[1]Steven L. Johnson, "Workplace Design Applications to Assembly Operations," in *Industrial Ergonomics: A Practitioner's Guide*, ed. David C. Alexander and B. Mustafa Pulat (Atlanta, Ga.: Industrial Engineering and Management Press, 1985), p. 204.

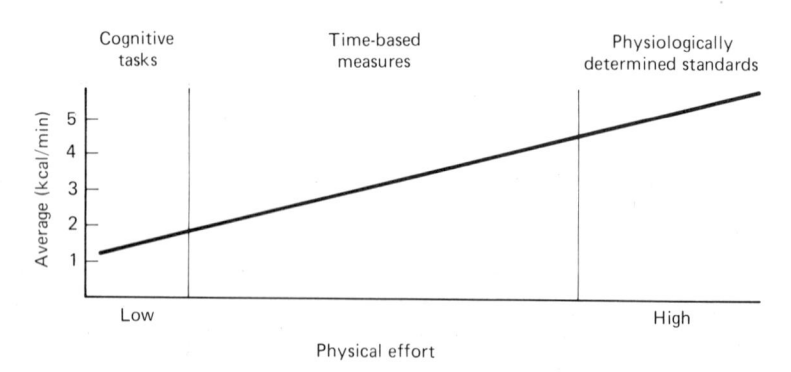

Figure 13-2 Determining a fair day's work.

The capabilities of people vary so much that the same work load forces some people to work rapidly, whereas others do the same job at a relaxed pace. This problem has faced industrial engineers for years while determining labor standards that are fair to both the organization and to each person on the job. Yet time-based measures are the easiest of the three methods to deal with. A fair day's effort for physiologically based jobs is 5 Kcal/min averaged over the 8-hour work shift. A discussion in Chapter 9 provides the background for determining physiologically based standards.

Assessing mental work loads is even more difficult. The rates for processing bits of information and/or decisions per unit time are possible criteria for measuring mental work loads. These are difficult to establish. Acceptable capabilities for people seem to vary quite widely.

In most cases, the answer to overutilization by people is to reduce the work load, slow the work pace, share the work load, select people who can meet the required work pace, increase the rest/recovery periods, or change the task. All of these have some impact on the productivity of the job, on the cost of performing the job, or they make management of the job more difficult by increasing administrative controls for that job.

Automation is often a good solution, especially as equipment becomes more versatile and can mimic human performance. Industrial robots replace strength, endurance, and manipulative functions, and other systems, such as bar codes, replace sensory input functions. Initial expenses may easily be justified, while the risk to human beings is reduced and performance enhanced.

13-2-3 Underutilization of People

While overutilization is the more common problem, underutilization of people can also result in reduced performance. Underutilization of the physical systems is not usually harmful, except where occasional heavy performance is required. For example, firefighters may have long periods of idle time interspersed with brief periods of intense activity. They need some activities, if only physical training, to help maintain necessary levels of fitness. Underutilization in cognitive areas shows up as operating errors resulting from boredom, lack of attention, or vigilance problems.

Increasing the work load with additional tasks can resolve these problems, as long as they do not interfere with the primary jobs of monitoring for infrequent problems and of providing rapid responses. One can blend in other tasks from another job to create additional work and reduce boredom. Routine tasks can be added such as logging information or making rounds for the purpose of providing additional stimulation to enhance alertness. False alarms can be used to avoid vigilance problems and to provide an understanding of the percentages and methods by which real alarms are responded to. Any additional tasks must be deferrable in order to maintain the original need, that is, a rapid response. One thing is clear—these types of jobs should not be automated too much. They are already devoid of stimuli and additional automation will simply accelerate the trend toward boredom and lack of attention. Complete automation, of course, is acceptable.

13-2-4 Solutions to Utilization Problems

The design of tasks can help avoid overutilization and underutilization problems of people at work. Task design includes changes to the work methods, fundamental changes to the product or process, or automation of the task.

The design of jobs provides a broader range of options. The work load can be shared or redistributed among more people. The workplace can be altered. Rest/recovery breaks can be added or redistributed to reduce fatigue buildup.

The most useful improvements to difficult jobs simply involve task reallocation. The jobs can be broken up so that the difficult tasks are shared among more people. Although it may be fatiguing to perform one task for an entire work shift, performing that same task for only a few hours may be suitable. An alternative approach, consolidation of many difficult tasks into one job is sometimes the best approach. By filling this job with a "stronger" person, potentially difficult tasks can be taken out of most of the jobs in the primary promotion sequence. This allows most people to proceed through the promotion sequence without having to perform these few difficult work tasks. Although this may present a few problems in filling one job, it eases the requirements for many other jobs. Often, through pay, work assignment, and supervisory methods, the more physically demanding job can be made desirable enough so that a stronger person will be willing to perform that job. Some physically demanding jobs are eagerly accepted by people wanting to "get into shape" or to build up muscle strength.

13-3 THE JOB DESIGN PROCESS

The job design process provides both a process for the overall design of jobs and a way to design tasks.

13-3-1 Overview of the Job Design Process

A conceptual model of the job design process must involve a careful examination both of the overall job and of its component tasks. Initially, the entire job is examined for possible improvement or elimination. Following that, each of the component tasks is defined and then examined for improvement or elimination. The remaining tasks are then reassembled into completed jobs.

In both cases, the redesigned jobs are evaluated to determine how well they fit in with other jobs in the promotion sequence. Pay rates are determined and "fair workday" issues are resolved. The job design model is shown in Figure 13-3. The major blocks of the model are discussed in the sections that follow.

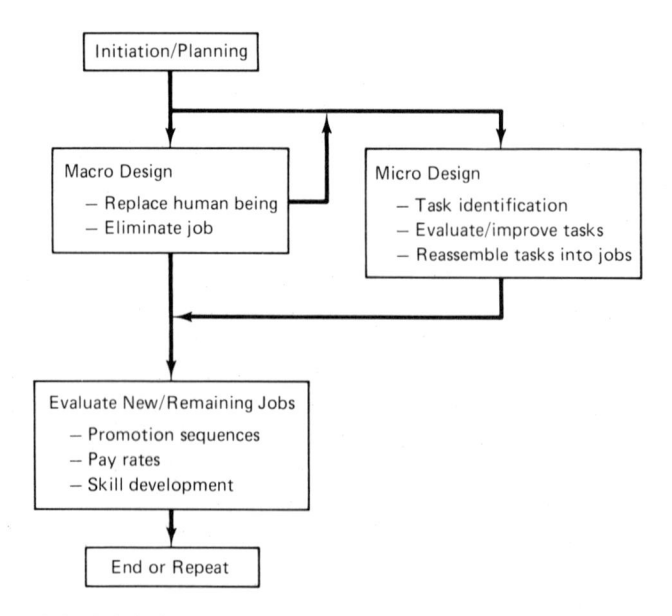

Figure 13-3 Job design model. *Source:* Adapted from David C. Alexander, "Job Design You Can Use," a continuing education seminar, sponsored by the Institute of Industrial Engineers, May 1982.

13-3-2 Macro Design

Macro design is a process of looking at the entire job to see if the job is even necessary. Or perhaps there is no need for a human being to perform the job. There are many situations when we use human labor inappropriately.

Jobs that require excessive amounts of human labor, or jobs that are difficult, dangerous, tedious, or monotonous are strong candidates for macro design. In these situations, replacing the person is appropriate. An example of macro design is to replace the human palletizing bags of material with an automatic palletizer or industrial robot. Spray painting done by an industrial robot is another way to remove the human being from a hazardous situation. In other cases, changing the manufacturing process is required for macro design. When fiberglass boats replaced hand-built wooden boats, the need for tedious and time-consuming human labor to build the product was eliminated.

Macro design is an excellent first step in the job design process because it keeps the ergonomist from improving jobs that simply are not necessary any longer. If one can eliminate the job, there is no need to provide a detailed analysis of each of the component tasks.

About 5 to 10% of the time, macro design is successful in either eliminating the job completely or in replacing the human requirements of the job. Usually, it is worthwhile to explore macro design. If the job is still required after macro design, then micro design is explored.

13-3-3 Micro Design

Micro design is that part of the process that involves looking at each individual task in detail. The first step in this task analysis is to identify each task.

Task identification. Task identification is discussed in Section 6-2-1. Table 13-1 shows the hierarchy of industrial work, which explains that a task is a separate identifiable part of the job.

Evaluation and improvements. Once the tasks are identified, each must be examined to determine ergonomic problems. Problems occur when the task requires abilities beyond those of the expected work population.

Task analysis, as it was described in Chapter 6, is a useful tool in identifying those tasks that result in ergonomic problems because of excessive requirements. Table 13-2 illustrates the process of listing the separate tasks in elements and then of determining the expected work population. Following that, the expected capability of this work population is determined. When the task requires abilities beyond those of the expected work population, a mis-

TABLE 13-1 Hierarchy of Industrial Work

Hierarchy of Work	Typical Measurement Units	Example
Function	Days/weeks/months	Shipping and receiving
Job	Hours/days	Local delivery; invoicing; warehousing
Task	Minutes/hours	Go to post office; load truck; unload parcel
Element	Seconds/minutes	Open truck door; get in truck; sign invoice
Movement	Microseconds/seconds	Reach for door handle; pull open door; grasp pencil

match is identified. Possible solutions to those problems are then listed, evaluated, and implemented.

The development of solutions to ergonomic problems was discussed extensively in Chapters 6 through 11. These chapters provide insight for specific types of ergonomic problems. Section 13-4 explores the use of automation to replace human beings performing tasks or jobs.

Assembling jobs from tasks. The final step in micro design is to reassemble jobs from component tasks. The analysis focused initially on the person's ability to perform the task one time. Now, the ability to perform the task repeatedly is evaluated. Also, evaluation of the performance of each task in combination with other tasks is required. It may be acceptable to perform a task one time, but the cumulative effect may exceed a person's capabilities.

An ergonomic goal, when assembling jobs from tasks, is that the ex-

TABLE 13-2 Task Analysis

Task/Element	Population	Population Capacity	Problem/Mismatch	Possible Solutions
Grasp knob at 76 in.	Male	95% can reach 76.8 in.	No	—
Grasp knob at 76 in.	All	95% can reach 73 in.	Yes	1. Step 2. Lower knob
Lift 15 lb, 10 lifts/min from 20 to 40 in.	All	19 lb at 12 lifts/min from 20 to 40 in.	No	—
Lift 25 lb, 12 lifts/min from 20 to 40 in.	All	19 lb at 12 lifts/min from 20 to 40 in.	Yes	1. Reduce weight 2. Reduce pace 3. Select people

pected work population would be able to perform each task and each job. This avoids barriers in the promotion sequence. Generally, the best approach is to spread out difficult tasks (using job assignment and/or job rotation) so that the cumulative effect is avoided.

In some cases, though, a task will exceed population capabilities. In this situation, rather than spreading that particular task through too many jobs, it should be consolidated into one job and removed from the normal promotion sequence. For example, color matching is a critical task in the paint, fabric, and plastics industries. Approximately 5% of the U.S. work population has defective color vision, so designing a wide range of jobs that require fine color discrimination will only lead to problems later. Color-related tasks should be consolidated into one job outside the normal promotion sequence. This will prevent eliminating people with color-defective vision from the majority of the jobs. A similar process can be used for physically demanding tasks that have proven unusually difficult to improve.

13-3-4 Evaluate

The final step in the job design process is an evaluation of the jobs to determine how well they fit with other jobs. Do they complement other jobs in the promotion sequence? Do they involve a "fair day's work" both for the organization and for the person on the job? Do they have unusual skill requirements, skills that are not developed in earlier jobs, or skills that are not needed in future jobs?

Whether changes were made in the macro or micro design steps, these items are important to evaluate for all the remaining jobs. At this point, pay rates are established and staffing levels are determined. If there are problems at this point in the process, the job may be redesigned by reassembling the component tasks.

13-4 AUTOMATION AND JOB DESIGN: A STRONG MARRIAGE

Earlier it was stated that a task must be performed by either a person or a machine. These two areas complement each other so well that it is difficult to discuss job design without also discussing automation. If a person cannot perform the task, a machine must perform it. If a machine cannot perform the task, a person must perform it.

13-4-1 People or Machines

Usually, when a task requirement is first identified, it is performed by a person, whether on the shop floor or in a laboratory or pilot plant. This is convenient, since the person is flexible and can perform other tasks if not fully utilized. Short term, the person is probably more economical to use than the machine, which must be amortized over longer periods. Eventually, however, it becomes desirable to automate the task or job for economical or safety and health reasons. Economics are based on larger manufacturing quantities or new equipment availability. Industrial robots, for example, are now taking over many jobs that were prohibitively expensive to automate.

The use of automation to complement or replace human labor is particularly apparent in those jobs that involve endurance or manual materials handling. Some of the standard ergonomic solutions to these problems require the selection of employees, reductions in the work rate, modifications to the equipment, or changes in the package sizes. All of these solutions involve reduced productivity on the job or extra administrative burdens such as ensuring that proper selection of employees takes place. Actually, a win-win solution to these problems is to automate the task. It keeps the person from performing a difficult or stressful task that can lead to injury. At the same time, it allows the organization to increase performance rates and enhances productivity. These types of solutions also upgrade the person from that of a manual laborer to an equipment operator, a trend that the ergonomist should favor since it utilizes the higher-order skills that a person brings to the job market.

The rest of this section centers on the evaluation of equipment to replace the human being in difficult and hazardous job situations.

13-4-2 The Innovation Process

The only way to make operational improvements is to match a need for improvement with the appropriate solution to that problem. For example, many ergonomists have talked about tasks being too difficult or hazardous for people to perform. Yet when confronted by management, they were unable to offer alternatives on how to perform the task in question.

Obviously, this problem should be corrected to have the best production operation. At the same time, design engineers may be aware of products that could resolve this dilemma, but are unaware of this application. This is where the marriage of automation and job design becomes critical. The need identified by the ergonomist must be married with the solution known by the designer to provide a practical innovation at the work site.

The innovation process is simply the linking of a needed task improvement with an invention capable of performing that task. The resulting combination of these two is the innovation at the work site. The process is shown in Figure 13-4.

When the innovation process is combined with job design, the process looks like Figure 13-5. The steps of the process parallel the steps used in the problem job model and in the job process.

The first step begins with definition of the task. The difficult elements

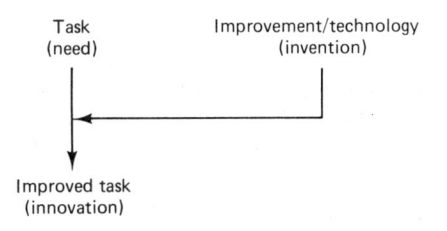

Chap. 13 Job Design   303

Figure 13-4 The innovation process. *Source:* Adapted from David C. Alexander, "Job Design: A Marriage of Ergonomics and Automation That Works in Industry," *Ergonomics News,* 19(3), Winter 1985, Figure 2. Reprinted by permission. Copyright © Institute of Industrial Engineers, 25 Technology Park/Atlanta, Norcross, GA 30092.

```
Task                    Improvement/technology
(need)                  (invention)
  │                           │
  │◄──────────────────────────┘
  ▼
Improved task
(innovation)
```

are highlighted and identified. Following that, possible improvements are determined. Specifically, many other people are contacted to determine if there is an alternative way to perform the difficult task or element that has been identified. This specific request of others is an important part of the innovation process, particularly since the requests are directed toward those having strong skills with industrial automation.

Ideally, a number of possible inventions or improvements are identified. These possibilities are evaluated to determine which are more suitable for this particular application. Some criteria for this evaluation are the cost of the improvement, convenience, how well it will work, and how practical it may

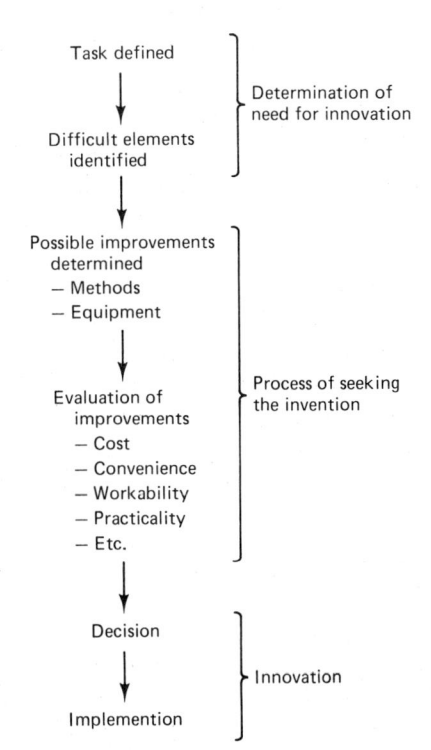

Figure 13-5 Innovation in job design. *Source:* Adapted from David C. Alexander, "Job Design: A Marriage of Ergonomics and Automation That Works in Industry," *Ergonomics News,* 19(3), Winter 1985, Figure 3. Reprinted by permission. Copyright © Institute of Industrial Engineers, 25 Technology Park/Atlanta, Norcross, GA 30092.

```
        Task defined              ⎫
             │                     ⎬  Determination of
             ▼                     ⎭  need for innovation
     Difficult elements
        identified
             │
             ▼
   Possible improvements           ⎫
        determined                  │
        — Methods                   │
        — Equipment                 │
             │                      │
             ▼                      ⎬  Process of seeking
      Evaluation of                 │   the invention
      improvements                  │
        — Cost                      │
        — Convenience               │
        — Workability               │
        — Practicality              │
        — Etc.                     ⎭
             │
             ▼
         Decision                  ⎫
             │                     ⎬  Innovation
             ▼                     ⎭
      Implementation
```

be. After the evaluation, a decision is made as to which improvements to pursue.

Following that, the improvement is implemented, resulting in an innovation at the work site.

13-4-3 Key Areas of Automation

The strengths of machines relative to those of humans point to the key areas of automation. Some of the strengths of machines for performing tasks are listed in Table 13-3.

Some of the key areas for automation, then, are to replace human muscle, to avoid excessive endurance requirements, and for sensory input. Since there are areas where the human being has definite limitations, using machines to perform in these areas is good ergonomic design. It allows people to be used in areas where we have relative strengths, such as making decisions, exerting judgment, and providing overall control to the process.

A word of caution in automating cognitive and monitoring tasks—automating too heavily can result in boredom and inattention on the part of the human user. Care must be exercised to avoid overautomation, with the result of jobs that are poorly designed for people.

13-4-4 Examples

This example will illustrate the use of an automatic palletizer to improve a production operation. A new facility was planned to manufacture and package small plastic chips. (The chips would be molded into products by other manufacturers.) The material was packaged into 25-kg (55.0-lb) bags which were palletized for transportation overseas. The bags were picked up from a conveyor line and manually placed on wooden pallets. Production rates required approximately 1500 bags to be palletized by each person during an 8-hour work shift. This obviously was a difficult job, since it required lifting, bending, stooping, and twisting to place the heavy bags. The job was machine-paced and often long periods of work were required between breaks.

The traditional approaches available for improving a job like this are:

1. Selection of employees, so that only the most physically strong people are put on that job
2. Rotation among other people working in a nearby area
3. A slower packaging rate
4. Smaller bag sizes
5. More frequent rest breaks
6. Developing a proper height for the package lift and package delivery points

TABLE 13-3 Capabilities of People versus Machines

People Excel In	Machines Excel In
Detection of certain forms of very low energy levels	Monitoring (both people and machines)
Sensitivity to an extremely wide variation of stimuli	Performing routine, repetitive, or very precise operations
Perceiving patterns and making generalizations about them	Responding very quickly to control signals
Detecting signals in high noise levels	Exerting great force, smoothly and with precision
Ability to store large amounts of information for long periods—and recalling relevant facts at appropriate moments	Storing and recalling large amounts of information in short time periods
Ability to exercise judgment where events cannot be completely defined	Performing complex and rapid computation with high accuracy
Improvising and adopting flexible procedures	Sensitivity to stimuli beyond the range of human sensitivity (infrared, radio waves, etc.)
Ability to react to unexpected low-probability events	Doing many different things at one time
Applying originality in solving problems (i.e., alternative solutions)	Deductive processes
Ability to profit from experience and alter course of action	Insensitivity to extraneous factors
Ability to perform fine manipulation, especially where misalignment appears unexpectedly	Ability to repeat operations very rapidly, continuously, and precisely the same way over a long period
Ability to continue to perform even when overloaded	Operating in environments which are hostile to people or beyond human tolerance
Ability to reason inductively	

Source: Wesley E. Woodson and Donald W. Conover, *Human Engineering Guide for Equipment Designers,* 2nd rev. ed. (Berkeley: University of California Press, 1966), pp. 1-23.

Each of these solutions had at least one problem associated with it. These problems correspond to the list above.

1. Establishing and validating a selection testing program
2. Administrative attention to assure proper rotation practices
3. Production constraints to meet necessary production demands
4. Marketing constraints that specify certain package sizes

5. Lower productivity and/or higher labor costs
6. Equipment redesign and associated cost of changing the equipment

A win-win solution was proposed, accepted, and implemented. The addition of an automatic palletizer to eliminate handling bags manually was the best solution. It clearly removed any ergonomic problems associated with manually palletizing material. The automatic palletizer concept was an expensive capital outlay, but it was entirely within the organization's objectives. It helped promote the strong safety and health program within the company and improved the status of jobs from laborer to machine operator. Other benefits were increased productivity, reduced packaging costs, and more uniform packages. In addition, the automatic palletizer would pay for itself entirely through labor saving. The palletizer was installed and the anticipated objectives were met.

Another example involved the addition of an industrial robot for a manual materials handling task. As the volume of this particular product increased, the requirements to package and palletize 50-lb bags of material increased. One unexpected result was an increase in the injury rate. As part of the plant's safety and health emphasis, a trained ergonomist was requested to evaluate the job for improvements.

Several traditional solutions were considered. These included:

1. Selection of people for the job
2. Changes in the pickup and delivery heights associated with the material
3. Rotation on the job
4. Increasing rest/recovery periods
5. Other redesigns of the equipment and layout in the bagging and palletizing area

Actual practice limited the options available. With 20 bagging stations, even a small cost per station amounted to a sizable expenditure. In addition, both bags and drums were packaged at each station, so traditional palletizing equipment was not suitable to handle this mix of products.

Each of the solutions initially developed would have resulted in some compromise in the overall objectives of the organization. To gain even marginal safety and health benefits would have required sizable expenditures, lower productivity, significant administrative control, or perhpas all three. At this point in the study, practical workable solutions were essentially nonexistent.

A broader view of the problem was proposed which involved the overall cost of packaging this material, coupled with the high costs of the injury rate. A "macro design" approach was taken, and other solutions now became apparent. An industrial robot could palletize both bags and drums. Cost prohibited furnishing each station with a robot, so consolidation was required. With high work rates, one robot could easily handle the volume from six to eight stations. (As a side benefit, consolidation of the stations made other improvements, such as dust control, much less expensive.) The results of this project were benefits in the areas of safety and health, ergonomics, economics, and productivity.

13-5 OTHER FACTORS IN JOB DESIGN

There are three factors that are important in an effective job. First, it must be a job that people can do. This involves the ergonomic aspects of job design. Second, it must be a job that people want to do. This involves the motivational aspects of job design. Third, it must be a job that has value to the organization. This involves the traditional industrial engineering aspects of job design.

Effective job design must integrate all three of these dimensions. Traditionally, ergonomics has been the last of the three areas to be considered and there are many cases where otherwise well-designed jobs have ergonomics problems.

Job design covers a wide range of technology and goes far beyond the scope of this book. The areas of motivation and organizational value will be discussed briefly. This is not a comprehensive discussion in either of these areas but should merely serve to provide a foundation for further exploration of these two critical areas.

13-5-1 Motivational Aspects of Job Design

The motivational aspects of job design cover the desire by the person to perform the job. The motivational aspects are important since, with a flexible and mobile labor market, people can easily quit a job that does not suit them. Second, jobs are becoming more discretionary. That is, the effort that can be put into a job ranges from the minimum effort to avoid being fired to the full effort required for truly high performance. This "discretionary effort" is controlled by the job holder, not the employer, and is highly correlated to the motivational aspects of job design.

This section follows closely the work of J. Richard Hackman and Greg R. Oldham, as discussed in their book *Work Redesign*. Their "job characteristics model" explains how the design of jobs can affect performance, motivation, and satisfaction outcomes at the work site. The model is shown in Figure 13-6. An explanation of the components of the model follow.

The desired performance, motivation, and satisfaction outcomes of a job are:

1. *High internal work motivation:* The person is internally motivated to perform the task, that is, being motivated for one's work because of the

```
                                CRITICAL
         CORE JOB             PSYCHOLOGICAL              OUTCOMES
      CHARACTERISTICS             STATES
```

Skill variety ⎤
Task identity ⎬──▶ Experienced meaningfulness of the work   High internal work motivation

Task significance ⎦   High "growth" satisfaction

Autonomy ──────────▶ Experienced responsibility for outcomes of the work   High general job satisfaction

Feedback from job ─▶ Knowledge of the actual results of the work activities   High work effectiveness

Moderators:
1. Knowledge and skill
2. Growth need strength
3. "Context" satisfactions

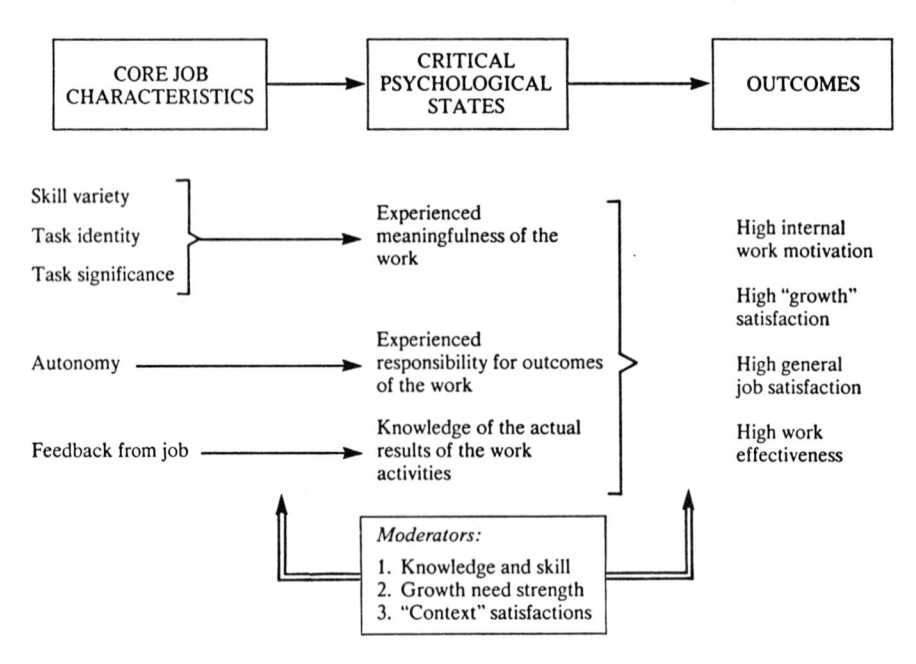

Figure 13-6 Complete job characteristics model. *Source:* J. Richard Hackman and Greg R. Oldham, *Work Redesign* (Reading, Mass.: Addison-Wesley Publishing Company, Inc., 1980), p. 90, Figure 4.6. Reprinted by permission.

positive internal feelings that are generated by doing well, rather than being motivated by external factors such as incentive pay.

2. *High growth satisfaction:* The job provides an opportunity for the person to grow on the job, and the people on the job find those opportunities personally exciting.
3. *High general satisfaction:* The job provides general overall satisfaction as measured by such questions as "Generally speaking, how satisfied are you with your job?" and "How frequently do you think of quitting your job?"
4. *High work effectiveness:* The person has high work effectiveness while performing the job as measured in both quantity and quality of the goods and services produced.

To have these outcomes, according to Hackman and Oldham, three critical psychological states are necessary. The person must find the work to be meaningful. (Is what I do important?) Second, the person must be responsible for the outcome of the work. (Have I personally provided an outcome that wouldn't be there otherwise?) Third, the person must have knowledge of the actual results of the work. (How did I do?) These three states are necessary to gain the desired outcomes, but unfortunately, these attributes cannot be designed into a job.

There is a relationship between some job characteristics and these psychological states, and fortunately, they can be designed into jobs. The five core job characteristics that relate closely to these states are skill variety, task identity, task significance, autonomy, and feedback from the job. Table 13-4 describes these job characteristics.

When the core job characteristics are provided, are the desired outcomes always there? Not necessarily, according to Hackman and Oldham. There are three "moderators," which may influence the degree to which the core job characteristics are translated into the desired outcomes. The first moderator, "knowledge and skill," involves those skills required to perform the task. Even in a well-designed job, if the person does not have the necessary knowledge and skill, that person will experience frustration, dissatisfaction, and poor job performance. The second moderator, "growth need strength," relates to the individual's desire to have a job that provides growth. A person who has low job growth needs will experience frustration if the job requires new learning and experience just to keep up. The third moderator, "context satisfactions," deals with the traditional "hygiene factors" of pay, coworkers, and supervision. Even in a well-designed job, a person will not have the desired outcomes if he or she has dissatisfaction with pay, with coworkers or supervision, with management practices, or with company policies.

In their book, Hackman and Oldham present diagnostic instruments to test for the strength of the core job characteristics and for the presence of the moderators. In addition to providing a quantitative assessment of these items, prescriptions for appropriate changes are also discussed.

The use of the "job characteristics model" is more useful for the design

TABLE 13-4 Core Job Characteristics

1. *Skill variety:* the degree to which a job requires a variety of different activities in carrying out the work, involving the use of a number of different skills and talents of the person
2. *Task identity:* the degree to which the job requires completion of a "whole" and identifiable piece of work, that is, doing a job from beginning to end with a visible outcome
3. *Task significance:* the degree to which the job has a substantial impact on the lives of other people, whether those people are in the immediate organization or in the world at large
4. *Autonomy:* the degree to which the job provides substantial freedom, independence, and discretion to the employee in scheduling the work and in determining the procedures to be used in carrying it out
5. *Feedback:* the degree to which carrying out the work activities required by the job provides the person with direct and clear information about the effectiveness of his or her performance

Source: J. Richard Hackman and Greg R. Oldham, *Work Redesign* (Reading, Mass.: Addison-Wesley Publishing Company, Inc., 1980), pp. 78-80. Copyright © 1980, Addison-Wesley. Reprinted with permission.

of jobs than for the design of tasks. Jobs are just a collection of tasks, and the same tasks assembled in different combinations can provide very different jobs. From the same tasks can come jobs that are very stimulating to ones that are motivationally bankrupt.

An example can illustrate this. During the assembly of automobiles, someone must put on the left rear tire of each automobile. A job can be designed so that one person puts the left rear tire on every automobile that comes by. This job provides little skill variety, and is deficient motivationally. The same task can be used to design a more varied job—mounting all four tires. Even more variety and autonomy can be provided by using a small team to assemble the entire drive train, including the task of mounting the left rear tire. By just regrouping required tasks, the job can be made to be highly motivating, or it can be designed to be limited and boring.

13-5-2 The Organizational Value of Job Design

Organizational value is the ratio of (1) the benefits to the organization of performing the job, and (2) the costs to the organization for the job. The greater the organizational value, the better for that organization. Value is normally calculated in dollars, but that is not always the proper measure, since with some jobs, the cost is easy to assess but the benefits are not. Using dollars may lead one to believe that there is no value for the organization, when, in fact, there really is.

Organizational value does not correlate positively or negatively with the ergonomic or motivational aspects of job design. The desired job can be strong on all three aspects.

Enhancing the organizational value is a matter of increasing the benefits to the organization or of reducing the costs for the organization. Benefits increase by gaining additional output or with a higher quality of output.

Costs are normally measured in dollars or in labor-hours. Improperly designed tasks are more expensive than properly designed ones. Extra idle time built into a job is more costly than if it were not there.

Usually, when labor costs are low, more of the tasks are done by people. As labor costs increase, there is increasing pressure to have those tasks performed by machines. This balance must be noted in the design of tasks and jobs.

Some types of job designs are more expensive than others, because of the additional administrative costs involved. For example, employee selection or job rotation seem to be reasonable solutions for some ergonomic problems. Yet these requirements can adversely affect the organizational value of those jobs by adding to the costs (hassle, not necessarily dollars) of those jobs.

The assembly of jobs from tasks can also affect the value. Some jobs may have idle time which increases the costs. By adding some deferrable tasks to the job, the value can be improved and a higher "yield" on the labor can be obtained.

The proper combination of related tasks can greatly enhance a job's value. For example, some organizations have a single person or work group operate, service, and maintain common equipment. The logic is that when the equipment is operating, a production operator is needed, yet when the equipment is down, someone is needed for service or repair but not for production. These tasks tend to be needed at different points in time, so having the same person perform them usually provides little conflict.

There are many ways to assess and improve the organizational value of jobs for the organization. The purpose of this section is not to provide all of those methods, but to introduce the concepts and ensure that the reader is aware of this important area.

13-5-3 Comprehensive Job Design

Comprehensive job design is difficult to do. It is unusual for one person to have the full skills in ergonomics, motivation, and industrial engineering to fully design jobs. The job designer must recognize his or her limitations and seek appropriate assistance. Developing the appropriate skills is important and should be done. However, on a short-term basis, a consultant may be preferable to a few articles or even a good book.

Since these three areas are so evenly balanced, designing a job with emphasis on only one or two of the areas is improper. The ergonomist certainly would not want jobs to be designed without the consideration of a skilled ergonomist. For that reason, the ergonomist must be sensitive to the design needs from the other areas. It is important to recognize one's own limitations in this area and to develop ways to augment one's skills.

13-6 SUMMARY

Job design represents a strong challenge for the ergonomist. The design of jobs for people can benefit from the skills of the ergonomist. In this key area, the ergonomist is forced to join forces with the industrial engineer and the behavioral scientist.

The proper design of jobs can enhance job performance and employee well-being at the same time. Well-designed jobs represent a "win-win" solution for everyone concerned.

ACKNOWLEDGMENTS

This chapter utilizes and builds on material from David C. Alexander, "Job Design You Can Use," a continuing education seminar, sponsored by the Institute of Industrial

Engineers, May, 1982; and David C. Alexander, "Ergonomics and Job Design—Keys to Improved Worker Performance," a continuing education seminar, sponsored by the Institute of Industrial Engineers, May, 1984.

Section 13-4 utilizes and builds on material from David C. Alexander, "Job Design: A Marriage of Ergonomics and Automation That Works in Industry," given at the 1981 Fall Industrial Engineering Conference in Washington, D. C., and David C. Alexander, "Job Design: A Marriage of Ergonomics and Automation That Works in Industry," *Ergonomics News*, 19(3), Winter 1985.

QUESTIONS

1. Define a job.
2. What are three critical elements of job design?
3. List and discuss the types of ergonomic problems that should be anticipated during job design. Give examples.
4. How is a "fair day's work" established for highly physical jobs? For jobs with high mental loads? For other jobs?
5. What are the solutions to underutilization of the person on the job? What precautions should be considered?
6. During job design, what is the value of macro design?
7. What are the steps in micro design?
8. Major job changes can result from the innovation process. What are the two elements that are required before an innovation can occur?
9. In the "job characteristics model" by Hackman and Oldham, high work effectiveness is an expected outcome. What additional outcomes also result?
10. What are the five core job characteristics that must be well designed in order to get the high work effectiveness of the job characteristics model?
11. What are the moderating influences between the core job characteristics and the desired outcomes of the job characteristics model?
12. What is the organizational value of a job to the organization? How is it commonly measured? How is it enhanced?

Chapter 14

ADMINISTRATIVE DECISIONS

The ergonomist can have a major impact on employee safety and health by influencing the policies and procedures within his or her organization. The policies influence procedures and designs, and ultimately affect the person on the job. This chapter shows those areas where the ergonomist can contribute to changes in administrative policy affecting people on the job.

14-1 INTRODUCTION

"I've been following some new research," a friend said, "which indicates that our schedules could be improved." He was right—the newly discovered information on sleep-wake patterns seemed helpful for designing shift schedules. In our investigation, we found that ergonomics was not the only factor involved in shift schedule design. There were federal and state legal requirements, operational needs, and the varying preferences of shift workers, just to name a few. In fact, ergonomics was only one of many criteria that had to be balanced in the design of shift schedules.

Shift work is a typical administrative decision area. There are many other situations where ergonomics will involve issues normally decided by administration. Ergonomics should influence many of the decisions involving people which come from the human resources area (also called personnel, employee relations, or industrial relations). Ergonomics can also influence many of the philosophical and policy issues facing an organization. Typically, however, ergonomics has been ignored in both of these areas, and only begins to have an influence from the most mature ergonomics programs. The administrative area is one of the keystones of a "second-generation ergonomics" program, since decisions are more complex, and its use only follows the successful application of many other ergonomic projects.

In the human resources areas, decisions on shift schedules, frequency and length of rest breaks, acceptable lifting limits, sound and light levels, limits and use of overtime, wage determination, job design, and the use of handicapped personnel are only a few areas where ergonomics can provide critical information and guidance for decisions.

In the organizational policy and philosophy areas, ergonomics can influence: the percentage of the population designed for, injury prevention versus correction after injuries have occurred, whether to pay extra for work in hazardous areas or to correct those situations, use of recovery breaks from difficult work, whether to use each person to his or her fullest potential, and whether to design jobs for people or to select people for jobs. Many of these issues are easier to resolve once one understands the degree of difficulty in carrying out the policy.

The characteristics of "administrative decisions" are:

1. Many issues are normally negotiated in employer–employee agreements.
2. The decision is not directly ergonomics, but affects its use and application greatly.
3. The decision involves the integration of many factors other than ergonomics.
4. The degree of quantitative measurement for the decision is limited.

The use of administrative policy is similar to the use of engineering design standards and purchasing specifications. They state the "rules" that one must use in that organization, and these "rules" are expected to be followed unless exceptions are agreed to up front.

Philosophies can affect policies as well as design standards and purchasing specifications, as noted in Figure 14-1. The philosophies are often a reflection of the organization's feeling toward its people and greatly affect whether the organization will even have an emphasis in ergonomics. Organizations that might be more sensitive in this area are the ones most sensitive to an ergonomics appeal based on social justifications. Smith and Smith[1] discuss this: "Whatever the firm can do to demonstrate its concern for the quality of the working life experienced by its employees will enchance its local public image and lead to improved labor relations and ability to compete for and

---

[1] Leo A. Smith and James L. Smith, "How Can an IE Justify a Human Factors Activities Program to Management?" *Industrial Engineering,* February 1982, p. 42. Printed with permission from *Industrial Engineering* Magazine, February 1982. Copyright Institute of Industrial Engineers, 25 Technology Park/Atlanta, Norcross, GA 30092.

```
Philosophy
(use all of population)
    ↓
Administrative policy
  (hire females)
Engineering standards
  (low lifting limits)
Purchasing specifications
  (buy 20-kg and 25-lb bags)
    ↓
Job design
  (for all people)
Equipment design
  (low heights and weights)
```

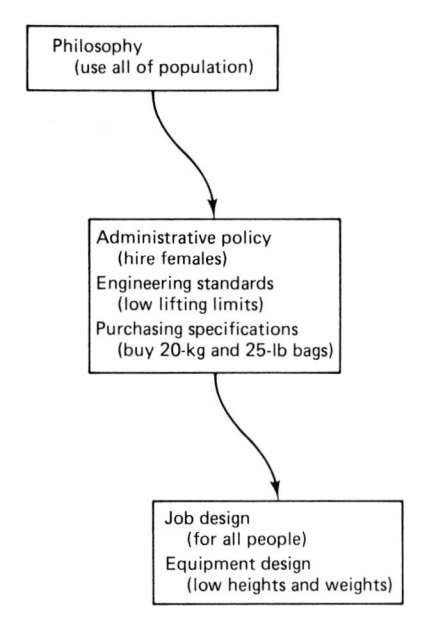

Figure 14-1 Influence of policy on job and equipment design.

retain qualified workers." They state that these circumstances are appropriate indicators of this sensitivity:

1. The firm employs nonunionized labor.
2. The firm has been in existence for a long time and is the principal employer in a particular area.
3. The firm is in a tight labor market and must compete actively for labor with surrounding companies.

Policy decisions are normally made in the human resources area and/or in conjunction with higher management. It requires credibility to have an influence in this area; perhaps more so than in the engineering design area, where decisions tend to be more quantitative. The credibility comes most easily from having strong contacts in the organization or from, in fact, actually being organizationally located there. Success with previous recommendations is also helpful. Support for influencing the policy area can come from NIOSH guidelines or from organizations such as the International Labor Organization. Many policy items will become, or have already been, the subject of legislation or contract negotiations (sometimes called "legislative ergonomics").

"Legislative ergonomics" can be contradictive, since the legislation can often run counter to proper ergonomics practice, while at the same time being based on ergonomics data. During the energy crisis, comfort standards were pushed to the limits, and safe working temperatures were violated (especially for hard work in hot environments) when the indoor temperature standards were set. Legal proposals for VDT workstations often call for adjustable desks, chairs, and footrests, when only two of these three are necessary.

The use and documentation of the policies and philosophies vary widely. Some are written and widely shared, others are written but not shared, and some are informal but widely understood. Policies on work times are usually well known. Policies on the use of handicapped are often planned but not widely known. Day-to-day activities such as lifting guidelines are informal and widely held, such as a "rule of thumb limiting lifting to 50-lb bags."

The policies often provide ease of use and clarity rather than effectiveness. For example, "lift no more than 50-lb" is easy to follow, but there are many situations where severe injuries could occur depending on posture, frequency of lift, and other conditions.

The rest of this chapter is devoted to the discussion of several situations where administrative decisions are used. The use of ergonomics to influence and shape organizational policies and philosophies is important for a balanced industrial ergonomics program. These are the major areas: work times and schedules, work task expectations, environmental issues, and pay and personnel policy issues. Figure 14-2 illustrates changes in the quantitative nature of decisions as one goes from "pure ergonomics" areas to the administrative decision areas.

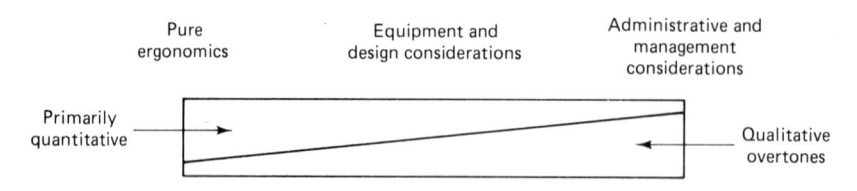

Figure 14-2 Pure ergonomic versus administrative decisions.

14-2 WORK TIMES AND SCHEDULES

Work times and schedules is one area where the value of ergonomics is quite visible. This area, like most other administrative decisions areas, is influenced from four major sources:

1. Legal and contractural obligations
2. Social concerns and preferences of the workers and tradition
3. Operational needs to run the business
4. Ergonomic concerns and human needs

These factors interact in different ways with the various decisions around work time and schedules (see Fig. 14-3). Work times are often negotiated or long standing and changes in this area are difficult to make and influence.

Chap. 14 Administrative Decisions 317

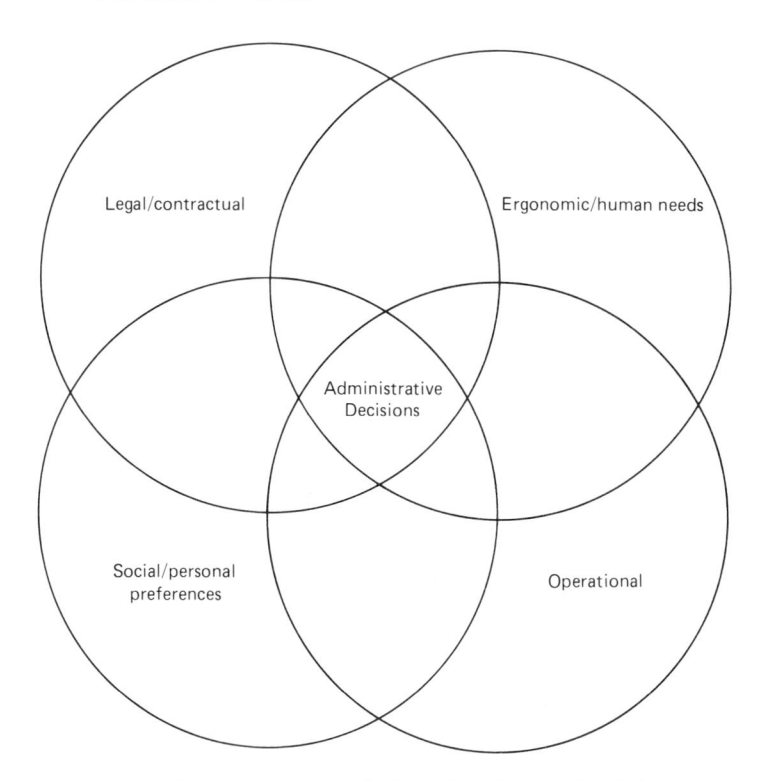

Figure 14-3 Interaction of major issues in administrative decisions.

Relative to work times and schedules, there are three major areas: (1) work and rest or recovery cycles, (2) overtime worked, and (3) work schedules. Each is discussed in a major section.

14-2-1 Work and Rest/Recovery Cycles

During paid time, there are work and nonwork periods. Nonwork periods range from 5% to well over 50% of the work day, depending on a variety of circumstances. Nonwork periods include coffee breaks, cleanup time, personal time, unavoidable delays, meal periods, or recovery periods. Only the recovery periods require analysis from the ergonomic standpoint.

Recovery periods are required when jobs are so difficult that they fully tax the people performing them. The analysis of work-recovery periods was detailed in Chapter 9. After first determining the length of recovery periods that are necessary, the workday should be examined for rest periods such as coffee and meal breaks or unavoidable delays. After these naturally occurring recovery periods are included, additional recovery periods should be added. These can be actual rest periods, or simply changes to easier work such as recordkeeping. From the physiological standpoint, shorter work/recovery cycles are far less tiring than longer work and rest periods. Operational factors and tradition, however, may require longer work periods than are physiologically desirable.

In one operation, the physiological cost ranged from 6 to 10 kcal/min worked. There were a few other work tasks that required less energy, but they represented a small percentage of the work day. Consequently, extensive recovery periods were required. The workday was designed to include 4.5 hours of heavy work, 1.5 hours of light work, and 2 hours of recovery. Only with critical physiological data was this schedule approved. The 1.5-hour blocks of heavy work were undesirable, but reflected a full unit of work that was hard to segment (see Table 14-1).

14-2-2 Overtime

Overtime is used as a buffer for unanticipated needs such as relief and temporarily high work volumes. Decisions on overtime are influenced by operational needs, followed by legal and contractual agreements, and to a lesser extent by social considerations and ergonomic needs.

The two major concerns with overtime are the amount of overtime worked, and when the overtime is scheduled. Ergonomics considerations are more valuable for scheduling than for determining the amount of overtime.

Amount of overtime. The amount of overtime is limited by social and legal/contractual agreements more than by ergonomic considerations. The workweek is approximately 40 hours based on tradition. From the ergonomics standpoint, the workweek could be longer. More important, the length of the workweek is what a person is adjusted to and the outside activities that have been established.

TABLE 14-1 Example of Work and Recovery Cycle

Work Time (min)	Task	Energy Cost *Kcal/min*	*Total kcal*
45	Set up	2	90
90	Heavy work	8	720
30	Rest	1	30
90	Heavy work	8	720
60	Rest and lunch	1.2	72
90	Heavy work	8	720
30	Rest	1	30
45	Clean up area	2	90
		Total	2472

For the short term, sleep loss and personal energy costs may limit the amount of overtime worked. Workdays of 12 to 16 hours may be worked, as may workweeks of 50 to 70 hours without serious long-term harm. Errors will be the first sign of fatigue, followed by physical tiredness.

For the long term, social, family and nonwork activities will limit the amount of overtime a person is willing to work.

These needs are reflected contractually and legally with overtime pay requirements. The overtime pay requirements influence operational needs, so that it becomes more economical to add staffing rather than incurring extensive overtime payments.

Scheduling of overtime. Overtime schedules can be determined to provide both the best quality work time and sleep time. A person's day is divided into four areas: sleep time, work time, personal time, and recreational time, as shown in Table 14-2. Recreational or "free" time is the time left over after other activities.

Overtime should be scheduled to utilize the person's recreational time rather than using sleep time. Sleep deprivation over a period of time will result in lower performance.

In one instance, a secretarial pool anticipated overtime of about 1 hour per day for one month. Questions of scheduling the time before or after the normal work shift were asked. Careful consideration of the two options revealed that the early work would result in sleep loss, while the later work would affect recreational time. To work the morning hour would mean getting up an hour earlier and losing an hour's sleep. Working the afternoon hour would mean using an hour's recreational time. Fatigue from the longer workday was evaluated, but since both options involved working 9 hours, the difference was not considered significant. Overtime was recommended to be scheduled after the normal work shift.

Overtime is needed for 24-hour work shifts, both for unanticipated relief needs when someone calls in sick and for periodic high volumes. With a three-shift operation, the scheduling of overtime has fatigue considerations. Figure

TABLE 14-2 Daily Activities (hours)

	8-Hour Workday	12-Hour Workday
Total day	24	24
Less:		
Work time plus travel	9	13
Sleep time	8	8
Personal time (meals, bathing, dressing, etc.)	3	3
Recreational time	4	0

```
Work shift   8 a.m.-4 p.m.
                                    Ergonomic impact:
  1. Call in person on day off        None
  2. Call person on 4-12 in           Person must get up early, some sleep loss
     early
Work shift   4 p.m.-12 p.m.
                                    Ergonomic impact:
  1. Call in person on day off        None
  2. Keep person on 8-4               Person goes to bed late, some sleep loss
     over late                           (generally uses recreational time)
Work shift   12 p.m.-8 a.m.
                                    Ergonomic impact:
  1. Call in person on day off        Person not adjusted to staying up all
                                         night; some sleep loss
  2. Keep person on 4-12              Person goes to bed later, some sleep loss
     over late
```

Figure 14-4 Overtime schedules for 24-hour operations.

14-4 shows some overtime schedules, while Figure 14-5 shows typical work and sleep patterns for different work shifts.

The first choice for all overtime is to have someone work on a scheduled day off. The second choice is to have someone work overtime during their recreational time. The next choice is to have someone move their sleep time, if possible, without losing that sleep time. The worst choice is to have someone work in lieu of sleep time.

14-2-3 Work Times and Schedules

Work times and schedules involve the length of the workday, the days of the week worked, the starting time, the regularity of the schedule, and so on.

For day-based workers, there are only a few major schedules used. The base schedule is 8 hours/day, 5 days/week. Common modifications include the days of the week worked and the starting times. The 4/40 schedule involves extended workdays in order to gain a longer weekend. Flextime is used to change starting times, providing a convenience to the worker if the operation can tolerate variable starting times. Ergonomically there are few improvements to day-based schedules.

For operations involving 24 hour/day staffing, the number of schedules and the ergonomics opportunities both increase. The criteria for establishing these schedules involve all four factors: legal/contractual, social, operational, and ergonomic. Depending on the criteria used most, some schedules seem to

Chap. 14 Administrative Decisions   321

	Work 8-4	Work 4-12	Work 12-8
12 p.m.	sleep	personal	work
1	sleep	free	work
2	sleep	sleep	work
3	sleep	sleep	work
4	sleep	sleep	work
5	sleep	sleep	work
6	sleep	sleep	work
7	personal	sleep	work
8	work	sleep	personal
9	work	sleep	sleep
10	work	personal	sleep
11	work	free	sleep
12 a.m.	work	free	sleep
1	work	free	sleep
2	work	free	sleep
3	work	personal	sleep
4	personal	work	sleep
5	free	work	personal
6	free	work	free
7	free	work	free
8	free	work	free
9	free	work	free
10	personal	work	free
11	sleep	work	personal

Figure 14-5 Typical work and sleep patterns for different work shifts.

be better than others. Some of the more important assessments of the 24 hour/day schedules are:

- Meeting operational needs
- Providing quality time off
- Providing quality sleep time
- Meeting all legal and contractual requirements

A few of the myriad of schedules available are shown in Figure 14-6, together with comments on their use.

Some key issues in designing shift schedules are discussed below. The direction of the rotation is critical for 8-hour shifts. The best direction is forward (for forward on the clock, e.g., after the day shifts of 8 a.m. to 4 p.m., begin the evening shifts of 4 p.m. until midnight), since this fits the natural pattern of human beings.

For continuous coverage with 8-hour work shifts:

1. "Southern Swing" — may be the most common schedule in U.S.; 4 crews; each crew works 7 "days" before scheduled time off; since all crews rotate, experience can be distributed among crews.
2. "Rapid Rotation" — popular in Europe; 4 crews; each crew works 7 "days" before scheduled time off; quick rotation of shifts minimizes cumulative fatigue; experience can be distributed among crews.
3. "Slow Rotation" — also widely used in U.S.; 3 crews with built-in relief on each crew; each crew works 4 weeks before rotation to another shift; provides time for adjustment to the current shift; experience can be distributed among crews.
4. "Fixed Shifts" — also widely used in U.S.; 3 crews, each on a fixed shift (may have smaller crews on later shifts); day shift tends to have more experienced people, third shift is less experienced.

For weekday coverage with 8-hour work shifts:

5. "5-day Shifts" — similar to the southern swing, slow rotation, and fixed shifts, except that the crews work only weekdays and have weekends off.

For continuous coverage with 12-hour work shifts:

6. "DuPont Schedule" — popularized by DuPont in U.S.; 4 crews; each crew works 3 or 4 "days" before scheduled time off; 7 consecutive days off each 4-week period; popular with shift workers; experience can be distributed among crews.
7. "EOWEO: Every Other Week End Off" — used in U.S. Gulf coast area; 4 crews; each crew works 2 or 3 "days" before scheduled time off; has the most weekends off; experience can be distributed among crews.

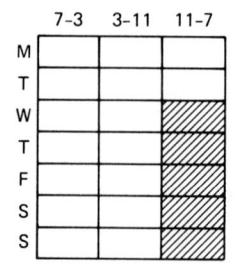

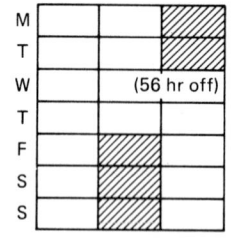

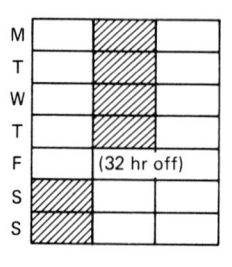

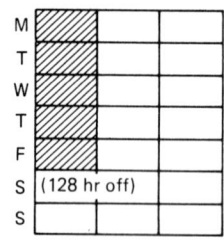

Figure 14-6 Shift schedules.

The length of the work shift traditionally has been 8 hours, although extended workdays are becoming more common. Usually, overtime is required, which greatly affects labor costs.

The number of consecutive days worked on the same shift ranges from only a few on rapid-rotation schedules to 7 for the southern swing. The longer the work periods, whether from extended workdays or extended workweeks, the longer the blocks of time off.

Chap. 14   Administrative Decisions   323

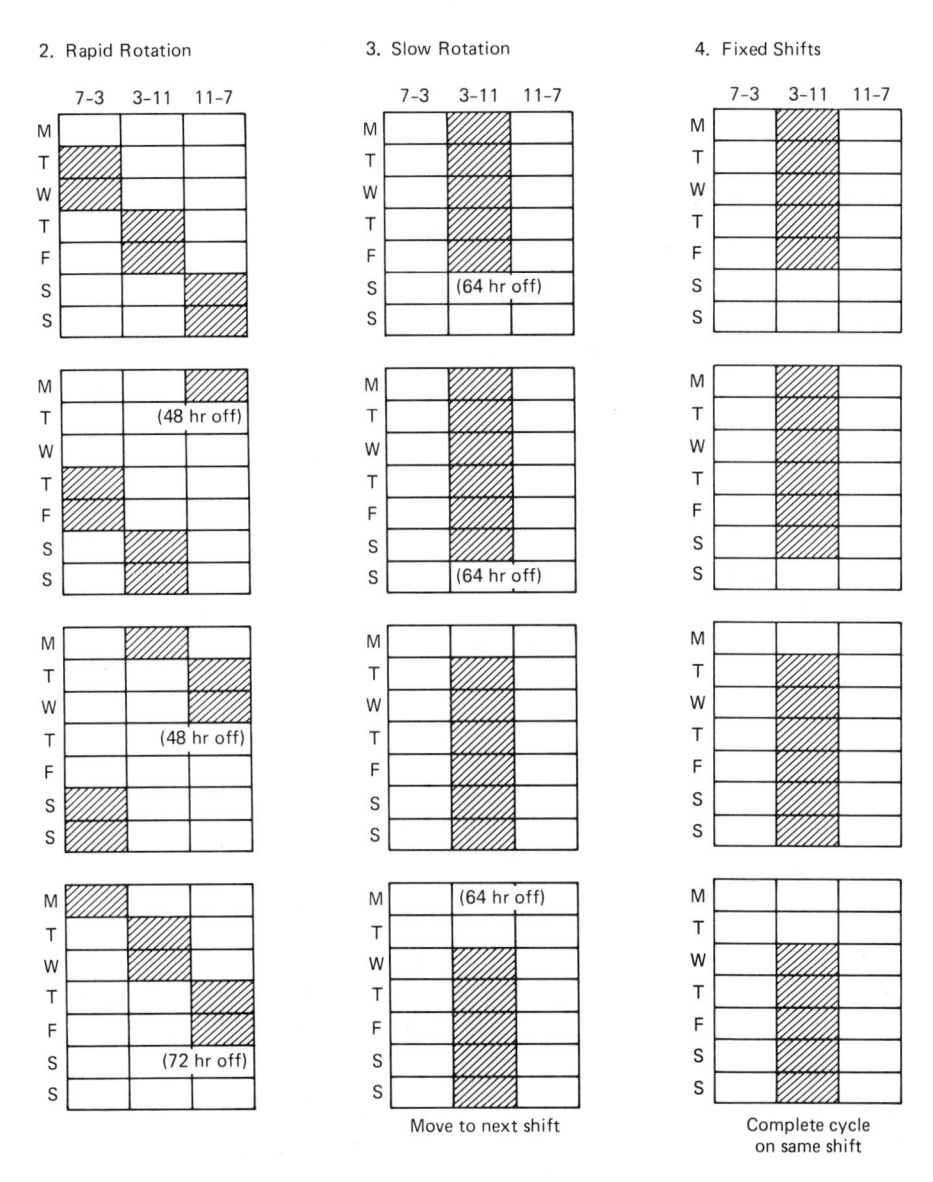

2. Rapid Rotation   3. Slow Rotation   4. Fixed Shifts

                                      Move to next shift   Complete cycle
                                                             on same shift

Figure 14-6 (*Cont.*)

The movement, or swing, from one shift to another is a topical question. Some researchers feel that a person can adjust to any schedule and that long blocks of time on a fixed schedule are preferable, whereas others feel that adjustment simply isn't ever complete and prefer a shorter time before rotating to avoid long periods of fighting a bad shift.

Time to be with family and friends is important. A good schedule will

324 Administrative Decisions Chap. 14

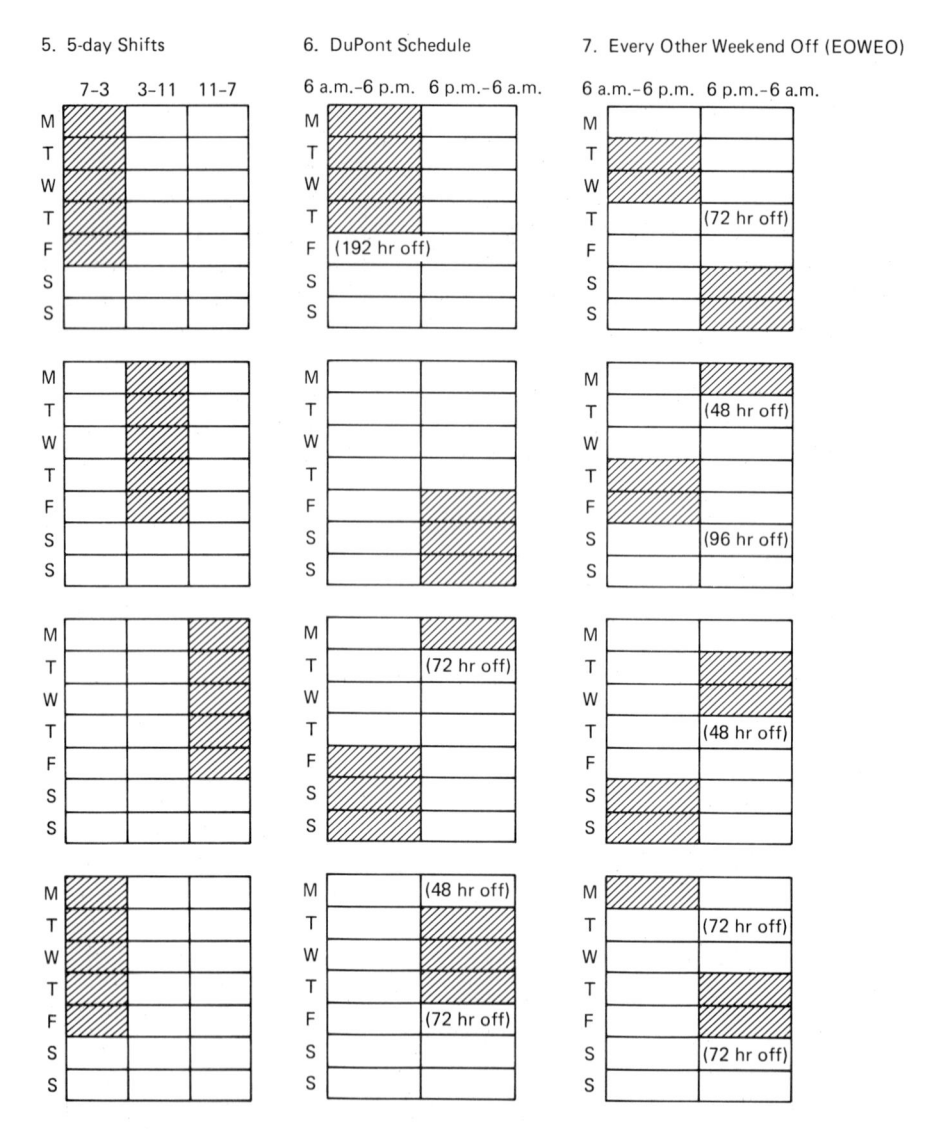

Figure 14-6 (*Cont.*)

provide nonwork and nonsleep time that fits the traditional recreational time periods of evenings and weekends off.

Schedule predictability is important to allow planning for nonworking members of the family. Predictability is possible from fixed schedules or from schedules that cycle on a short time period such as 4 weeks.

14-3 WORK TASK EXPECTATIONS

The underlying assumptions about the work force greatly influence the design of jobs and tasks. If, for example, one designs a job in a foundry environment, the expectations of heavy materials handling could influence the tasks required. The designer would probably make no special provisions to deliver a 50-lb box of preprinted forms required in the department. Yet the same designer might make an accommodation for handling 50-lb boxes of copying paper used in a secretarial operation.

These unwritten work task expectations may not be documented or even clearly outlined, but they affect the design of jobs. It will be useful to clarify these expectations prior to designing jobs. Deviations from these expectations require deliberate effort and usually require approval from management. Changes to the traditional expectations may, however, be initiated by the administation to overcome Equal Employment Opportunity or other problems.

These expectations are determined by tradition, culture, and experience. The expectations influence the work population, and in turn the work population influences the expectations, as noted in Figure 14-7. Therefore, direct intervention is often necessary to change the expectations before a different work population can be rapidly and safely introduced to the job.

14-3-1 Expectations about Job and Task Assignments

Are all jobs designed for the entire work populaton, or will some people be excused from some jobs? If everyone is expected to perform every job, the jobs must be designed to the lowest common denominator, that is, for the person with the lowest strength, with the least endurance, or with the poorest dexterity. These jobs, then, can be performed by everyone, but they will utilize only a small proportion of the more capable person's capabilities. The tendency may be to design jobs not for the tenth or fifteenth percentile, but for the fortieth or fiftieth percentile, and then simply to select people who meet only those criteria. Having a variety of assignments allows a wider variety of

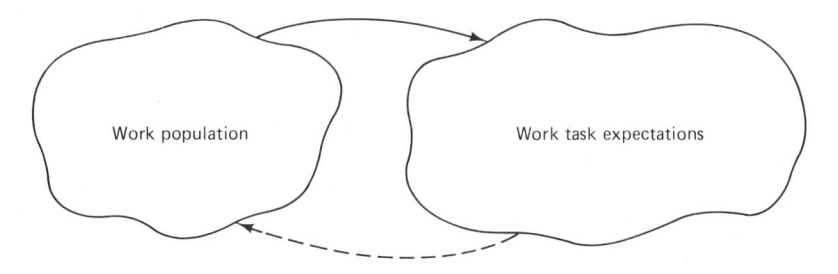

Figure 14-7 Relationship of work population to work task expectations.

people to be employed, although the placement and movement of personnel may be complicated somewhat by who is capable of performing each job.

Rotation of jobs or tasks is a related issue in the design of jobs. Rotation can provide the moderating influence for difficult jobs that will make them acceptable to a wider variety of people. Certainly, there are costs associated with job rotation, such as extra job knowledge and work skills. Yet these costs may be small in comparison with the opportunity to broaden the work force or reduce the injury rate.

14-3-2 Expectations about Physical Capacities

Most plants have either written or unwritten limitations on the amount of weight handled. These limitations are usually far reaching and influence internal operations as well as the weights of materials purchased or shipped. These limitations will control the work population more directly than most other work task expectations.

If the weight limitation is set at 60 lb, each person in the work population will have to lift that amount or risk an injury. The jobs designed and the equipment used will reflect this limitation. With people handling weights of 60 lb, equipment will be used and designed into the system only for material weighing more than 60 lb. If attempts are made to change the work population later, it will be difficult to add the necessary equipment because of space limitations. Without the required changes, the work population will remain the same and frustrate any desires to broaden the work force.

During the introduction of metric standards to the chemical industry, the "standard bag" was being changed from 50 lb to either 20 kg (44 lb) or to 25 kg (55 lb). The standard bag became 25 kg. This decision will have a far-ranging impact on the makeup of the work force in this industry for many years to come. Those people not capable of handling the 25-kg bags either will not be employed, will leave in frustration, or will have higher risks of injury.

14-3-3 Expectations on the Use of Physical Assists

The use of assists to help with manual materials handling tasks is another expectation that influences the design of jobs and work force. Assists are carts, hoists, buggies, spring-mounted tools, and so on, that ease the difficulty of specific tasks.

Are these assists used at all and, if so, for what percentage of the work population? Are they provided for only the most difficult jobs, or are they routinely required whenever job difficulty is anticipated? Are they required to be used, or is their use optional? The answers to these questions will help determine the design of jobs.

If the use of the assists is not required, they may not be used by all the people who need them. When the risk of injury is perceived to be small, people will not willingly use the assist. Yet even a small (less than 1%) injury risk will result in many injuries each year. Table 14-3 lists the factors affecting the acceptability of these assists by workers. Of course, operator participation in determining the need for the tool and in the development of the tool will aid in the acceptability of the tool. And with operator participation, the tool should adequately reflect the actual needs of the operators.

One last issue concerning physical assists is when they are placed into use. Are they provided before or after injury occurs? It is common to find assists installed in response to an injury, yet a better philosophy is to anticipate these needs and install the assists prior to injuries.

14-3-4 Expectations about Work Allowances and Work Standards

How are work allowances set? Are they provided for every job, or are they used only when they are necessary? Are they scientifically set, or are they determined by "best guess"?

Work allowances are a combination of tradition, legal and contractual obligations, and physiological needs. The physiological needs can be determined scientifically based on a fair day's work and actual effort required by the job.

Pacing for machine-controlled work cycles is another issue. To set the pace, one must know the distribution of times required to perform a unit of work. Setting the cycle for the average, then, forces 50% of the operators to

TABLE 14-3 Operator Acceptance of Assist Devices

The following factors tend to affect the acceptability of assists and tools by operating personnel.

Increases Acceptability	Decreases Acceptability
Remove steps from the operation	Add extra steps to the operation
Reduce time to perform the operation	Increase time to perform the operation
Remove an awkward or difficult step	Add an awkward step or make the operation awkward or difficult
Change from a previous, more difficult method	Change from a previous "easier" method
An assist that is universally used by all employees	An assist that is only needed or used by part of the work force
A "penalty" for not using the assist (e.g., a strain from trying to lift something that is too heavy instead of using a chain-fall)	A "penalty" for using the assist (e.g., if using the assist takes longer than if the employee doesn't use the assist)

work too fast and 50% to work too slow. The impact on those working too fast is far more serious than the impact on the other workers. However, setting the slower cycle necessary to accommodate the slower workers further compounds productivity problems. The use of operator-controlled cycles is recommended where possible.

14-4 WORK ENVIRONMENT

The work environment is largely determined by administrative decisions, made with information about operational requirements, costs to modify the environment, injury or illness rates, and legal or contractual requirements.

Since ergonomics is one of several criteria used to make these decisions, the work environment is another appropriate issue for this chapter. For illustration, several work environment issues are discussed.

14-4-1 Noise

Based on hearing loss, noise levels required by OSHA have an 8-hour average of 90 dBA. This is one of the most specific legal requirements for environmental workplace limits. Fortunately, it is well supported ergonomically. Some organizations have specified lower dBA limits, either based on their own evidence or in anticipation of a change in the legal limits.

Noise also affects communications and mental concentration, and some organizations have set lower noise levels where these are required. Offices, in particular, often have lower noise levels established. The ergonomics literature can be a big help in determining these noise levels.

The use of music or of background (masking) noise is generally an administrative decision. Ergonomic research is mixed in its support of these, and often other criteria overshadow ergonomics in these decisions.

14-4-2 Illumination Levels

The quantity and quality of light has been the subject of numerous ergonomic studies. However, with lighting, ergonomics is not a single issue (and may not even be the most important of several issues) in determining illumination levels. The significant lowering of actual illumination levels based on increasing energy costs illustrates the dominance of cost criteria on lighting.

Certainly, ergonomics will have a larger impact on the quality of illumination than on the quantity of illumination. As long as certain minimum standards are met, it is doubtful that ergonomics problems will result from the quantity of illumination. Ergonomics has proven an acceptable reason for changes to the quality of light, as illustrated in the inspection literature. Thus, where it is important, ergonomics can influence illumination levels.

14-4-3 Color

The colors used in the work environment can affect both safety and habitability. Wall and equipment colors, however, are probably influenced more by personal choice than by known data. Economic criteria such as repainting and light reflection are also strong factors.

Operational safety is enhanced when colors distinguish the equipment from its background. Yellow, orange, and red signify danger or moving parts (see Table 14-4).

Habitability is enhanced by nondistracting colors. Some colors excite, while others soothe. The colors that are particularly good from the human standpoint are shown in Table 14-5.

14-4-4 Temperature and Humidity Considerations

Most inside work areas have established temperature and humidity ranges. Air velocity may be controlled as well, although it is not as noticeable. The established limits, which may vary from summer to winter, are used in the design of the facilities. The limits usually are set by management, although legal guidelines have been used since the energy crisis. Some temperature limits are set to accommodate electronic and computer equipment, even though the same comfortable temperature cannot be justified for the computer operator alone.

Even within these bounds, though, ergonomic data can help to establish appropriate work climates which consider the work load and the climate together. Ergonomic data may even form the basis for an exception to the legislated temperature limits, since the work climate must be safe (i.e., not help to promote heat stress) as well as being reasonably comfortable.

TABLE 14-4 OSHA Safety Color Code

Color	Designation
Red	Fire: fire protection equipment, hoses, extinguishers
	Danger: flammable liquids, danger signs
	Stop: emergency stops, stop signs
Orange	Dangerous equipment: machines parts that cut, crush, and shock
Yellow	Caution: physical hazards for stumbling, falling, tripping, striking, caught between
Green	Safety: first-aid equipment
Blue	Warning: caution against starting, using, or moving equipment under repair
Purple	Radiation: x-ray and others

Source: W. H. Weiss, "Surround Your Workers with COLORS," *Production Engineering,* July 1984, p. 73. Excerpted from *Production Engineering.* Copyright Penton/IPC, Inc., July 1984.

TABLE 14-5 Desirable Workplace Colors

Plant Environment	Suggested Colors
Ceiling	White, almost without exception
High-temperature areas	Cool colors such as light green, aqua
Vaulty, chilly spaces	Warm tones such as peach, soft yellow, beige
Casual spaces	Lighter, cleaner hues, such as light blue, coral
For critical seeing tasks	Soft colors, such as pale gold, fern green, colonial green, smoky blue, taupe
Storage areas	White, pastel tints
Stairwells, corridors	Bright tones of soft yellow
Large expanse of unimportant equipment, such as bins	Medium gray
Ordinary objects to be highlighted	Yellow, yellow-green, orange, red orange
Traffic control, housekeeping	White, gray, black
Safety	OSHA colors

Source: W. H. Weiss, "Surround Your Workers with COLORS," *Production Engineering,* July 1984, p. 72. Excerpted from *Production Engineering,* Copyright Penton/IPC, Inc., July 1984.

The use of ergonomic data can help to establish work pace criteria for use in hot and/or humid environments. These environments may be the result of process heat, such as in a bakery, or of high external temperatures. When work in these environments can result in severe heat stress, the pace should be adjusted to compensate for the increased temperature. The Air Force uses such a system to curtail outside drills above 90°F and to reduce them to half pace at 88°F. Similar limitations, an ability to self-pace the work, or self-regulated recovery breaks are all suitable responses to hot work environments. Although these are management decisions, each can be initiated and supported based on current ergonomic literature.

14-4-5 Clothing

The required use of special clothing items, and who purchases those items, are clearly administrative decisions. Special work clothing includes safety items such as hard hats, safety glasses, and gloves; personal protective items such as respirators, impermeable suits, and earmuffs or earplugs; task-required items such as coveralls, steel-toed shoes, or gloves; and temperature-related items, such as reflective suits, cooling vests, and vortex suits for hot areas and thermal suits, coveralls, gloves, and hats for cold work areas.

The use of most of these items is determined largely by tradition and contractual arrangements. Legislation will require the use of some safety items. Ergonomics, however, can and should influence decisions with the temperature-related items. There are many ways of dealing with either heat or cold stress, and clothing is one of those ways. The ergonomist should be able to evaluate all possible alternatives, and when clothing changes are the most effective and economical, those recommendations should be made. For example, the use of personal cooling systems may be the most effective way of repairing hot process equipment without a lengthy cool-down period.

Requiring that specific items be used is an administrative decision. Decisions to share the equipment will also depend on the price per unit and how often it is needed. Ergonomic criteria, such as whether it will be equally effective for different-sized people, can also affect this decision.

14-4-6 Workplace Contaminants

Contaminants at the workplace are primarily the domain of the industrial hygienist, and threshold limit values (TLVs) for chemical exposures are set by regulatory agencies and/or by company management. Ordinarily, ergonomics has little to do with this process. However, with high work paces or long work durations, ergonomics can help to assess and set safe TLVs.

When the work pace is heavy, the person will inspire larger amounts of air per unit of work time. This increases the exposure for the person with the heavier job. The ergonomist can determine, for any given job, the work pace and the volume of inspired air. The results of this analysis coupled with current TLV data can have these results:

1. Lower the work pace to match the current TLV.
2. Adjust the TLV to match the heavier work pace.

The fitness of the work force is also important. If less fit people perform the task, their respiratory rate will be higher, increasing their exposure. If the current exposure level is close to the TLV, this additional level of exposure could be enough to cause a work illness.

Work duration also affects exposure levels. Extended workdays, such as a 10-or 12-hour work schedule, increase the exposure proportionately over the 8-hour work schedule (which is the basis for TLVs). The design of work schedules must consider exposure to workplace contaminants and their TLVs. Rotation on critical jobs can lessen the exposure time, as can the grouping of critical tasks to form those jobs.

14-5 PAY AND PERSONNEL

Ergonomics, by definition, is concerned with personnel at work. In this section, the use of ergonomics specifically to enhance traditional personnel functions is explored. It is useful to demonstrate where it is appropriate for decisions to be influenced by the use of ergonomics information or techniques.

14-5-1 Pay

When pay systems involve a quantitative assessment of several work factors, ergonomics can provide reliable and repeatable evaluation methods. Some ergonomic-related factors that are used in these pay systems are physical difficulty, environmental stressors, and required work skill levels.

Physical difficulty can be assessed on strength requirements or on physiological costs (kcal). Environmental stressors can be measured and compared to desirable levels. When the environmental and physiological factors interact, this can also be determined. Skill levels can be based on performance curves and workplace errors.

In each of these situations, the ergonomic techniques allow scientific, quantitative measurements that produce consistent and reliable results. This is clearly preferable to the qualitative assessments that are commonly used for these systems.

14-5-2 Personnel

The primary personnel function is the procurement of qualified people for job openings. Ergonomics is helpful in assessing and matching people with jobs.

Specifically, ergonomics is used as a basis for the selection of personnel for physically demanding jobs. This involves an assessment of the requirements of the job and an assessment of the capabilities of the person. When these two are compared, a prediction of acceptable job performance is made.

Selection for specific, physically demanding jobs has limited use due to its difficulty. However, these same techniques are used on limited basis for initial hiring decisions. The requirements of the entry-level jobs are known, and job applicants are often compared with these requirements. Although this is a very crude system, the decisions can be improved with quantitative data on the entry-level jobs.

Job-matching techniques have proven useful in the placement of people with temporary or permanent disabilities. After an injury or illness, ergonomic measurement techniques can assess a person's ability to perform specific tasks. This information, coupled with job requirements, can allow a person to successfully perform a required job. The huge medical costs associated with disabilities make this a very economical proposition. Further details are provided in Chapter 15.

In each of these placement situations, ergonomics can contribute measurement techniques. The ergonomist can assist with the actual assessments or can establish processes and procedures to be used by others to perform the assessments.

14-5-3 Use of Personnel Information

Many organizations have extensive systems to capture personnel data related to in-plant injuries, accidents, lost time for various reasons, medical visits, and the like. The ergonomist should influence this information, its use, and its analysis.

Information like this is very useful when it is used to anticipate problems and to prevent their initial occurrence or reoccurrence. The ergonomist should therefore, have easy access to the information. Since the ergonomist knows the types of problems that occur at the work site and on the job, the ergonomist looking at the data firsthand can see the clues that identify problem areas. Problems that affect people based on gender or age are often strength or endurance related. Wrist and hand injuries point toward excessive movements or work stresses. Higher frequencies for specific jobs may only point out the job, not the problem.

The ergonomist should also help determine when and how the information is analyzed. For example, gross injury rates are of little value in diagnosing specific jobs requiring improvements. At the same time, injury rates for individual jobs may not provide enough information to even show a trend. Injuries by job categories may prove to be the most useful way to determine trends.

This information can also be used to demonstrate the effectiveness of an ergonomics program. Long term, the injury rates should have a downward trend, especially in areas where there has been extensive use of ergonomics.

When planning the use of this information, the ergonomist should consider how often the information should be analyzed, how difficult the analyses are to perform, whether they have proven useful in identifying problem areas, and whether management will use this information in approving job studies or job changes.

14-6 SUMMARY

This chapter showed the critical impact that the ergonomist must have on the administrative policies affecting the person on the job. In the first place, the use of ergonomics information is critical to making the correct decision on many of these policy and procedure issues.

In the second place, and perhaps more important long term, the policies of an organization are like an automatic pilot in a vehicle. If the policies strongly encourage an "ergonomics approach" in those decisions affecting people on the job, ergonomics will be implemented easily. On the other hand, if the organizational policies do not encourage the use of ergonomics, each job and equipment design decision will have to be won individually. The mature ergonomics program should have an influence on organizational policy, especially in the areas discussed in this chapter.

QUESTIONS

1. List several of the areas where administrative decisions should be affected by knowledge from the field of ergonomics, but typically have not in the past.
2. List four typical characteristics of "adminstrative decisions" that could involve ergonomics.
3. Why are "administrative decisions" so important to the ergonomist?
4. What are the ways that policies and philosophies are known within an organization? What implications does this have for the ergonomist?
5. List the four major sources of information that affect administrative decisions.
6. How does the current work force have a subtle but powerful influence on the make-up of the future work force?
7. When machine pacing is required, what problems occur if the pace is set for the average operator?
8. Which colors signify danger or moving parts?
9. How can a physically demanding job increase exposure to workplace contaminants?

Chapter 15

THE DISABLED EMPLOYEE

Designing for people with functional disabilities is a normal function of the ergonomist. With increasing medical and disability costs, the interest in reemploying injured workers will increase. This chapter provides an overview of the processes and techniques used in the design of jobs, equipment, and facilities to accommodate people with disabilities.

Accommodations like this are good business practice. The economics shown in this chapter illustrate the favorable costs associated with employment of the handicapped.

15-1 INTRODUCTION

"So you can see that I've got quite a problem," my client said. She had just reviewed a personnel situation she was facing. It seemed that a clerk had been involved in an automobile accident, and the result was a leg with severely reduced functions. My client knew that a replacement was necessary. As she had said: "The job requires delivering reports, filing, obtaining office supplies from central storage, and other walking. With a cane, it's impossible to carry much." The question was not whether to get a replacement—that had already been decided—but what to do with the incumbent. Since I was familiar with most of the jobs in the plant, my client wanted me to help find another job for the clerk.

Now I knew that I was supposed to be thinking of new jobs, but I couldn't help asking a few questions about the old job. "Did she consider using a small cart? Could any tasks be reassigned? What about a leg brace? How was the employee managing similar tasks at home? What did the physician have to say about the injury—would it strengthen with use?" It didn't take long to make my point. Not only could the clerk return to the old job, but there would be no training time for the replacement. It was decided to buy a small cart. Later I found that the clerk was doing fine on the job, and that the cart was in constant use—others were borrowing it because they could carry more and get their own jobs done faster. And another cart was on order.

The industrial ergonomist should be interested in the employment of disabled workers for two reasons. First, there is an economic advantage to employing disabled workers rather than continuing to incur the expense of long-term disability costs. Second, the ergonomist, of all people, must realize that accommodating someone with a disability is usually no more difficult than accommodating any other change in the work population.

According to the National Safety Council, there are some 2.3 million disabling work injuries each year, of which approximately 80,000 are viewed as permanent impairments. This represents a very large pool of disabled employees to be brought back to the work force. A breakdown of the types of disabilities found in the working-age population is shown in Table 15-1.

It is not uncommon to find that the number of disabled people receiving disability benefits in an organization can run as high as 2 to 3% of the current work force. This represents an unusually large cost to the organization and it is a cost that the industrial ergonomist can reduce.

Accommodation of disabled employees is perceived to be prohibitively expensive, yet the facts tell a completely different story. The costs of accommodation are often found to be "no big deal," according to "the Berkeley Study."[1] In this study of 2,000 federal contractors, it was found that 80% of the accommodations cost less than $500 each.

On a macro scale, the economics of employment go far beyond the individual employer-employee relationship. One should look beyond these costs in determining policy regarding the hiring of people with disabilities. A person who becomes disabled represents a substantial cost to federal, state, and local governments for compensations and welfare funds if that person is not employed. Employment, on the other hand, provides additional tax revenues to these same governments. Employment turns a tax drain into a tax asset. At the same time, the social costs are quite large since the self-esteem and well-being of disabled people without employment are substantially lower than those with employment. This large national viewpoint certainly suggests that a conscientious employer seeks to employ people with disabilities whenever it is possible and appropriate at the work site.

---

[1]"A Study of Accommodations Provided to Handicapped Employees by Federal Contractors," Contract No. J-9-E-1-0009; prepared for the U.S. Department of Labor, Employment Standards Administration, by Berkeley Planning Associates, Berkeley, Calif.

TABLE 15-1 Types of Disabilities Found in the Working-Age Population

Disability	Approximate Percent Affected
Musculoskeletal system	40
Heart	25
Respiratory system	13
Special senses	6
Cancer	3
Metabolic, gastrointestinal, and urinary disabilities	14

Source: Thomas J. Armstrong and Dev S. Kochhar, "Work Performance and Handicapped Persons," in *Industrial Engineering Handbook,* ed. Gavriel Salvendy (New York: John Wiley & Sons, Inc., 1981), p. 6.7.7.

There is also federal legislation regarding the employment of people with disabilities. The Vocational Rehabilitation Act of 1973 requires employers with federal contracts in excess of $2500 to take affirmative action to employ qualified handicapped individuals. The term *qualified* is taken to mean that the handicapped person has an impairment that substantially limits one or more of his or her major life activities, but that the person would be able to perform a particular job if reasonable accommodations were made. Accommodations are considered reasonable unless it can be shown that they impose undue hardships due to business necessity and financial cost and expenses.

15-2 OVERVIEW OF EMPLOYER CONCERNS

In speaking with employers about the employment of people with disabilities, several issues always seem to arise. These issues are presented here.

15-2-1 Types of Disabilities

When discussing disabilities, it becomes necessary to qualify the disability. This provides a framework for determining possible choices of employment. Three general areas are certain to arise, as noted below.

Temporary versus permanent disability. The approach to handling temporary and permanent disabilities can be vastly different. With temporary disabilities, the objective may be simply to provide a quicker return to work for the employee. The classic example of this situation involves a lifting injury, where the employee should not return to that task until the injury has completely healed. To facilitate a quicker return to work, an alternative job that does not require lifting can be assigned temporarily.

Permanent disabilities, on the other hand, often require a completely different approach. Substantial accommodations may have to be made to the work side in order for the person to perform his or her old job. If these accommodations are not possible, an alternative job with training may be needed.

The disability may be stable or progressively deteriorating. Stable disabilities—from injuries or polio—are easier to design for and the condition may even improve with therapy or job duties. Progressive disabilities such as arthritis or heart disease may be aggravated by job performance.

Current employee versus a new hire. The process of returning a current employee to his or her old job or even to a new job is quite different from placing a new employee. One incentive to accommodate a disabled employee is the extended disability payments for that person. Many organizations find it expedient to reemploy previous employees before they begin to hire new personnel with disabilities.

The current employee is also a "known" commodity, with dependable work habits, friends, and a knowledge of the organization and how it operates. It is easier for an employer who is unfamiliar with hiring the disabled to take a risk in these circumstances.

What is interesting, however, is that employers begin to gain confidence in their ability to hire and fully utilize someone with a disability through this process. It then becomes much easier to interview and employ a new person with a disability.

Three types of disabilities. There are three general types of disabilities: physical, sensory impairment, and developmental.

Physical disabilities often come to mind first. These impairments take the form of movement and mobility disorders, such as paralysis, amputations, and coordination or other musculoskeletal disorders. They also include cardiovascular and respiratory disabilities.

Another general type of disability is sensory impairment. People who have such impairments may experience partial or total blindness, hearing losses, or other forms of sensory deprivation.

The third general type of disability is called developmental disabilities. These involve mental retardation and emotional disabilities.

Table 15-2 lists the frequency of these disabilities. There are specific methods of dealing with each of these general types of disabilities.

15-2-2 The Economics of Employment

The economics of employment are not as simple as they seem. Initially, employers think only of the expenses of accommodation. Fortunately, there are substantial cost benefits as well. At the plant level, there are all the economic considerations noted below.

TABLE 15-2 Frequency of Disabilities

Disability	Frequency of Occurrence (Percent of U.S. Population Affected)
Difficulty in interpreting information	Approx. 7
Limitation of sight	Less than 1
Limitation of hearing	Approx. 3
Limitation of speech	Approx. 4
Susceptibility to fainting, dizziness, seizures	Approx. 2
Lack of coordination	Approx. 1
Limitation of stamina	Approx. 3
Difficulty in moving head	Approx. 1
Limitation of sensation	Approx. 1
Difficulty in lifting and reaching with arms	Approx. 6
Difficulty in handling and fingering	Approx. 1
Inability to use upper extremities	Approx. 2
Difficulty in sitting	Approx. 5
Difficulty in using lower extremities	Approx. 4
Poor balance	Approx. 2

Source: James Mueller, *Designing for Functional Limitations* (Washington, D.C.: The Job Development Laboratory, The George Washington University Rehabilitation Research and Training Center, 1980).

Medical and disability costs. Some of the fastest-rising costs are associated with medical expenses. These include both medical benefits costs and disability payments. Locating these costs can be difficult. Either they are buried under a general account item and hidden from view, or they are spread out so thinly that they are difficult to accumulate. Since these costs are so large, it is worth the time for the ergonomist to uncover and study them.

The number of people on the disability roles can range from 1 to 3% of the current work force. One financial institution had 120 of 4000 employees on the disabled roles. A manufacturer had approximately 400 of 20,000 employees on a disability list.

Tax incentives. The government has recognized the potential burdens of employment of people with disabilities and has chosen to offset some of those expenses with tax deductions. In the early 1980s, up to $25,000 could be deducted per year for qualified architectural and transportation barrier removal expenses needed to accommodate handicapped workers. There were also targeted job tax credits which gave credit for 50% of the first year's wages up to a maximum of $3000 and 25% of the second year's wages to a maximum of $1500 when a qualified handicapped applicant was employed.

Current tax benefits can be obtained from the Internal Revenue Service, the Department of Labor, and state rehabilitation agencies. These tax benefits, coupled with the reduced disability costs discussed earlier, can make it quite attractive for an employer to seek job accommodations.

Accommodation expenses. Expenses may be required to redesign a work site or work area. However, the cost of these expenses is often overestimated. The Berkeley Study, mentioned earlier, reveals some interesting facts about accommodation costs. The study involved 2000 federal contractors and took 20 months to complete. Table 15-3 presents a summary of accommodation costs.

A summary of this study[2] stated:

The reported accommodations took many forms:

- Adapting the work environments and location of the job
- Retraining or selectively placing the worker in jobs needing no accommodation
- Providing transportation, special equipment or aides
- Redesigning the worker's job, and
- Re-orienting or providing special training to supervisors and co-workers.

TABLE 15-3 Total Cost of Accommodation to Company

Cost	Frequency of Response Number	Percent
None	458	51.1
$ 1–99	169	18.5
100–499	109	11.9
500–999	57	6.2
1,000–1,999	39	4.3
2,000–4,999	35	3.8
5,000–9,999	9	1.0
10,000–14,999	8	0.9
15,000–19,999	6	0.7
20,000 or more	15	1.6
Total for which cost was reported	915	100.0

Source: Frederick Collignon, Mary Vencill, and Linda Toms Barker, "A Study of Accommodations Provided to Handicapped Employees by Federal Contractors," Contract No. J-9-E-1-0009, p. 29, Table 6a. Prepared for the U.S. Department of Labor, Employment Standards Administration, by Berkeley Planning Associates, Berkeley, Calif., 1982.

[2]*Virginia Bulletin,* The Governor's Overall Advisory Council on Needs of Handicapped Persons, 2(3), July 1983.

No particular type of accommodation dominates and most workers received more than one type of accommodation.

Overall work improvements. It is not unusual for changes made specifically to accommodate a disability to make the job easier for other, ablebodied persons. For exmple, a ramp to accommodate a wheelchair will also accommodate a dolly used to move parts. Rarely will an accommodation hinder another worker. These changes can have a positive economic impact, although these improvements are rarely measured.

Summary of expenses. Job accommodation is usually a paying proposition for most firms and should be regarded as simply good business practice. Although most accommodations cost little to nothing, even expensive accommodations can be cost-effective. One national computer company spent $7000 to return a spinal cord–injured worker to full-time employment and at the same time saved $1400 per month in disability payments. This type of return on investment would be envied by any astute businessperson.

15-2-3 Employing Disabled Persons

The ergonomist will find that the process of employing someone with a disability is not substantially different from the process of employing any new person. The initial concern is with finding a successful job match. The abilities and skills of the employee should be matched with jobs the organization has. Part of this will involve making necessary accommodations. Secondary issues will prove critical later. These issues include access outside the work area, emergencies, and promotional opportunities.

Access to and from the job site, and transportation from the home to the work building, are obvious needs. Access to employee services such as the credit union, benefits, and medical facilities must also be considered. Safety during emergencies can easily be planned ahead and the special needs of the handicapped considered—assistance down stairs, guides for sight impairments, and special alarms for the hearing impaired.

The reactions of current workers to a new worker (with or without a disability) can range from acceptance to curiosity to rejection. Initial preparation of the work group—both peers and supervisors—will smooth the entry. Often the work group can be a substantial asset in helping any new person perform the job easier and better.

Promotional opportunities are not often considered in the haste to make the initial placement. Yet possible future jobs should be noted and an assessment of needed accommodations made. Ideally, the original accommodations will continue to be useful, and job skills and work knowledge will build from one job to the next.

15-2-4 Multidisciplined Approaches to Job Placement

With severe disabilities, a multidisciplined approach to job placement and accommodation is necessary. Ergonomists can specify task and equipment changes, medical and rehabilitation personnel can determine functional abilities, engineering personnel can design actual equipment changes, and EEO/personnel specialists know personnel policies, benefits, and so on. Each person on the team brings a strength to the process of designing for the severely disabled.

Smaller plants or companies may not have access to these resources, yet there is help available from local agencies and area resources. The social services agencies in city or county government are familiar with rehabilitation, and nearby universities will have professional staff who are able to help with the design of the accommodations. Other resources, such as the President's Committee on Employment of the Handicapped or the American Foundation for the Blind, have accommodation information available by mail. Even local resources such as high school shops can be used to build special jigs or fixtures. Creativity and ingenuity are often the best resources that one has in implementing appropriate accommodations for the handicapped.

15-2-5 Job Placement and Transitional Work Centers

Some industries have found it convenient to have a central coordinator for handicapped placements. These vary in sophistication and whether the placement is temporary or permanent. The coordinator always seeks to find suitable jobs that use the abilities of that person in a productive and cost-effective manner for the organization.

The process for employees with temporary disabilities such as broken bones or sprains and strains involves an assessment of the disability, followed by temporary jobs that do not affect the injury. The jobs may be in a "sheltered work environment" where small jobs are brought in. An alternative is to loan these employees to areas requiring help, thus avoiding the need to hire temporary help from outside the company. Some jobs are almost clichés, like stuffing envelopes for stockholders' meetings, sorting parts, minor repairs, delivering parts, and the teardown and reclamation of scrap, yet these jobs are regularly performed and someone must do them.

Placements for permanent disabilities are more complex than this. The disabilities may be physical, sensory, or developmental. Nemeth[3] discusses placement programs for permanent disabilities so that a match is made between employee restrictions and the physical requirements of the job. Once

---

[3]Susan E. Nemeth, "Job Documentation for Employee Placement," *Ergonomics News,* 19(2), Fall 1984, p. 1. Reprinted from *Ergonomics News,* Fall 1984 Newsletter of the Institute of Industrial Engineers, 25 Technology Park/Atlanta, Norcross, GA 30092.

an injured employee is determined to be unable to return to the former job, a detailed review of the employee restrictions are made. The placement coordinator, ergonomist, industrial engineer, and plant physician are all involved in the determination of job suitability. Nemeth states: "Job placement systems range from informal systems, where the Placement Coordinator may work purely from his/her knowledge or from a list of a few jobs that usually meet most restrictions, to formally documented systems where nearly all jobs on the manufacturing floor are included in a computer file." One assessment system is provided in Table 15-4. Regardless of which system is used, the requirements assessed on the job should match the abilities assessed for the workers. Like most task analyses, there are no specific rules as to the amount of detail required, so several iterations may be needed to get the correct information detail.

TABLE 15-4 Job Documentation System

Physical Requirement Categories for Jobs

1. Use of the hands: right, left, either, both, or not required
2. Manipulative finger movements
3. Use of the wrists: flexing or rotating movements
4. Standing or walking: percent of job that requires standing or walking
5. Use of feet: foot-pedal operation
6. Use of the arms: right, left, either, both, or not required; use above and below shoulder level
7. Lifting requirements: weight and frequency, above and below waist level
8. Pushing or pulling: maximum force required
9. Bending, stooping, or twisting movements: degrees of bend, stoop, or twist required
10. Vision: minimum sight requirements, depth perception, color vision
11. Hearing: minimum auditory requirements, noise level
12. Environmental conditions: temperature, atmospheric contaminants, air-conditioned, skin irritants
13. Paced or unpaced task

Physical Ability Categories for People

1. Use hands, fingers, wrists
2. Stand, walk, use feet
3. Use arms above and below waist level
4. Lift above and below waist level
5. Push and pull
6. Bend, stoop, twist
7. See
8. Hear
9. Work in particular environments
10. Work on paced jobs

Source: Susan E. Nemeth, "Job Documentation for Employee Placement," *Ergonomics News,* 19(2), Fall 1984, pp. 1-2. Reprinted by permission. Copyright Institute of Industrial Engineers, 25 Technology Park/Atlanta, Norcross, GA 30092.

15-3 THE ACCOMMODATION PROCESS

In Chapter 6 a process was introduced for the redesign of jobs and equipment that resulted in ergonomics problems. Designing for new work populations or known work limitations is a specific subset of that process. This section discusses that specific process for the design of jobs for people with disabilities. It can be used for temporary or permanent disabilities, although any costly accommodations will probably be reserved for long-term disabilities.

Job accommodation is very compatible with industrial ergonomics. Once a mismatch in a work system (person-task-environment) is discovered, traditional ergonomics principles state that the preferred approach is to redesign the task to eliminate the problem. Secondary approaches include enhancing the person's ability to perform the task and reallocation of tasks to others who can perform them.

For job accommodation of the disabled employee, there are three successive and preferred steps. The first step is to remove excessive task demands by redesign of the equipment or facility so that everyone can perform the job. The second step is to enhance personal performance through job aids such as phone amplifiers, electric carts, or adaptive devices such as protheses and braces. The third step is to transfer activities to other workers who can perform them. Figure 15-1 illustrates this process. It is typical to use several of these methods concurrently.

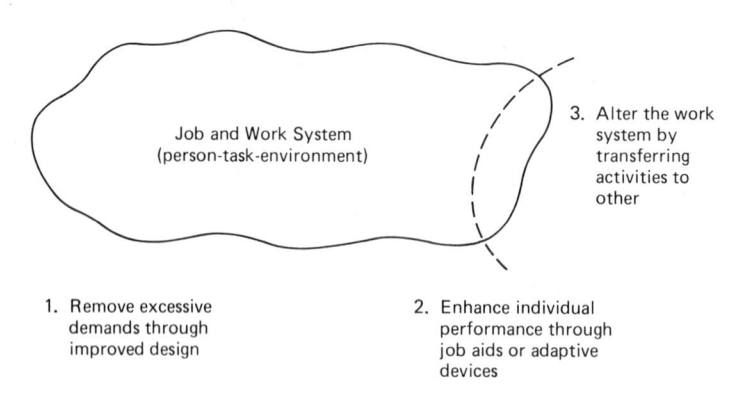

Figure 15-1 Job accommodation alternatives.

15-3-1 The Process for Job Accommodation

The accommodation process involves several different steps, each outlined below, and shown in Figure 15-2.

Step 1. The initial step ensures that the person is qualified for the job. The person must have the basic skills and job knowledge required to perform

Chap. 15 The Disabled Employee 345

Step 1. Individual qualified for job

Step 2. Define disability
 Functional capabilities
 Limitations

Perform task analysis for proposed job

Comparison

Acceptable — Mismatch

Step 3. List redesign accommodations needed
 Job — Change methods, processes, redesign tasks
 Equipment — Fixtures, jigs, special tooling, location
 Facilities — Accessibility, parking
 Training — Job skills, methods
 Environment
Explore job aids and enhancements
Consider — Costs
 — Productivity enhancements
 — Aid to others with disabilities
 — Aid to able-bodied workers

Select/implement alternatives

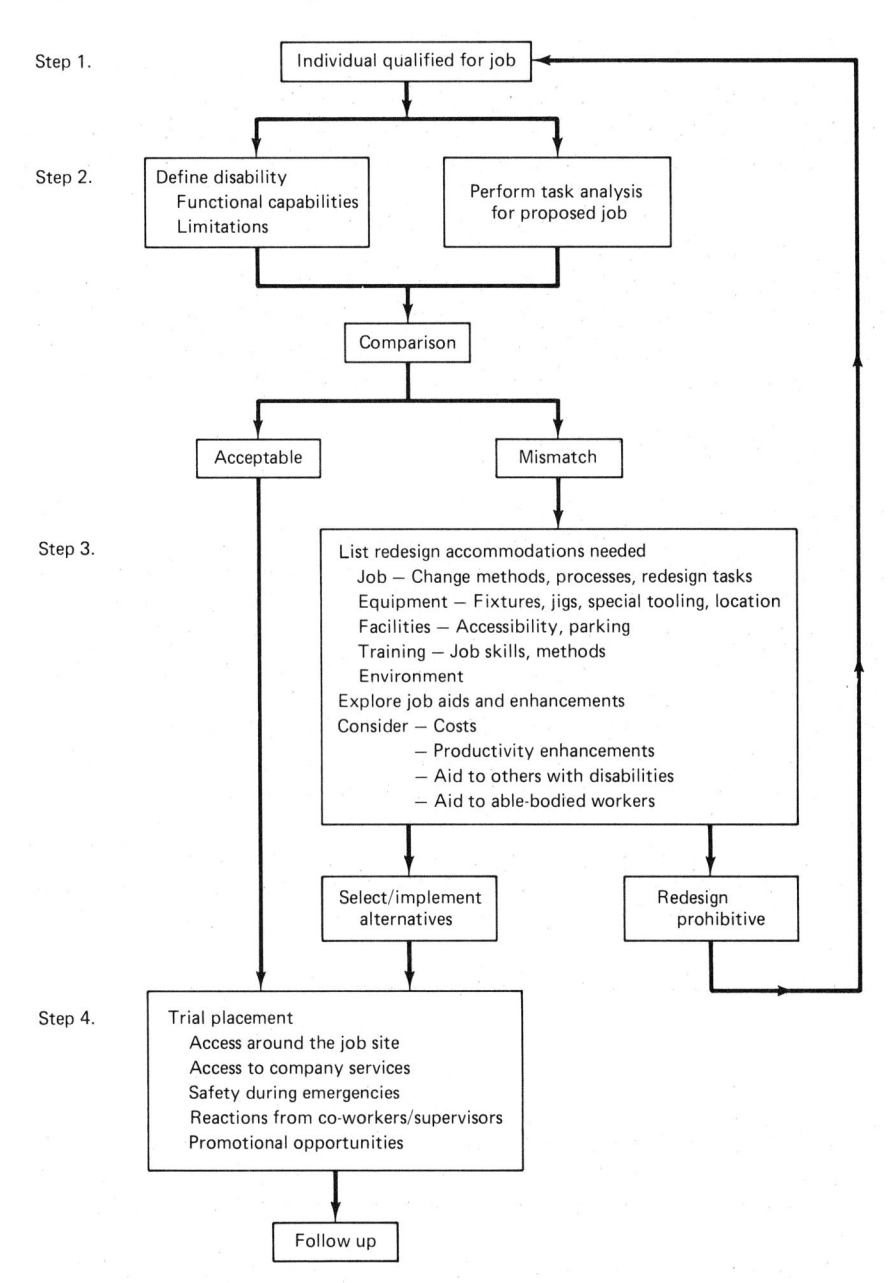

Redesign prohibitive

Step 4. Trial placement
 Access around the job site
 Access to company services
 Safety during emergencies
 Reactions from co-workers/supervisors
 Promotional opportunities

Follow up

Figure 15-2 Process for job accommodation. *Source:* Adapted from David C. Alexander. "Job Design You Can Use," a continuing education seminar, sponsored by the Institute of Industrial Engineers, May 1982.

the job successfully. If this is not true, another job must be found, or a job training process is required. This, of course, is the first step in any employment process.

Step 2. The second step in the accommodation process is assessment of both the requirements of the job and the abilities of the person. For the person, there are a number of assessment methodologies available to determine reach, strength, and aerobic capabilities. Usually, however, a complete assessment is unnecessary—only the specific tasks required by that job must be assessed. Ergonomists and medical and rehabilitation personnel can perform this assessment.

Task analysis is effective for analyzing the job. Task analysis was discussed in Chapter 6, and more detailed discussions are available elsewhere. Industrial engineering personnel are excellent resources to perform task analyses.

A comparison of the job requirements and the person's capabilities is then made. If the person is acceptable with no accommodations on the job, then the person is placed. Step 3 discusses needed job accommodations.

Step 3. With a mismatch between the job requirements and the abilities of the person, job accommodations are needed. The changes may be to the facilities, to the equipment used, to the work area and environment, to the job, or even to the person.

The changes to facilities and equipment represent the hardware changes. These include jigs, fixtures, and tooling for special tasks. Additional changes are raised desks, lower water fountains, wider doorways, and so on. Special equipment for mobility needs or for sensory enhancement may be required. Facility changes can provide needed access within and around the work site.

Job redesigns are often cheaper to implement. These involve changes such as rearranging tasks, omitting difficult tasks, and changed work methods and/or processes.

Training can enhance skills to perform additional tasks or to perform the old tasks in new ways. Training can be effective by itself, although it is often used in conjunction with other changes.

Other changes, such as modifying a work schedule, assigning a nearby parking spot, and changing the work site to the ground floor can be effective and inexpensive. An able-bodied "buddy" may be assigned to handle difficult lifts, excessive reaches, and other infrequent and unusual situations.

Job aids can enhance limited functions. Electric carts help mobility, phone amplifiers help the hearing impaired, and braces aid stability. All of the accommodations should be evaluated for costs/benefits, the impact on productivity (they may enhance or reduce productivity), and the aid to others with disabilities and/or to able-bodied workers.

If the redesigns are acceptable from an economic standpoint, they are implemented. If not, the process of finding another possible job begins again.

Step 4. Once a suitable job is found, a trial placement is made. Such factors as access around the job site, access to company services, safety during emergencies, reactions from coworkers and supervision, and promotional opportunities and plans are all addressed at this time.

Close contact should be made during the early part of the trial to assess its success. Is the job being performed to expectations? Are the accommodations working as anticipated? Is output acceptable? Are personal relationships developing? Have the emergency procedures been tried? As time on the job increases, the follow-ups will become less frequent. The process for job accommodation is shown in Figure 15-2.

15-3-2 Making the Accommodation Process Easier

Many people think of the accommodation process as burdensome and difficult. It is not usually that difficult, with the process outlined above as a guide. Even so, there are some ways to make the process even easier.

By involving the disabled person in the process, workable ideas often surface earlier. That person may have made the same accommodation at home or elsewhere, or may know someone else who has made a similar accommodation. For practical accommodations, this is an unusually good source of information.

Let one person become the focus for these placements. Experience is a great teacher, and what is learned on one placement or accommodation will be useful later. That person can also become the keeper of accommodation literature and design standards. These resources should be collected in advance and cataloged. Then, when it is needed, accommodation information is readily available. Useful resources include "Specifications for Making Building and Facilities Accessible to and Usable by Physically Handicapped People" (published by the American National Standards Institute, Inc.), the Vocational Rehabilitation Act of 1973, as well as specific pamphlets and design information from the President's Committee on Employment of the Handicapped, the American Council of the Blind, the American Foundation for the Blind, the American Heart Association, the National Easter Seal Society, the United Cerebral Palsy Association, and other societies.

The use of preplacement discussions with the work group, both peers and supervisors, can dispel unnecessary apprehension. An understanding of the disability and how it may affect performance will help the work group accept the new worker. Let the work group help with on-the-job adjustments—when minor problems occur, involve them in the solution. Establish clear expectations about the success of this job placement. And do not let others provide unneeded production assistance—however helpful it may seem at first, it will become a sore point later.

Publicize successful accommodations, both within your work organization and within the community. This will help to dispel myths of difficult placements and will ease your next placement. At the same time, it will begin to develop a network of others interested in accommodations for handicapped employees. This local network can prove valuable for solving problems in the future.

When possible, set the tone of strong top management commitment to the accommodation of disabled workers. Successful placements are the best way to develop this commitment. Lower disability payments further reinforce this commitment.

15-4 PREDESIGN FOR ACCOMMODATIONS

Many job placements are easy if the work site and facilities are already designed to meet the needs of people with disabilities. Predesign must occur both to ensure accessibility to the work site, as well as usability within the work site.

Many times an accommodation is prohibitively expensive simply because of earlier design decisions. For example, narrow aisleways, if they are part of the structural integrity of the building, can limit use of that building by people in wheelchairs. Similarly, excessive stairs at a building's entry will block entry by mobility impaired persons. At the work site, lower light switches cost nothing during initial construction, but later prove to be very expensive to relocate.

Design of facilities to accommodate people with disabilities need not be prohibitively expensive. In fact, given the trends in disabilities and the age of the population, predesigning for these eventual needs is simply good engineering and business practice.

15-4-1 Design Standards

Design standards are available from the American National Standards Institute in the form of "Specifications for Making Building and Facilities Accessible to and Usable by Physically Handicapped People." These standards provide dimensions that ensure accessibility for mobility impaired and visually handicapped people.

These standards do not tell you how to design a work site, but rather provide raw information that can be utilized to design the work site. In that sense they are much like anthropometric data—very valuable if you know how to use the information, but of little value if you want a finished design.

Information is provided in these and other areas:

Inside accessibility
    Minimum space requirements
    Reaches and clearances
    Ramps, stairs, and elevators
    Doors and aisles
Outside accessibility
    Parking
    Curb cuts
    Ground surfaces
    Entrances
Bathrooms
Work areas
    Storage
    Seating, tables, and work surfaces
    Assembly areas
Services
    Water fountains
    Telephones

Some examples of the type of information found in the ANSI Standard are shown in Figure 15-3 and in Table 15-5.

15-4-2 Integrated Work Site Designs

The ergonomist and others may be interested in a "complete work site" that integrates all of the necessary factors. Two model work sites, designed by Jim Mueller of the Job Development Laboratory at George Washington University, are shown in Figures 15-4 and 15-5. Figure 15-4 is an office-type setting and Figure 15-5 is a more industrial setting. These design guidelines can aid a large percentage of the disabled employees and can benefit the nondisabled employee as well.

In addition to these general design guidelines, the Job Development Laboratory also provided detailed design recommendations for specific disabilities such as poor balance, difficulty in lifting, difficulty in sitting, limitations of hearing, and so on.

Others have also developed consolidated designs. The Appliance Information Service of Whirlpool Corporation has written "Designs for Independent Living—Kitchen and Laundry Designs for Disabled Persons." It provides the same "finished design" information that is so usable for the practitioner.

15-5 ACCOMMODATIONS AND AIDS

There are many accommodations and job aids available and workable accommodations are still increasing dramatically. One reason is the increasing use

(a) Minimum clear width for single wheelchair

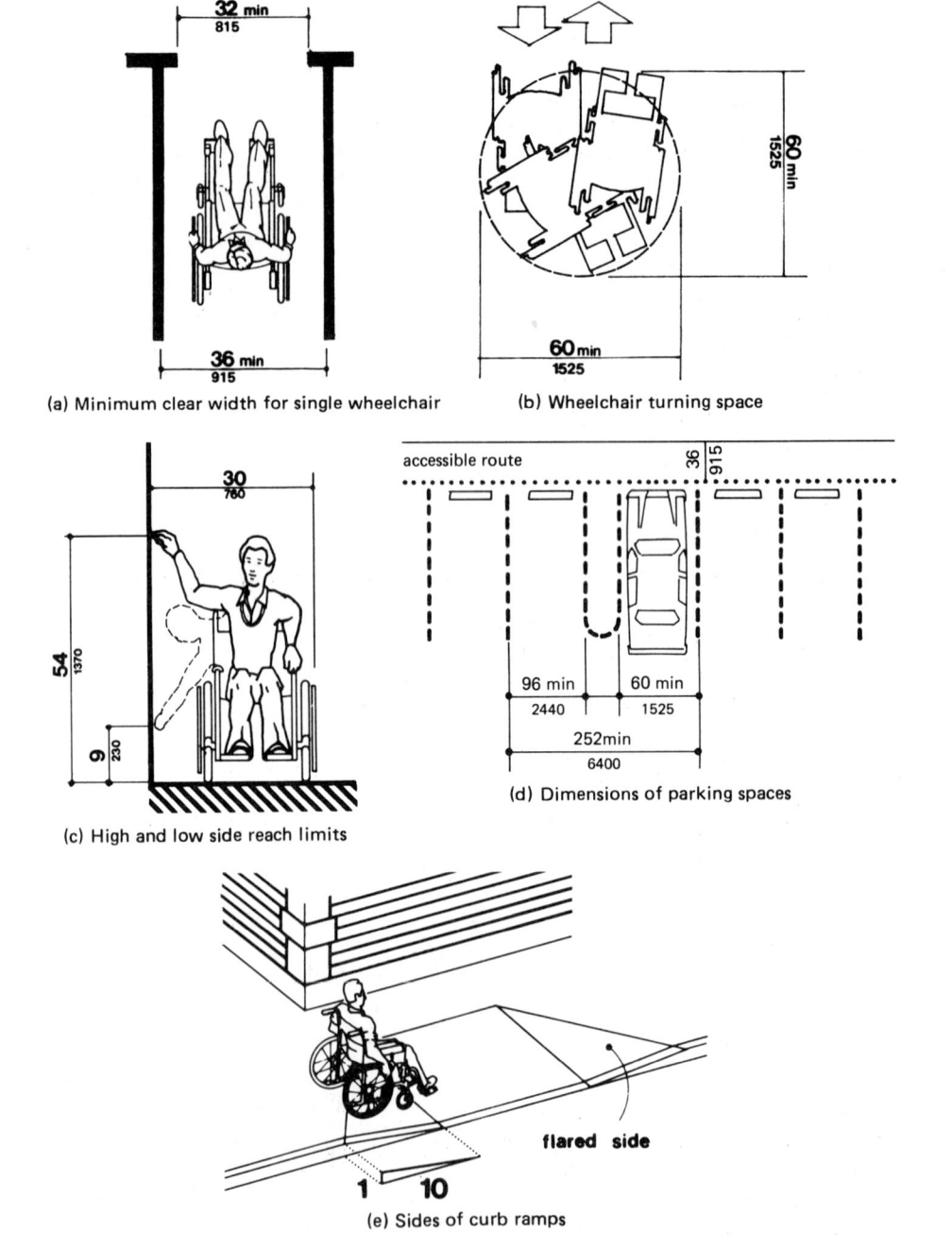

(b) Wheelchair turning space

(c) High and low side reach limits

(d) Dimensions of parking spaces

(e) Sides of curb ramps

Figure 15-3 ANSI design specifications. *Source:* American National Standards Institute, *Specifications for Making Buildings and Facilities Accessible to and Usable by Physically Handicapped People, ANSI A117.1-1980.* Reprinted by permission. Copyright 1980 by the American National Standards Institute. Copies of this standard may be purchased from ANSI, 1430 Broadway, New York, NY 10018.

350

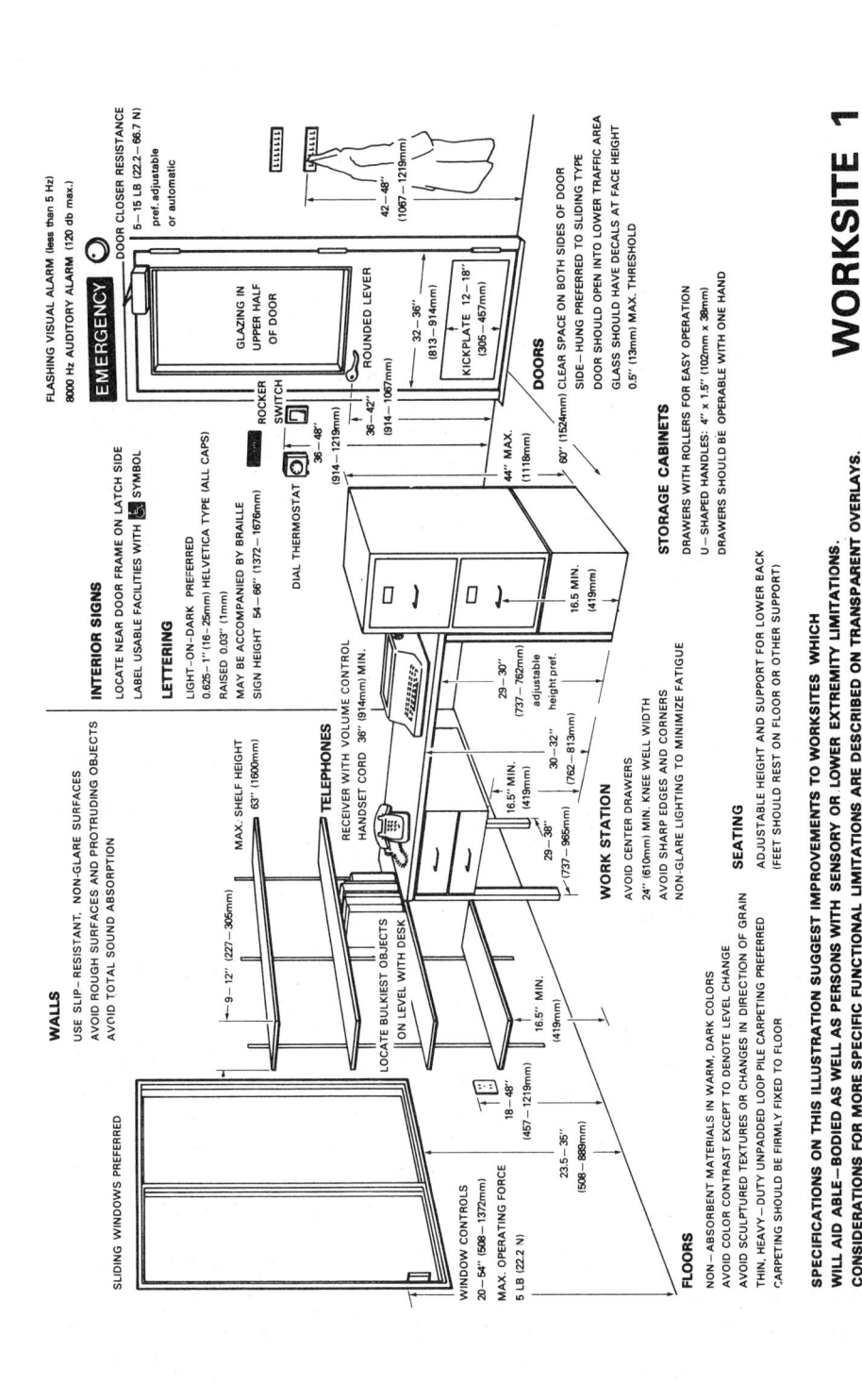

Figure 15-4 Worksite 1: the office setting. *Source:* James Mueller, *Designing for Functional Limitations* (Washington, D.C.: The Job Development Laboratory, The George Washington University Rehabilitation Research and Training Center, 1980), p. 5.

351

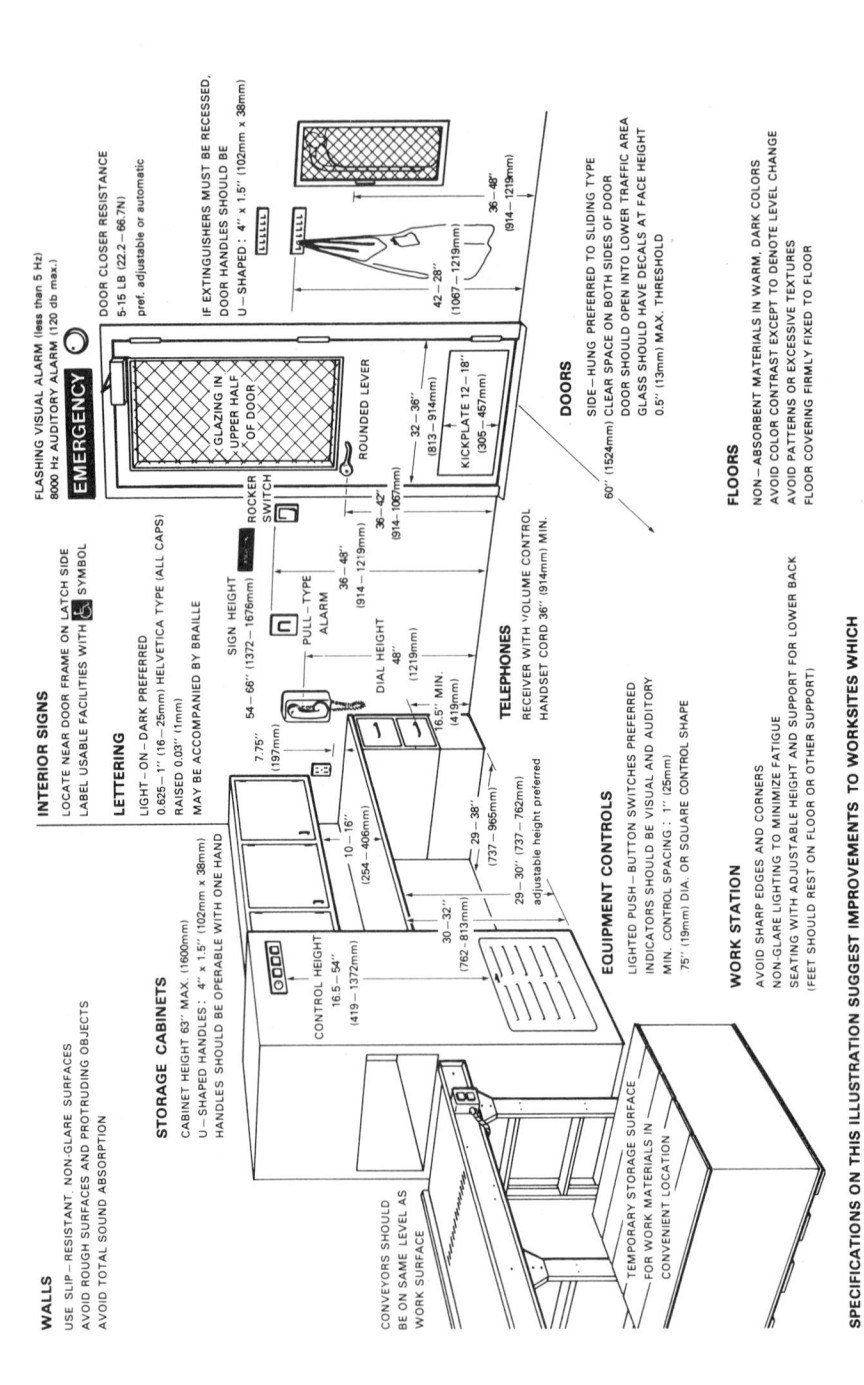

Figure 15-5 Worksite 2: the industrial setting. *Source:* James Mueller, *Designing for Functional Limitations* (Washington, D.C.: The Job Development Laboratory, The George Washington University Rehabilitation Research and Training Center, 1980), p. 7.

Chap. 15 The Disabled Employee

TABLE 15-5 Convenient Heights of Work Surfaces for Seated People[1]

	Short Women		Tall Men	
Conditions of Use	*in.*	*mm*	*in.*	*mm*
Seated in a wheelchair:				
Manual work:				
Desk or removable armrests	26	660	30	760
Fixed, full-sized armrests[2]	32[3]	815	32[3]	815
Light, detailed work:				
Desk or removable armrests	29	735	34	865
Fixed, full-size armrests[2]	32[3]	815	34	865
Seated in a 16-in. (405-mm)-high chair:				
Manual work	26	660	27	685
Light, detailed work	28	710	31	785

[1] All dimensions are based on a work-surface thickness of $1\frac{1}{2}$ in. (38 mm) and a clearance of $1\frac{1}{2}$ in. (38 mm) between legs and the underside of a work surface.

[2] This type of wheelchair arm does not interfere with the positioning of a wheelchair under a work surface.

[3] This dimension is limited by the height of the armrests; a lower height would be preferable. Some people in this group prefer lower work surfaces, which require positioning the wheelchair back from the edge of the counter.

Source: American National Standards Institute, *Specifications for Making Buildings and Facilities Accessible to and Usable by Physically Handicapped People,* ANSI A117.1-1980 (New York: American National Standards Institute, 1980, p. 67. Copyright 1980 by the American National Standards Institute and reprinted with permission. Copies of this standard may be purchased from the American National Standards Institute, 1430 Broadway, NY, NY 10018.

of electronics technology to mimic human functions and to transport data. For example, the Kurzweil reader is a device that will actually read printed text aloud. Jobs that once had to be done "at the office" can now be done anywhere there is a computer terminal and a phone line. Even developmental disabilities can be enhanced through computer memory storage and through decision rules placed in programs. Many accommodations are used in combination, especially for more severely and multiply disabled persons.

Armstrong and Kochhar discuss accommodations: "Accommodations include either enhancement of worker performance by means of an adaptive device or modification of the work and workplace so as to reduce the performance requirements. A guiding principle is to remove completely or to reduce to a tolerable level the job demands that a disabled worker cannot meet. When such demands cannot be eliminated or reduced to a tolerable level, the work should be reorganized so that the more difficult activities are transferred to those workers who are capable of performing them."[4]

Obviously, accommodations work best when they are focused on clearly defined tasks that the person must perform. Seek information on accommodations and aids from a variety of sources. The accommodation need not be elaborate to be effective—in fact, some of the more functional accommodations are the result of common sense and ingenuity.

15-5-1 Types of Accommodations

As noted in Section 15-3, job accommodations take several forms:

1. Removing excessive demands through improved design
2. Enhancing individual performance through job aids or adaptive devices
3. Altering the work system by transferring activities to others or changing job assignments

The President's Committee on Employment of the Handicapped lists some common accommodations, as noted below.[5]

Common job site accommodations
- Building modifications: ramps, washroom remodeling, wider doors and aisles, elevators
- Providing training in disability to supervisors and coworkers
- Providing convenient parking and carpool or van transportation
- Providing flexible working hours, rest periods, or work at home

Common job task accommodations
- Adjusting the work space: lowering workbenches, providing a private office
- Adjusting the work area: moving a job to a ground floor location
- Retraining or placing the disabled employee in jobs that require no accommodation
- Providing special assistance: readers for blind workers, sign language interpreters and TTY equipment for deaf workers, talking calculators, and reading machines

---

[4]Thomas J. Armstrong and Dev S. Kochhar, "Work Performance and Handicapped Persons," in *Industrial Engineering Handbook*, ed. Gavriel Salvendy (New York: John Wiley & Sons, Inc., 1981), p. 6.7.7.

[5]*Employer Guide—Simple Steps to Job Accommodation* (Washington, D.C.: The President's Committee on Employment of the Handicapped).

- Redesigning elements of the job to fit the disabled worker
- Redesigning written examinations

The experience gained in the Berkeley Study is useful in providing information on the types of accommodation actually used in industry. Their experience indicates that a variety of accommodations are used, and the stereotyped accommodations, such as ramps and expensive reading machines, do not prevail. Table 15-6 provides data on the accommodations found. The authors state: "There is no typical pattern of accommodations provided to handicapped workers, even when divided according to disability groups."

15-5-2 Work Aids and Enhancements

Work aids enhance a person's abilities. The aids can be complex, like the Kurzweil reading machine, or simple, like a pivoting work stool. The aids are not normally specific to a job but can enhance performance in many situations. The aids are also good ergonomic design, since they often make a job easier for able-bodied employees as well.

Aids are used to enhance or replace a sensory input, or to enhance or replace a mobility or strength requirement. Aids seem to be determined by need, creativity, and imagination. Descriptions of a few of the many aids follow.[6,7]

- Talking calipers and other tools allow blind machinists to use hearing instead of vision to read these tools.
- Talking calculators allow the user to hear the numbers entered and calculated instead of seeing them.
- TTY equipment sends a written message so that a deaf person can use vision instead of hearing.
- Sit-stand workstools can ease the burden of standing for long hours, yet provide the mobility that hairdressers and barbers need for their work.
- Rotating bookcases/racks for convenient storage of heavy reference books can reduce the need to lift and handle heavy books.
- Small carts provide the needed support and carrying capabilities for people with balance/stability problems.
- Wrist supports ease shoulder and arm stress for people using typewriters and data-entry devices.

---

[6]Adapted from Gerd Elmfeldt and others, *Adapting Work Sites for People with Disabilities: Ideas from Sweden* (Bromma, Sweden: The Swedish Institute for the Handicapped, 1981).

[7]Adapted from James Mueller, *Designing for Functional Limitations* (Washington, D.C.: The Job Development Laboratory, The George Washington University Rehabilitation Research and Training Center, 1980).

TABLE 15-6 Types of Accommodations Provided

Category 1: Adaptations of the work environment and location, including the following types of accommodations: removed barriers, adjusted work environment, adjusted table/desk, other rearrangements, relocated work site.

Removing architectural barrier for individual	5.7%
Adjusting the work environment (heat, light, ventilation)	2.8%
Adjusting table, desk, bench, etc.	6.4%
Other rearrangements of work site	4.6%
Relocating work site	1.5%

Category 2: Provision of special equipment and assistance, including the following types of accommodations: modified telephone, typewriter, etc., provided audiovisual aids such as microfilm or dictaphone, provided other special equipment, tools, or devices, provided transportation or other mobility assistance, and assigned aides, readers, etc. (These are accommodations which enable an employee to perform the assigned job functions by providing "something extra.")

Modifying telephone, typewriter, etc.	2.8%
Providing microfilm, dictaphone, audiovisual aids	1.1%
Providing other special equipment, tools, or devices	4.7%
Providing transportation or other mobility assistance while on job	3.2%
Assigning aides, readers, etc.	3.2%

Category 3: Accommodations involving job modifications, including: reassigned tasks, modified work hours, and other modifications of work procedure.

Assigning tasks to other workers	8.8%
Modifying work hours or schedules	5.2%
Other modification of work procedures	8.9%

Category 4: Training and transfer included the following accommodations: provided additional training and transferred employee to another job. These are grouped together because they are both one-time efforts which may make further accommodation unnecessary by minimizing the work limitations imposed by the disability. They involve adjustments on the part of the individual accommodated worker.

Providing additional training	5.2%
Transferring employee to another job	8.7%

Category 5: Orientation consisted of orienting coworkers and supervisors to provide special assistance. This was the single most frequently cited form of accommodation. Though it happens in a variety of ways, both formal and informal, it basically involves special consideration or efforts on the part of other individuals in the work situation.

Orienting supervisors and coworkers to provide necessary assistance	18.0%
Other	8.9%

Source: Frederick Collignon, Mary Vencill, and Linda Toms Barker, "A Study of Accommodations Provided to Handicapped Employees by Federal Contractors," Contract No. J-9-E-1-0009; prepared for the U.S. Department of Labor, Employment Standards Administration, by Berkeley Planning Associates, Berkeley, Calif., 1982, pp. 23-24.

Chap. 15   The Disabled Employee   357

- Electric staplers reduce heavy forces.
- Phone amplifiers enhance sound for hearing-impaired persons.
- Pen/pencil holders molded to fit the hand and large pencils/pens enhance grip strength for arthritics and others with low grip strength.
- Automatic storage/retrieval units ease lifting and carrying requirements for people with low back strength and low lifting strength.
- Electric carts enhance mobility.
- Lifting hoists enhance low lifting capabilities.
- Single-unit faucet controls reduce the need for hand–eye coordination and hand stability.
- Lever door handles reduce the need for grip strength and coordination.

Obviously, there are thousands of other work aids and enhancements. This list is more of a stimulator than anything else—it is intended to show how the aids either enhance or replace a normal human function, and how many aids there might be.

Useful aids mimic typical lost human functions, such as sensory input (sight, hearing, feeling), standing/stability, lifting, holding, manipulating, moving or carrying, coordination, stamina, and learning/interpreting information. Typical aids, therefore, are with communications and writing devices, workshop equipment and tools, transportation aids, lifting devices, and work furnishings and office aids.

15-5-3 Examples of Accommodations

David Rhyne tells of a successful accommodation: "... a manual press operator who was permanently disabled by an automobile accident.... Foot pedals were used to compensate for the loss of strength in his arms; a locking system was utilized so that tools could be placed in openings by feel; a container of specific capacity eliminated the need to count; and ear plugs were utilized to reduce shop noise and ensuing headaches. The total job accommodation package cost approximately $250."[8]

In an article in the *Washington Post*, Caroline Miller discusses another situation: "Currently, 3M is spending about $6,000 to install a mechanical lift and redo a restroom for a worker who was seriously injured in an automobile accident two years ago. Now a quadraplegic who gets around in a motorized wheelchair, the employee will be working at a computer terminal nearby."[9] The article further discusses a construction worker who suffered orthopedic injuries in a fall who is now an administrative coordinator for a construction

---
[8] David M. Rhyne, "IEs Can Play Vital Role in Bringing the Disabled into Economic Mainstream," *Industrial Engineering,* April 1984, p. 63.
[9] *The Washington Post,* November 25, 1984, p. F4.

group; an ironworker suffering from neurological disease who is now operating a toolroom; a production operator with an injured wrist who now works with audiovisual equipment; and a material handler with a dislocated shoulder who is now drawing detailed design layouts.

Some excellent accommodations are presented by the Swedish Institute for the Handicapped.[10] The disabilities are all serious. Some are highlighted below.

- A television repairman working successfully with a double-hook prosthesis, due to an amputation resulting from an explosion.
- A forestry worker who operates an articulated tractor after a tumor forced the amputation of an arm and shoulder. The tractor was modified to place all controls on one side and to use foot pedals.
- A sanitation worker with disc degeneration that limited lifting of workshop refuse. A hoist and carts with wheels eliminated the need for lifting.
- A clerical worker with rheumatoid arthritis was transferred to a job that involved less constant typing and was provided an electric typewriter with wrist rests, a pushbutton phone, a height-adjustable chair, and a few other office tool changes.
- A factory worker who suffered reduced arm and leg strength after a stroke is now working as an expediter, a job that requires movement about the plant on a small scooter.

15-6 SUMMARY

The design of jobs and equipment to accommodate people with handicaps is a natural function for the industrial ergonomist. This chapter discussed the importance of the issues and the concerns that most employers seem to have.

A model for assessing the need for accommodations and for developing accommodations was presented. Examples of many successful (they work and are cost-effective) accommodations are discussed. The ergonomist should be proactive in encouraging employers to employ people with disabilities.

QUESTIONS

1. Why should the ergonomist be interested in the employment of disabled workers?
2. Which disabilities are the most common? What general types of disabilities should the ergonomist be most prepared to accommodate?

[10]Elmfeldt and others, *Adapting Work Sites for People with Disabilities: Ideas from Sweden.*

Chap. 15   The Disabled Employee   **359**

3. What is an estimate of the percentage of people receiving disability benefits in an organization?
4. According to the Berkeley Study, what percentage of the accommodations cost less than $500?
5. List an organization's financial concerns associated with the re-employment of someone who incurs a disabling injury off the job.
6. Give two examples of accommodations for disabled workers that benefit able-bodied workers.
7. How can employees with temporary disabilities be brought back to work sooner?
8. What is required to successfully bring a permanently disabled employee back to work?
9. List the types of physical abilities assessed in the Job Documentation System discussed by Nemeth.
10. What are the three methods, in order, for providing job accommodations?
11. What is the one essential requirement before accommodation begins?
12. List and briefly discuss the four steps in the process for job accommodations.
13. How can the ergonomist make the accommodation process easier?
14. What are some of the design features that should be considered for new facilities to ensure that disabled employees can be easily and inexpensively accommodated?

Part 5
KEEPING IT GOING

Chapter 16

SUSTAINING THE ERGONOMICS EFFORT

Part 5 (Chapter 16) acknowledges the fact that an ergonomics program will have a short-term effect on the organization. Only by sustaining the ergonomics effort long term will it have maximum impact.

The ergonomics effort will be more effective if it becomes more than just another program. The ergonomist must manage the effort so that it will be sustained over the long term. The ergonomist will be more effective when the program is well accepted rather than spending time and effort fighting for survival. Greater effectiveness also occurs when there are more people using ergonomics. Therefore, the ergonomist needs to win converts to the practice of ergonomics.

16-1 INTRODUCTION

I had a call the other day from a friend in another part of the country. "How do you keep the momentum going?" he asked. "We have been pushing our ergonomics program for 15 months now, and people have quit asking me to come speak to their work groups about ergonomics. I think they must be losing interest. How do I keep their interest up?" We discussed several programs with other companies that had failed and why. Not managing the ergonomics effort was a big contributor.

The ergonomics effort must grow and spread for it to become truly effective. There are corrections to existing designs to make as well as new designs to influence. If we want to have designs that utilize the human being well, the ergonomics effort must be sustained over a long period.

Managing an ergonomics effort to sustain its growth is substantially different from simply practicing ergonomics. The perspective must shift from just providing ergonomics information and data to that of a critical manager who wants to end the ergonomics program. By convincing that manager, you will satisfy the skepticism of the organization management.

Managing the ergonomics program to sustain its existence and growth is essential, even if the program is very small. The alternative is to have the program die out.

To ensure the growth of the ergonomics effort, one must:

1. Observe the natural stages of growth for an ergonomics program.
2. Provide more benefits than costs.
3. Win converts to the strengths of ergonomics.

16-2 THE NATURAL GROWTH OF AN ERGONOMICS PROGRAM

Many industrial ergonomics efforts begin in response to a current problem in the organization expected to be resolved through the use of ergonomics principles. That problem may be a rash of carpal tunnel syndrome injuries, complaints of eyestrain or backache at VDTs, or rising medical and health care costs. With success at this initial endeavor, other problems are tackled. Their successful resolution leads, in turn, to more ergonomics projects.

At some point, the ergonomist also begins to work on the prevention of problems and injuries. This second stage emphasizes effective design, and ergonomic design information is widely shared. During this stage, proper ergonomic design becomes the responsibility of the designer rather than of the ergonomist.

A third stage may emerge with more advanced industrial ergonomics efforts. That stage is one of influencing the organization's philosophy, policies, and design standards and specifications, all relative to the use of people.

In the third stage, for example, the ergonomist might sponsor or contribute to engineering design standards that allow people with all major handicaps to work in any facility. Or a policy might be established that requires that all equipment and job design possibilities be exhausted before employee selection is utilized. A policy on the use of VDTs could be established that ensures the adjustability of equipment. In this stage, shift work systems are designed from a knowledge of physiological principles, rather than only on business needs and/or on personnel preferences.

The correction of an existing problem certainly has an immediate impact on the operation and the people involved. It is important to correct these problems. Yet the solution of that particular problem or the improvement of the design of that workplace has little impact on other people or on other jobs. The work that goes into the influence of design standards or policies, on the other hand, rarely has an immediate impact on people at the job site. The design standard has to filter down and be used in actual designs before it helps the person on the job. Yet that standard can eventually have a greater impact on more designs than can the results from any single project. These three stages are shown in Figure 16-1, together with information regarding the impact of these stages.

The evolution from correction to prevention of problems is natural and will occur as existing problems are corrected. This provides time to train others in the use of ergonomics and for the development of design guidelines. The evolution to the third stage is slower, and may have to be forced to occur. The third stage is characterized by the integration of ergonomics with other information into the establishment of organizational policy or design standards. Note that there is a substantial difference between design guidelines (stage two) and design standards (stage three). Design standards must be followed, whereas guidelines are usually only suggested design practices.

Shift work illustrates well the stage three concerns. Shift work is an area where ergonomics should have an influence, but often does not. Most existing shift work systems were designed for operational needs, for personal preference of the shift workers, or to meet legal constraints for overtime. With an effective third-stage ergonomics program, ergonomics is used actively in designing and evaluating shift schedule alternatives.

The use of ergonomics from the initial design stages is another indication of an effective third-stage ergonomics effort. When the person is fully integrated into the design, just like the equipment and facilities, the person will be utilized better. When to begin to consider designing for the person, however, is largely a policy issue. By influencing the policy to encourage the use of ergonomics early in the design process, there will be a much larger impact.

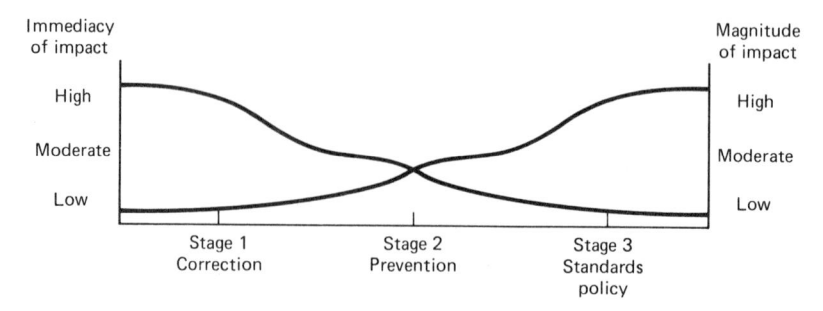

Figure 16-1 Stages of an industrial ergonomics effort.

Simply having organizational policies and design standards that clearly state that the designs will accommodate the full range of the work population can prevent more problems than any ergonomist can resolve in a lifetime. By limiting the weight lifted, for example, there will be far fewer manual materials handling problems for the ergonomist to resolve.

Growth of the ergonomics effort parallels the growth of the ergonomist. No matter how thrilling it was to solve the first problem, answering that same question over and over will get tiresome. There are only so many times that using the NIOSH lifting guidelines to determine work pace will continue to hold your interest. The growth of the ergonomics program to stage three results in fewer mundane and repetitive questions being asked of the ergonomist. At the same time, it will allow the ergonomist to grow personally and professionally by encouraging the application of ergonomics to:

- Broader issues, such as rising medical costs
- More complex questions, such as those involving shift work systems
- Bigger projects, such as workplace and job design for a new facility
- New applications, such as the use of ergonomics for product design

16-3 BENEFITS GREATER THAN COSTS

For long-term survival of any organizational function, the benefits of that function must be greater than the costs to the organization. Ergonomics is no exception. Those who claim that it is unnecessary to justify an ergonomics program are misinformed.

The justification, however, does not have to be completely economic. Justification is simply a measure of benefits relative to costs. Smith and Smith[1] discuss three methods of justification: increased productivity, less nonproductive time, and social/legal responsibility. They point out that justification without clear economic benefits is not unusual and is a feasible means of promoting ergonomics.

Justification is certainly easier when there is an economic basis. Comparing the costs and the benefits is more straightforward when dollars are used. The ergonomist may find it difficult to measure such items as lost time, absenteeism, or employee safety. Yet measuring (assessing) these key variables is what management does when they evaluate your proposals.

Safety is one "intangible" area frequently used to justify ergonomic proposals and ultimately, the ergonomics effort itself. Although the cost of an

---

[1] Leo A. Smith and James L. Smith, "How Can an IE Justify a Human Factors Program to Management?," *Industrial Engineering*, 14(2), February 1982, p. 40. Copyright © Institute of Industrial Engineers, 25 Technology Park/Atlanta, Norcross, GA 30092.

injury is not easy to measure, it is possible. By seeking historical information about injuries, one can develop reasonable predictors of future injuries. Take, for example, back injuries. The ergonomist can find the medical costs to treat the previous injuries, the money paid to the employees (disability and makeup wages) during time off, and rehabilitation expenses for the employees returned to work. These expenses are normally large and will often justify the expenditure of funds to study a job with a high injury rate and then to implement necessary changes.

Absenteeism, too, is measurable. There are usually large costs associated with relief staffing. By examining these costs, one can develop a strong case for those changes that will lower the absentee rate.

Turnover is measurable as well. The ergonomist should add the costs to employ a new person (recruiting, selection, orientation, etc.) with the costs to train a person on that job. For complex jobs, the time to reach proficiency will be measured in months, not days. These costs are unusually large and therefore are desirable to avoid.

If measurement is impossible, carefully compile a list of the benefits that are expected. Point out those benefits that are intangible and those that are tangible but difficult to calculate. Many benefits are simply not easy or convenient to calculate, although they will have a definite financial impact on the organization.

Regularly projecting and measuring the results of ergonomics projects is good management for the ergonomics effort. If there is a backlog of projects, this will allow the better projects to be selected first. With the limited resources of many ergonomics programs, everyone should be striving to have the largest gain for the resources used.

Measuring the results of ergonomics projects also allows case studies to be explained better. The case studies simply sound better when there are dollars associated with the results. An example with a savings of $20,000 relative to a cost of $1500 is easier to appreciate than one that resulted in two fewer back injuries per year at a cost of $1500. The case studies and stories are more likely to get retold when there are dramatic returns and simple facts.

16-4 WINNING CONVERTS TO ERGONOMICS

One of the easiest ways to destroy an effective ergonomics effort is to maintain a small group of technical specialists waiting for the phone to ring. Designers will not regularly call this group because they forget, they feel it is unnecessary, or they get tired of calling. In any event, ergonomics will not be used. A more effective approach is to win converts to ergonomics, so that more and more people are willing and able to practice the basics of ergonomics.

16-4-1 Building Commitment

People who work in the organizational development area realize there is a definite process for building commitment to a new concept. There are a series of steps that one goes through in developing this commitment. These steps are presented in Figure 16-2. Building commitment to the use of ergonomics follows these steps. A person progresses from one step to another. It is important to understand where a potential convert may be on this chart, since each step has a different intervention strategy.

The first step is to simply become aware of ergonomics. The person must learn about ergonomics—what it is and what it can do. Only after the awareness phase can that person accept ergonomics as a useful tool. Asking someone to accept ergonomics before he or she knows what it is will only result in confusion.

Acceptance occurs when the person has a positive image of ergonomics. Acceptance of ergonomics is followed by a willingness to give it a try—to use it to solve a problem or in a design. Of course, a problem must also be there before the trial can take place.

If the trial is successful, regular use of ergonomics can occur. Regular use will occur faster when the initial applications are more successful. If the initial trials are unsuccessful, regular use will probably not occur. Special attention to ensure successful trials is worthwhile.

With regular use, ergonomics will be incorporated into operating and design procedures. At this stage, the use of ergonomics will be mandated by the organization and using ergonomics will be easier than not using ergonomics. Developing procedures based on ergonomics principles will occur only after repeated successful applications.

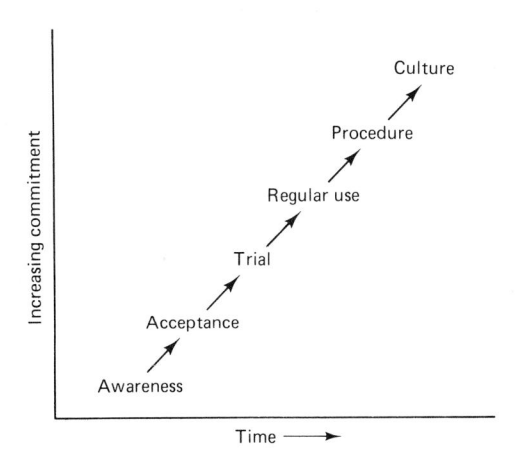

Figure 16-2 Building commitment to the use of ergonomics.

The highest commitment is "culture based." The culture of using ergonomics occurs when people ask why ergonomics was not used in a design, rather than asking why ergonomics should be used. When the "culture" dictates that ergonomics be used to design equipment, facilities, jobs, and products, it becomes very difficult to overlook. The wider this cultural acceptance of ergonomics, the more likely that designers will freely include ergonomics in their designs, and the more likely managers will insist on its use. Building ergonomics into the culture is a long-term effort, yet the results will be the most effective in sustaining the ergonomics effort.

16-4-2 Demonstrating Its Use

In the stages of building commitment, each stage is enhanced when there is a history of successful use of ergonomics. Using success stories while creating awareness adds to a positive acceptance. Similarly, success during a trial will help encourage long-term use of ergonomics.

Tell stories and encourage their retelling. This is a good way to develop an image of ergonomics as positive, productive, and profitable.

16-4-3 The Three Audiences

There are three unique groups that can influence the spread of ergonomics use in your organization. Each deserves special consideration. These groups are the managers who allocate resources for the ergonomics efforts, the designers who must use ergonomics principles, and the actual users—the operators, the mechanics, the clerks—who directly benefit from ergonomic design.

Management. Management is most pleased when ergonomics does what you said it would do, that is, to reduce operating costs and improve performance. This group should be aware of success stories that demonstrate reductions in injuries, better and safer performance, and lower costs.

Commitment is enhanced when the use of ergonomics provides benefits that outweigh its costs. Full commitment is obtained when ergonomics becomes part of the human resources efforts in the organization.

Designers. Designers prefer the situation where their designs are better after including ergonomics than before. The designer should see the use of ergonomics as a win-win situation rather than a compromise. For example, when ergonomics is included in the design of a piece of equipment, there should be no need to impair other necessary functions.

Commitment is enhanced when the use of ergonomics provides a better design. The inclusion of ergonomics when developing standards is further evidence of strong commitment.

The Actual Users. Although these people may not seem to have much influence on the use of ergonomics, they, in fact, can have a substantial influence. When this group is familiar with ergonomics and can identify problems, ergonomics use will increase. This group can also ask questions about ergonomics in the design of new facilities and equipment. Commitment will be enhanced when ergonomically designed equipment and facilities are easier to use and they realize that ergonomics contributed to that ease of use.

16-5 SUMMARY

For the ergonomics effort to be fully successful, it must be sustained over a long period. Creating the "culture" that requires everyone to consider the use of the human being from the initial part of the design is the best way to accomplish this goal.

QUESTIONS

1. Why is it important to sustain the ergonomics effort?
2. What are three stages in the natural growth of an ergonomics effort?
3. What are some of the "Stage Three" areas that the ergonomist might help develop?
4. What are the general methods of justifying an ergonomics program?
5. Describe the difference between intangible and incalculable savings. Are the following areas intangible or incalculable: safety? absenteeism? turnover?
6. List the steps in building commitment to ergonomics and the appropriate interventions at each step.

INDEX

A

Absenteeism, 364
Accident and medical costs, savings on, 25-26
Accommodation of disabled employees:
  available accommodations and aids, 349, 353-55, 357-58
  costs of the, 336, 338, 340-41
  design standards for the, 348-49
  overall work improvement resulting from the, 341
  predesign for the, 348-49
  process for the, 344, 346-48
Action limit (AL) for lifting, 144-49, 150, 169, 176, 177
Adjustability, 129, 281, 282, 285
Administrative solutions:
  cold stress and, 239
  hard jobs and, 213-15
  heavy jobs and, 176-80, 185, 326
  job design technology, 188-89, 292-311
  job redesign, 176-77, 214-15
  lifting and, 144-45, 150, 169, 176-80, 185, 326
  philosophy of, 313-16
  recovery periods, 189, 209, 212-13, 215, 296, 297
  rotation of personnel, 94, 189, 215, 321-23
  selection of personnel, 94, 144-45, 150, 177-79, 215, 281, 296, 297, 302, 306
  training of personnel, 93-94, 144-45, 150, 179-80, 185, 215
  versus design solutions, 89-90, 92-93
  when not to use, 176
  when to use, 176
  work times and schedules, 316-24
Aerobic work, definition of, 209
American Conference of Governmental Industrial Hygienists (ACGIH), 219
American Foundation for the Blind, 342
American Industrial Hygiene Association, 30, 195
American Industrial Hygiene Association Journal, 34
American National Standards Institute, 348

369

Anaerobic work, definition of, 209
Analysis:
   of cold stress, 238-39
   of hard jobs, 206
   of heat stress, 218-20, 225-27
   of human error, 259-60
   of lifting methods, 144
   link, 131
   major task, 78-80
   predesign, 98-99
   task, 86-89, 299-301
   of work/recovery cycles, 317-18
Anthropometric problems:
   changing work population and, 97-98, 229-30
   definition of, 12
   identification of, 79
   job design and, 294
   purchase of ergonomically designed products and, 281-82
   questions regarding, 82
Anthropometry:
   average person and, 132
   data ranges and, 132
   definitions of, 12, 102, 131
   design aids and, 138
   importance of human body dimensions, 131-32
   use of data in, 134-36
   whom to design for, 132-34
Applications of industrial ergonomics:
   equipment and facilities design, 23, 102-3, 226-29, 245-46, 278-80
   importance of knowledge in, 16-17, 293
   in-plant studies, 22
   product design, 23-24, 278-80, 288-91
   work times and schedules, 362
Applications of industrial ergonomics, philosophy of:
   best solution not always chosen, 6
   importance of criteria other than ergonomics, 5-6
   importance of making firm recommendations, 6
   technical and nontechnical solutions, 5
   timely solutions, 5
   workable solutions, 4-5
Applied Ergonomics, 34
Area specialists, 47
Armstrong, Thomas J., 353-54
Assembly, errors resulting from, 273-74
Assessment of hard jobs:
   using comparison tables, 196-97
   using heart rate, 199-200, 201-2, 207, 208, 210-11, 212
   using oxygen consumption, 197, 199, 201, 202, 207, 208-9, 212
Assessment of work loads, 295-96
Astrand, Per-Olof, 195, 208, 209, 212, 213
Auburn University, 31
Automation:
   examples of, 304-7
   job design and, 294, 296, 297, 299, 301-7
   key areas of, 304
   manual materials handling and, 172, 306
   savings from, 172, 296, 302, 306
   selection of personnel and, 302, 306
   work/recovery cycles and, 306
Average person, anthropometric concept of the, 132
Ayoub, M. M., 151

B

Beevis, D., 23, 36
Berkeley Study, the, 336, 340, 355
Bioengineering, 3
Biomechancial problem(s):
   changing work population and, 97
   examples of, 96, 283
   identification of, 14, 96
   job design and, 294
   purchase of ergonomically designed products and, 282-83
Biomechanics, 3
Blum, Max A., 36-37, 46
Brouha, Lucien, 195, 201-2, 210-11
Bursitis, 188

Index

C

Cardiovascular system, jobs that are tough on the. *See* Hard jobs
Cardiovascular system problems:
  excessive work load and, 207, 212
  heat stress and, 13, 216, 217–18, 226–27, 284
  job design and, 294
  overstress and, 194–95
  peak work loads and, 208–12
  purchase of ergonomically designed products and, 284
Carpal tunnel syndrome, 188, 361
Carrying, 165–67
Causes:
  of cumulative trauma disorders, 188, 189–92
  of human error, 256–60, 262, 265, 267–68, 268–69, 270–71, 272–73, 273–74, 319
  of injuries, 141, 143, 189–92
  of manipulative problems, 283
  of strength problems, 282
Changes:
  engineering design, 89–90, 306
  of environment, 92, 173, 214, 227–30
  of equipment, 92, 306
  in job, 91–92, 214–15, 230, 302, 306
  of methods, 91
  of the people performing the job, 215, 306
  of processes, 92
  of product, 92, 302
  of tools, 92
  of workplace, 92, 171, 172, 214, 297
  in work population, 97–98, 229–30
  in work/recovery cycles, 215, 230, 232, 306
  in work task expectations, 325
Chemical exposures, 331
Clearances:
  visual, 119
  workplace design and, 106
  work space design and, 120–21
Clothing, special work, 330–31
Cognitive problems:
  difference between motivational problems and, 16
  examples of, 15–16
  job design and, 294
  purchase of ergonomically designed products and, 282
Cold stress:
  analysis of, 238–39
  effects of, 237–38
  examples of, 235, 240
  solutions to, 239–40
  work/recovery cycles and, 239
Color, 329
Comfort, employee, 11, 12
Consolidated designs, 114–15
Consultant:
  ergonomist as a, 17–18
  comprehensive job design and a, 311
Control and display tasks, errors resulting from, 260, 262, 264
Correction versus prevention of problems, 10, 362
Cost-effectiveness of industrial ergonomics, 9–10, 11–12, 228
Costs:
  of absenteeism, 364
  of accommodation of disabled employees, 336, 338, 340–41
  of equipment, 277–78, 280, 306
  of injuries, 141, 186–87, 306, 363–64
  of job design, 310
  of job rotation, 326
  of lighting, 328
  of maintenance, 278, 281
  medical and disability, 336, 339, 361
  operating, 278, 280–81, 284
  of overtime, 319
  of service, 278, 281
  of turnover, 364
  vendors and, 278, 280
Costs, savings on:
  accident and medical, 25–26
  design and development engineering time and costs, 26–27
  production, 25
Cumulative trauma disorders (CTDs):
  bursitis, 188
  carpal tunnel syndrome, 188, 361

Cumulative trauma disorders (*cont.*)
  causes of, 188
  correction of, 188-89
  definition of, 188
  hand tool design and, 189-92
  prevention of, 188-89
  tendinitis, 188
  tenosynovitis, 188

D

Design(s):
  for accommodation of disabled employees, 348-49
  aids, 138
  author's principles of, 189
  changes, 89-90, 214, 306
  consolidated, 114-15
  cumulative trauma disorders and, 188-89
  and development engineering time and costs, 26-27
  dynamic, 119-21
  equipment and facilities, 23, 102-3, 226-29, 245-46, 278-80
  guidelines, 124, 126, 185-86, 362
  hand tool, 189-92, 281, 282
  information, 49-52
  job, 188-89, 292-311, 325-28
  macro, 299, 306
  maintenance needs and, 120-21
  manual materials handling and problems of, 141, 143, 167, 185-86
  micro, 299-301
  product, 23-24, 278-80, 288-91
  production needs and, 120
  service needs and, 120, 278, 283
  standards, 50, 104-6, 185-86, 348-49, 361, 362, 363
  static, 119-21
  trade-offs, 114-15, 120, 126-28
  vendor knowledge of ergonomics and, 278-80
  workplace, 101-6, 114-16, 121, 123-24, 126-31, 188-89, 214
  work space, 101-4, 116, 118-21, 123-24, 126-31, 188-89

  of work times and schedules, 320-24
  See also Problems
Designer(s):
  comprehensive job design and a, 311
  concept of the average person and the, 132
  engineering solutions and, 169, 171
  ergonomist as a, 17
  and evaluation of equipment for proper ergonomic design, 280-81, 284-86
  importance of human body dimensions to the, 131-32
  needs of the, 46-47
  quantitative data and, 140
  training of, 47-48
  as users of ergonomics, 43, 121, 123-24, 278, 361, 366
Design person, definition of, 138
"Designs for Independent Living—Kitchen and Laundry Designs for Disabled Persons," 349
Design solutions:
  adjustability, 129, 281, 282, 285
  automation, 172, 294, 296, 297, 299, 301-7
  changing environment, 92, 173, 214, 227-30
  changing equipment, 92
  changing jobs, 91-92, 230
  changing methods, 91
  changing processes, 92
  changing products, 92, 173
  changing tasks, 91, 230
  changing tools, 92
  changing workplace, 92, 297
  cold stress and, 239
  container redesign, 171-72
  controlling the user population, 129
  design process, 130, 288-91
  eliminating workers, 92
  equipment redesign, 172-73
  handling process redesign, 173
  job design technology, 188-89, 292-311
  job redesign, 171, 173
  mechanization, 172, 294

overdesigning, 129
redesigning methods, 173
savings and, 173, 176
versus administrative solutions, 89–90, 92–93
win-win solutions, 130, 302, 306
workplace redesign, 171, 172, 214, 297
Diffrient, Niels, 138
Dimensional specifications, 286
Disabilities:
  current employee versus a new hire, 338
  developmental, 338
  physical, 338
  sensory impairment, 338
  temporary versus permanent, 337–38
Disabled employee(s):
  accommodation process for, 344, 346–48
  available accommodations and aids for, 349, 353–55, 357–58
  costs of accommodating, 336, 338, 340–41
  costs of medical and disability benefits for, 336, 339
  example of a, 335–36
  job placement and transitional work centers for, 342–43
  multidisciplined approaches to job placement for, 342
  overall work improvement resulting from accommodations for, 341
  predesign for accommodations for, 348–49
  process of employing, 341
  reasons for employing, 336
  tax incentives for employing, 339–40
  types of disabilities, 337–38
Display and control, errors resulting from, 260, 262, 264
Dreyfuss, Henry, 138
Drum handling, 169
Dynamic design versus static design, 119–21

E

Eastman Kodak Company, 29
Efficiency rates, 281
Electrocardiogram (EKG), 202
Employee:
  comfort, 11, 12
  health, 11–12, 207
  productivity, 11–12, 296, 302, 306, 310–11
  safety, 11
  See also Personnel; Work time and schedules
Endurance:
  job design and, 294
  problems, 13, 194–95, 207, 284, 294
  questions regarding, 82
Engineering:
  design changes, 89–90, 214, 306
  psychology, 3
Engineering solutions, lifting problems and, 150, 169, 171–73, 176
Engineers:
  training of, 47–48
  as users of ergonomics, 43
Environment:
  changes in the, 92, 173, 214, 227–30, 306
  job design and, 294
  the problem job model and the, 83
Environmental problems:
  cold stress, 235, 237–40
  color, 329
  definitions of, 14–15, 236
  examples of, 235, 240
  heat stress, 13, 31–32, 99, 216, 217–22, 224–32, 237, 284, 329–30, 330–31
  job design and, 294
  lighting, 240–41, 243–46, 328
  noise, 246–50, 252, 280, 283, 284, 328
  purchase of ergonomically designed products and, 283–84
  solutions to, 236–37, 246
  special clothing, 330–31
  temperature and humidity, 329–30

Environmental problems (*cont.*)
  vibration, 252-53
  workplace contaminants, 331
Equal Employment Opportunity, 325
Equipment:
  costs of, 277-78, 280-81, 306
  and facilities design, 23, 102-3, 226-29, 245-46, 278-80
  fatigue and, 281
  heat stress and, 31-32, 222, 329-30, 330-31
  manual materials handling and, 172-73, 176, 306
  redesign, 172-73
  repair of, 283, 311
  used in an ergonomics effort, 31-32, 34, 222
  used in work physiology, 201-2
  vendors and, 278-80
Equipment, purchase of ergonomically designed:
  difficulties in the, 277-80
  endurance problems and the, 284
  environmental problems and the, 283-84
  evaluation of, for proper ergonomic design, 280-81, 284-86
  human error problems and the, 282
  importance of the, 276-77
  manipulation problems and the, 283
  operating cost increases and the, 280-81
  physical size problems and the, 281-82
  product design and the, 288-91
  selection of personnel and the, 281
  specifications for the, 286-87
  strength problems and the, 282-83
  vendors and the, 278-80
Ergonomics:
  definition of industrial, 2
  goal of, 1-2, 295
  legislative, 315-16
  See also Industrial ergonomics effort, planning an
Ergonomics, 34
"Ergonomics Guide to Assessment of Metabolic and Cardiac Costs of Physical Work," 195
"Ergonomics Guide to Assessment of Physical Work Capacity," 195, 202
Ergonomics program:
  benefits of an, 363-64
  effectiveness of an, 360-61
  how to ensure the growth of an, 361
  how to win converts to an, 364-67
  stages of growth of an, 361-63
Ergonomics specialists, 43, 364
Ergonomist:
  comprehensive job design and an, 311
  engineering solutions and the, 169, 171
  and evaluation of equipment for proper ergonomic design, 280-81, 284-86
  part-time, 18
  personal growth of the, 363
  responsibilities of the, 187
  role of the, 17-18
  selection of personnel and the, 178
Error. *See* Human error
Evaluation:
  of equipment for proper ergonomic design, 280-81, 284-86, 289
  of the job design process, 299-300, 301
Excessive work loads, 207, 212, 295-96
Expectations. *See* Work task expectations
Experience. *See* Knowledge, importance of in industrial ergonomics
External problems, 14-15, 283-84, 294

F

Facilities and equipment design, 23, 102-3, 226-29, 245-46, 278-80
Fair day's work:
  concept of a, 194-95
  job design evaluation and a, 301
  methods of determining a, 295-96

Index

Farr, Donald E., 49
Fatigue:
   equipment and, 281
   excessive work loads and, 207
   overtime and, 319
   peak work loads and, 208
   prediction of, 210-11
First-line supervisors as users of ergonomics, 41-42
Fitness training, 215
Forms and paperwork, errors resulting from, 270-71
Functional specifications, 286-87

G

General Motors Corporation, 29, 36
George Washington University, Job Development Laboratory of, 349
Guidelines:
   design, 124, 126, 185-86, 362
   design process solutions, 130
   difference between standards and, 362
   manual materials handling, 164-69, 185-86
   NIOSH lifting, 143-50, 169, 171, 176, 179, 315, 363

H

Hackman, J. Richard, 307, 308, 309
"Handtool Design," 190
Hand tools:
   cumulative trauma disorders and design of, 189-92
   design and use of, 190-92, 281, 282
Hard jobs:
   analysis of, 206
   definition of, 204
   reasons for concern regarding, 206
Hard jobs, assessment of:
   using comparison tables, 196-97
   using heart rate, 199-200, 201-2, 207, 208, 210-11, 212
   using oxygen consumption, 197, 199, 201, 202, 207, 208-9, 212
Hard jobs, problems with:
   excessive demands for an entire work shift, 207
   high work loads for a few muscle groups, 212
   peak work loads, 208-12
   work/recovery cycles, 209, 212-13, 306
Hard jobs, solutions to:
   changes in the job, 214-15, 302, 306
   changes in the people performing the job, 215, 306
   changes in the work/recovery cycle, 215, 306
   criteria for choosing, 213-14
   redesign of the physical environment, 214, 227-30, 306
   selection of personnel, 215
Hawthorne effect, 84, 85
Health, employee, 11-12, 207
Heart rate (HR), 199-200, 201-2, 207, 208, 210-11, 212, 217, 226
Heating, ventilation, and air-conditioning (HVAC) engineers, 227, 228, 229, 230, 232
Heat stress:
   analysis of, 218-20, 225-27
   cardiovascular problems and, 13, 216, 217-18, 284
   ergonomist's equipment and, 31-32, 222
   predesign analysis and, 99
   prevention of, 227-30, 329-30, 330-31
   solutions to, 230-32
   threshold limit values for, 219
   wet-bulb globe temperature and, 219-22, 224-27
Heat Stress Index (HSI), 219, 226
Heavy jobs:
   administrative solutions for, 176-80, 185, 326
   automation and, 172
   cumulative trauma disorders and, 188-92

Heavy jobs (*cont.*)
  engineering solutions for, 169, 171–73, 176
  injuries and, 140, 141, 143, 186–87, 306
  lifting, 141, 143–64, 169, 171–73, 176, 179–80, 185, 326
  manual materials handling (other than lifting), 141, 143, 164–69, 172, 173, 176, 179–80, 185, 306
Hot jobs:
  analysis of, 218–20, 225–27
  examples of, 216
  prevention of heat stress and, 227–30, 329–30, 330–31
  problems with, 217–18, 284
  wet-bulb globe temperature and, 219–22, 224–27
Human engineering, 3
Human error:
  analysis of, 259–60
  causes of, 256–60, 262, 265, 267–68, 268–69, 270–71, 272–73, 273–74
  costs of, 254–55
  detection of, 255–56
  error-producing situations, 256–60
  examples of, 254–55, 282
  identifying problems of, 79
  overtime and, 319
  questions regarding, 82
  rates of, 255, 278, 281
  solutions to, 262, 264, 265–66, 267–68, 268–69, 270–71, 272–73, 273–74, 282
Human error, tasks that produce:
  assembly, 273–74
  control and display, 260, 262, 264
  forms and paperwork, 270–71
  information processing, 268–69
  inspection and monitoring, 266–68
  instruction, 272–73
  labels and signs, 265–66
Human factors, 3
Human factors engineering (HFE), 3, 46
Human Factors Journal, 34
Human Factors Society, 22, 30–31
"Humanscale 1/2/3," 138

Humidity and temperature considerations, 329–30

I

Identification:
  of a problem job, 78
  of problems, 14, 54–56, 64–69, 71–74, 79, 206, 225–27, 250
Illumination. *See* Lighting
Implementing an industrial ergonomics effort:
  creating interest in, 44–46
  developing broad-based support for, 40–41
  factors to consider in, 39–40
  lasting change and, 52
  making it work, 46–51
  user needs and, 41–43
Improvement(s):
  employee comfort and, 11, 12
  employee health and, 11–12
  employee productivity and, 11–12, 296, 302, 306, 310–11
  employee safety and, 11
  intangible method for measuring, 4
  overall, resulting from employing disabled employees, 341
  savings on accident and medical costs, 25–26
  savings on design and development engineering time and costs, 26–27
  savings on production costs, 25
  tangible method for measuring, 3–4
Industrial Engineering, 34
The Industrial Environment—Its Evaluation and Control, 222, 253
Industrial ergonomics. *See* Ergonomics
Industrial ergonomics effort, planning an:
  beginning, 20–21
  factors to be considered in, 21
  opportunities for improvement, 25–27
  organization, 27–29
  resources, 30–32, 34

selling the concept to management, 34–37
types of applications, 22–25
Information:
  design, 48–52
  sources of for an ergonomics effort, 34
Information processing, errors resulting from, 268–69
Injuries:
  causes of, 141, 143, 189–92
  costs of, 141, 186–87, 306, 363–64
  cumulative trauma disorders, 188–92
  heavy jobs and, 140, 141, 143, 186–87, 306
  lifting, 141, 186–87, 326
  manual materials handling, 141, 172, 173, 176, 186–87, 306
  prevention of, 140, 186–87, 189–92, 227–30, 361, 363–64
  rates of, 278, 306, 326, 364
  See also Employee
Innovation process, job design and the, 302–4
Institute of Industrial Engineers, 30
Instruction, errors resulting from, 272–73
International Labor Organization, 315

J

Jiang, B. C., 151
Job, definition of a, 293
Job, an effective:
  definition of, 293
  factors important in, 307
Job design:
  automation and, 294, 296, 297, 299, 301–7
  comprehensive, 311
  costs of, 310
  definition of, 293
  effective job and, 307
  evaluation of, 299–300, 301
  importance of, 292–94
  innovation process and, 302–4
  macro design, 299, 306
  maintenance and, 311
  manipulative problems and, 294
  mechanization and, 294
  micro design, 299–301
  motivational aspects of, 307–10
  organizational values of, 310–11
  process of, 298–301
  repair and, 311
  service and, 311
  technology of, 188–89, 292–311
  utilization of personnel and, 294–97
  work task expectations and, 325–28
Job placement for disabled employees:
  multidisciplined approaches to, 342
  transitional work centers and, 342–43
Job redesign, 171, 173, 176–77
Job rotation, 214–15, 326
Jobs, assessment of difficult:
  using comparison tables, 196–97
  using heart rate, 199–200, 201–2, 207, 208, 210–11, 212
  using oxygen consumption, 197, 199, 201, 202, 207, 208–9, 212
Job Severity Index (JSI):
  examples of the use of the, 160–64
  formula for the, 151
  NIOSH Guides and the, 151
  purpose of the, 150–51
  using the, 151–60
Job types:
  hard, 204, 206–15
  heavy, 140–41, 143–69, 171–73, 176–80, 185–92, 326
  hot, 216–22, 224–32
Johnson, Steve, 295
Journal of the American Medical Association, 34

K

Kinesology problems, 14, 283, 294
Knowledge:
  importance of in industrial ergonomics, 16–17, 293
  tiers of, 29, 47
  vendor, 278–80
Kochhar, Dev S., 353–54
Kurzweil reader, 353, 355

L

Labels and signs, errors resulting from, 265-66
Legislative ergonomics, 315-16
Lifting:
 action limit (AL) and, 144-49, 169
 administrative solutions and, 144-45, 326
 analysis of methods for, 144
 cumulative trauma disorders, 188-92
 design solutions and, 171-73, 176, 188-89
 engineering solutions and, 150, 169, 171-73, 176
 injuries, 141, 186-87, 326
 Job Severity Index (JSI) and, 150-64
 Lift School, principles of the, 179-80
 maximum permissible level (MPL) and, 145-49, 169
 NIOSH guidelines for, 143-50, 151, 169, 171, 176, 179, 315, 363
 physical training for, 185
 selection of personnel and, 144-45, 150
 skill acquisition for, 180, 185
 solutions, 150
 training for, 179-80
 See also Manual materials handling
Lift School, the, 179-80
Lighting:
 costs of, 328
 problems, 240-41, 243-46, 283-84, 328
 solutions, 246
Line supervisors as users of ergonomics, 42, 225
Link analysis, definition of, 131
Lucas, Richard, 22, 44-45

M

Macro design, 299, 306
Maintenance:
 costs of, 278, 281
 design and, 120-21, 278
 job design and, 311
Major task analysis, 78-80
Management:
 and allocation of resources for an ergonomics effort, 366
 and evaluation of equipment for proper ergonomic design, 284
 planning of an ergonomics proposal to, 34-37
 presentation of an ergonomics proposal to, 37
Managers as users of ergonomics, 42-43
Manipulative problems:
 causes of, 283
 definition of, 14
 examples of, 14, 283
 solutions to, 283
 job design and, 294
Manual materials handling (MMH):
 automation and, 172, 306
 carrying, 165-67
 cumulative trauma disorders, 188-92
 design problems and, 141, 143, 185-86
 drum handling, 169
 ergonomist and, 187
 guidelines for, 164-69, 185-86
 injuries and, 141, 172, 173, 176, 186-87, 306
 management of, 185-87
 mechanization and, 172
 physical training for, 185
 pulling, 168
 pushing, 167
 shoveling, 168-69
 skill acquisition for, 180, 185
 training for, 179-80
 See also Lifting
Maximum permissible level (MPL) for lifting, 145-49, 150, 169, 176
"Measure of Man, The," 138
Mechanization:
 definition of, 214
 job design and, 294
 manual materials handling and, 172
 physical assists, 326-27
Medical and disability costs, 336, 339, 361

Meister, David E., 49
Micro design, 299–301
Miller, Caroline, 357
Mini-Guide for Lifting, A, 151
Monitoring and inspection, errors resulting from, 266–68
Motivation, job design and, 307–10
Motivational problems, difference between cognitive problems and, 16
Mueller, Jim, 349
Musculoskeletal system, jobs that are tough on the. *See* Heavy jobs
Musculoskeletal system problems, 141, 186–87, 194–95

N

National Institute for Occupational Safety and Health (NIOSH), 74
  lifting guidelines of, 143–50, 151, 169, 171, 176, 179, 315, 363
National Safety Council, 336
National Safety Council Journal, 34
Natural selection, 179
Nemeth, Susan E., 190, 342, 343
NIOSH Guides, 143–50, 169, 171, 176, 179, 315, 363
  Job Severity Index and the, 151
Noise:
  difference between sound and, 246
  hearing loss and, 247–48, 250, 252, 328
  interference with work performance and, 249, 250, 252, 328
  limits, 280, 283, 328
  problems, 246–50, 252, 280, 283, 284, 328
  solutions, 250, 252
  speech interference and, 248–49, 250, 252
North Carolina State University, 31

O

Off-site problem identification:
  data bases and, 67–68
  intuition and, 69
  professional contacts and, 68–69
Oldham, Greg R., 307, 308, 309
On-site problem identification:
  interactive discussions and, 65
  on-site studies and, 66–67
  physical inspections and, 65
  setting priorities and, 67
  walk-through inspections and, 65–66
Operating costs, 278, 280–81, 284
Operators:
  as users of ergonomics, 41–42
  control of work pace by, 189
Optimum utilization, definition of, 295
OSHA, 328
Overdesigning, 129
Overtime:
  amount of, 318–19
  costs of, 319
  fatigue and, 319
  scheduling of, 319–20
Overutilization of personnel:
  definition of, 295
  solutions to, 296, 297
  work/recovery cycles and, 296, 297
Overwork, 207
Oxygen consumption, 197, 199, 201, 202, 207, 208–9, 212, 226

P

Paperwork and forms, errors resulting from, 270–71
Pareto principle, 29
Part-time ergonomist, 18
Payoff rates, 281
Pay systems, ergonomics and, 332
Peak work loads, 208–12
Personnel:
  ergonomics and, 332–33
  people as resources in an ergonomics effort, 30–31
  rotation of, 94, 189, 215, 321–23
  training of, 93–94, 144–45, 150, 179–80, 185, 215
  See also Employee; Work times and schedules

Personnel, selection of:
  automation and the, 302, 306
  hard jobs and the, 215
  lifting and the, 144-45, 150
  natural selection, 179, 215
  negative aspects of the, 177-79
  overutilization of personnel and the, 296, 297
  positive aspects of the, 177
  problem job model and the, 94
  purchase of ergonomically designed equipment and the, 281
  self-selection, 179
  underutilization of personnel and the, 297
Personnel, utilization of:
  job design and, 294-97
  optimum, 295
  over-, 295-96
  solutions to problems of, 296, 297
  under-, 296-97
Philosophy:
  of administrative solutions, 313-16
  of applications of industrial ergonomics, 4-6
  of problem identification, 72-74
Physical:
  assists, 326-27
  capacities and work task expectations, 326
  disabilities, 338
  training, 185
Physical size problems, 12, 281-82
  changing work population and, 97-98
  identification of, 79
  job design and, 294
  questions regarding, 82
Physiology in Industry, 195, 202
Pine, Steven M., 27
Planning an industrial ergonomics effort, 20-32, 34-37
Policies and procedures. *See* Administrative solutions
Predesign:
  accommodation of disabled employees and, 348-49
  analysis, 98-99

President's Committee on Employment of the Handicapped, 342, 354
Prevention:
  of cumulative trauma disorders, 188-89
  evaluation of equipment for proper ergonomic design, 280-81, 284-86
  of heat-related problems, 227-30, 329-30, 330-31
  of injuries, 140, 186-87, 189-92, 227-30, 361, 363-64
  versus correction of problems, 10, 362
Problem job model, the:
  administrative changes and, 89-90, 92-94
  anthropometric questions and, 82
  changing work population and, 97-98
  clarification questions and, 82
  design solutions and, 91-92
  determining what needs to be corrected, 86
  determining whether the problem has been corrected, 92
  determining whether the problem is real or perceived, 81-83
  determining whether the problem needs to be solved, 80-81
  determining whether this is the right problem to solve, 80-81
  endurance questions and, 82
  environmental questions and, 83
  Hawthorne effect and, 84, 85
  human error questions and, 82
  identification of a problem job and, 78
  major task analysis and, 78-80
  perceived problems and, 83-85
  pinpointing specific problems with, 96-97
  predesign analysis and, 98-99
  redesign and, 89-90
  resolution of the problem and, 94
  selection of personnel and, 94
  strength questions and, 82
  task analysis and, 86-89, 299-301

Problems:
- cognitive (thought), 15-16, 282, 294
- difference between cognitive and motivational, 16
- endurance (cardiovascular), 13, 194-95, 207, 216, 217-18, 284, 294
- environmental (external), 14-15, 235-41, 243-50, 252-53, 283-84, 294
- excessive work loads, 207, 212, 295-96
- fatigue, 207, 208, 210-11, 281, 319
- with hard jobs, 207-13
- high work loads for a few muscle groups, 212
- with hot jobs, 217-18
- human error, 79, 282, 319
- lifting, 141, 143-64, 169, 326
- lighting, 240-41, 243-46, 283-84, 328
- manipulative (kinesiology), 14, 283, 294
- manual materials handling and design, 141, 143
- noise, 246-50, 252, 280, 283, 284, 328
- overutilization of personnel, 295-96, 297
- peak work loads, 208-12
- perceived, 83-85
- physical size (anthropometric), 12, 79, 82, 97-98, 281-82, 294
- prevention versus correction of, 10, 362
- strength (biomechanical), 14, 96-97, 282-83, 294
- underutilization of personnel, 296-97
- vibration, 252-53, 283-84
- ways to avoid ergonomics, 76
- work/recovery cycles, 209, 212-13, 306

Problems, identification of:
- advanced methods, 64-69
- checklists for the, 56, 64
- engineering solutions and, 169-70
- importance of, 54-55
- off-site, 67-69
- on-site, 65-67
- philosophy of, 72-74
- problem prevention and, 69, 71, 227-30
- technology and, 72
- traditional method, 56, 64

Problems, solutions to. *See* Solutions
Product design, 23-24, 278-80, 288-91
Production:
- design and, 120
- savings on costs of, 25

Productivity, employee, 11-12, 296, 302, 306, 310-11
Purchasing. *See* Equipment, purchase of ergonomically designed
Purdue University, 31
Pushing, 167

R

Rates:
- efficiency, 281
- heart, 199-200, 201-2, 207, 208, 210-11, 212, 217, 226
- human error, 255, 278, 281
- injury, 278, 306, 326, 364
- payoff, 281

Redesign. *See* Design solutions
Repair of equipment, 283, 311
Resources used in an ergonomics effort:
- equipment, 31-32, 34, 222
- information, 34
- people, 30-31

Rest periods, 189, 209, 212-13, 225, 296, 297, 306, 317-18
Rest/recovery cycles. *See* Work/recovery cycles
Reynaud's syndrome, 192
Rhyne, David, 357
Robots. *See* Automation
Rodahl, Kaare, 195
Rotation:
- of jobs, 214-15, 326
- of personnel, 94, 189, 215, 321-23

S

Safe Practices for Manual Lifting, 74
Safety, employee, 11

Safety, employee (*cont.*)
  color and, 329
  special work clothing and, 330
Savings:
  on accident and medical costs, 25–26
  from automation, 172, 296, 302, 306
  on design and development engineering time and costs, 26–27
  design solutions and, 173, 176
  from mechanization, 172
  on production costs, 25
Schedules. *See* Work times and schedules
Schulz, Kenneth A., 27
Selan, J. L., 151
Self-selection of personnel, 179
Service:
  costs of, 278, 281
  design and, 120, 278, 283
  job design and, 311
Shift work. *See* Work times and schedules
Shoveling, 168–69
Signs and labels, errors resulting from, 265–66
Singleton, W. T., 28, 30
Slade, I. M., 23, 36
Smith, James L., 35–36, 314–15
Smith, Leo A., 35–36, 314–15
Solutions:
  adjustability, 129, 281, 282, 285
  administrative, 89–90, 92–94, 144–45, 150, 169, 176–80, 185, 188–89, 209, 212–15, 239, 313–34
  to cold stress, 239–40
  controlling the user population, 129
  design, 91–92, 128–30, 171–73, 176, 188–89, 214, 227–30
  design process guidelines, 130
  engineering, 150, 169, 171–73, 176
  to environmental problems, 236–37
  to hard jobs, 213–15
  to heat stress, 230–32, 329–30, 330–31
  to heavy jobs, 169, 171–73, 176–80, 185, 326
  to human error, 262, 264, 265–66, 267–68, 268–69, 270–71, 272–73, 273–74, 282
  to lifting problems, 150, 326
  to lighting problems, 246
  link analysis, 131
  to manipulative problems, 283
  to noise problems, 250, 252, 280, 328
  overdesigning, 129
  to problems of overutilization, 296, 297
  to problems of underutilization, 297
  technical and nontechnical, 5
  timely, 5
  use frequency, 131
  use sequence, 131
  win-win, 120, 302, 306
  workable, 4–5
Soule, Robert D., 253
Specialists:
  area, 47
  ergonomics, 43, 364
Specifications:
  dimensional, 286
  functional, 286–87
  how to determine, 287
"Specifications for Making Building and Facilities Accessible to and Usable by Physically Handicapped People," 348
Standards:
  design, 50, 104–6, 185–86, 348–49, 361, 362, 363
  difference between guidelines and, 362
  work, 327–28
Static design versus dynamic design, 119–21
Strains & Sprains: A Worker's Guide to Job Design, 42
Strength:
  job design and, 294
  questions regarding, 82
Strength problems:
  causes of, 282
  changing work population and, 97
  examples of, 96, 283

identification of, 14, 96
job design and, 294
purchase of ergonomically designed products and, 282-83
Stress:
cold, 237-40
excessive work requirements and, 207
heat, 13, 31-32, 99, 216, 217-22, 224-32, 237, 329-30, 330-31
Studies, in-plant, 22
Supervisors:
first-line, 41-42
line, 42, 225
Swedish Institute for the Handicapped, 358

T

Task analysis, 86-89, 299-301
Tax incentives for hiring disabled employees, 339-40
Technology, problem identification and the use of, 72
Temperature and humidity considerations, 329-30
Tendinitis, 188
Tenosynovitis, 188
Testing. *See* Evaluation
Texas Technological University, 31
Textbook on Work Physiology, 195
Thought problems:
difference between motivational problems and, 16
examples of, 15-16
job design and, 294
purchase of ergonomically designed products and, 282
Threshold limit values (TLVs):
for chemical exposures, 331
for heat stress, 219
Tiers of knowledge concept, 29, 47
Tools:
changes of, 92
cumulative trauma disorders and design of, 189-92
design and use of, 190-92, 281, 282

Trade-offs, design:
affecting a single person, 127-28
affecting two different populations, 127
consolidated designs and, 114-15
types of, 126-28
within a population, 127
work space design and, 120
Training:
of designers, 47-48
of engineers, 47-48
fitness, 215
lifting and, 144-45
major limitation of widespread, 48
major strength of widespread, 48
manual materials handling, 179-80
of personnel, 93-94, 144-45, 150, 179-80, 185
physical, 185
skill acquisition, 180, 185
Turnover of personnel, 364

U

Underutilization of personnel, 296-97
University of Michigan, 31
Use frequency, definition of, 131
User population, controlling the, 129
Users of ergonomic information:
actual, 367
designers, 43, 278, 361, 366
engineers, 43
ergonomics specialists, 43, 364
first-line supervisors, 41-42
line supervisors, 42, 225
managers, 42-43
operators, 41-42
Use sequence, definition of, 131

V

Vendors:
equipment costs and, 278, 280
and evaluation of equipment for proper ergonomic design, 280-81, 284-86

Vendors (*cont.*)
  knowledge of ergonomics and, 278–80
Vibration, 252–53
"Vibration," 253
Virginia Polytechnic Institute and State University, 31
Visual clearance, definition of, 119
Vocational Rehabilitation Act of 1973, 337

W

Weiman, Novia, 24
Wet-bulb globe temperature (WBGT):
  definitions of, 218–19
  use of in industry, 220–27
Whirlpool Corporation, Appliance Information Service of, 349
Win-win solutions, 130, 302, 306
Work:
  aerobic, 209
  anaerobic, 209
  concept of a fair day's, 194–95, 295–96, 301
Work allowances, 327–28
Work loads:
  assessment of, 295–96
  excessive, 207, 212, 295–96
  peak, 208–12
Work pace, control of by operators, 189
Work physiology:
  assessment of difficult jobs, 196–97, 199–200
  definition of, 195
  methods and equipment, 201–2
  questions answered by, 196
  using the assessment techniques, 200–201
Workplace contaminants, 331
Workplace design, 101–4, 214
  anthropometry and, 131–36, 138
  clearances and, 106
  consolidated designs and, 114–15
  cumulative trauma disorders and, 188–89
  difference between work space design and, 119–20
  guidelines, 124, 126
  principles of, 104–6
  processes and concerns, 116, 121, 123–24, 126–31
  reaches and, 106
  solutions, 128–30
  trade-offs and, 126–28
Work population, changes in the, 97–98, 229–30
Work Practices Guide for Manual Lifting, 143–50
Work/recovery cycles:
  analysis of, 317–18
  automation and, 306
  changes in, 215, 230, 232, 306
  cold stress and, 239
  overutilization of people and, 296, 297
  peak work loads and, 209
  problems with, 209, 212–13, 306
Work redesign, 307
Work requirements, excessive, 207, 212, 295–96
Work space design, 101–4
  anthropometry and, 131–36, 138
  clearances, and, 120–21
  cumulative trauma disorders and, 188–89
  design trade-offs and, 120
  difference between workplace design and, 119–20
  elements of, 116, 118, 120–21
  examples of, 121
  guidelines, 124, 126
  processes and concerns, 121, 123–24, 126–31
  solutions, 128–30
  trade-offs and, 126–28
  visual clearance and, 119
Work standards, 327–28
Work task expectations:
  changes in, 325
  job design and, 325–28
  about job and task assignments, 325–26

about physical capacities, 326
about the use of physical assists, 326-27
about work allowances and work standards, 327-28
Work times and schedules:
designing of, 320-24
ergonomics applied to, 362
factors that influence, 316
overtime, 318-20
work/recovery cycles, 317-18